BIRDS
of
OHIO

James S. McCormac
Gregory Kennedy

with contributions from Chris Fisher & Andy Bezener

LONE
PINE

Lone Pine Publishing International

Distributed by Lone Pine Publishing
1808 B Street NW, Suite 140
Auburn, WA, USA 98001

Website: www.lonepinepublishing.com

National Library of Canada Cataloguing in Publication
McCormac, James S., 1962–
    Birds of Ohio / James S. McCormac, Gregory Kennedy.

Includes bibliographical references and index.
ISBN-13: 978-1-55105-392-9
ISBN-10: 1-55105-392-6

    1.-Birds—Ohio.   2.-Birds—Ohio—Identification.   I.-Kennedy, Gregory, 1956–
II.-Title.

QL684.O3M33 2004   598'.09771   C2004-901086-7

Cover Illustration: Wood Duck by Gary Ross
Illustrations: Gary Ross, Ted Nordhagen, Ewa Pluciennik

PC: P16

CONTENTS

ACKNOWLEDGMENTS

Successful completion of a project of this scope is never a singular labor; many people make critical contributions. I am indebted to my fellow members of the Ohio Bird Records Committee for generously allowing me the use of the seasonal abundance graphs included within. In particular, David Dister, Joseph Hammond, Robert Harlan, Bernie Master, and Bill Whan, whose meticulous research crafted these graphs. Also, Bob Royse made helpful suggestions, which I adopted. I owe a special debt of gratitude to Bill Whan, Editor of *The Ohio Cardinal*, who graciously reviewed much of my work and improved it greatly. The Division of Wildlife of the Ohio Department of Natural Resources was very helpful in providing data on selected species, and biologist Kathy Shipley was especially forthcoming with information about endangered species. Laura Busby of Ohio Audubon was the catalyst that made this project possible. Topnotch editor Gary Whyte managed to shape my scrawlings into a decipherable form, and teach me what professional editing is all about.

I began manifesting a keen interest in birds at a very early age, and certain people were vital in encouraging me to explore ornithology. Foremost are my parents, John and Martha, who always supported my interest and even drove me to see rarities before I had a driver's license. My 4th grade teacher, Debora Moon, strongly encouraged me and made a big impression. Jim Fry, one of Ohio's top listers, was an early and important mentor, as was Bruce Peterjohn, Ohio's premier field ornithologist of modern times. Larry Rosche and the late Dave Corbin had a big, positive impact on me in my early years of birding. In recent years, I've had the good fortune to become friends with Bernie Master, Peter King, Bill Thompson and his much better half, Julie Zickefoose, spending many an hour afield with them. All have taught me much in their own way. Very special thanks, too, to all of Ohio's birders who share their sightings, support birding and habitat conservation and make that special effort to encourage the new birder.—*Jim McCormac*

Thanks are also extended to the growing family of ornithologists and dedicated birders who have offered their inspiration and expertise to help build Lone Pine's expanding library of field guides. Thanks go to John Acorn, Chris Fisher, Andy Bezener and Eloise Pulos for their contributions to previous books in this series. Thanks also go to Marcus C. England for his preliminary research on this book. In addition, thank you to Gary Ross, Ted Nordhagen and Ewa Pluciennik, whose skilled illustrations have brought each page to life.

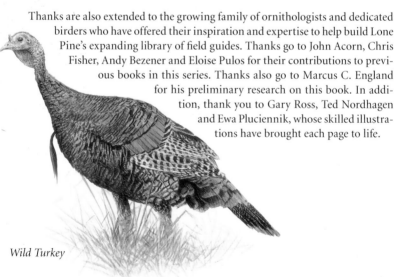

Wild Turkey

Greater White-fronted Goose
size 30 in • p. 38

Snow Goose
size 30 in • p. 39

Canada Goose
size 35 in • p. 40

Brant
size 26 in • p. 41

Mute Swan
size 60 in • p. 42

Tundra Swan
size 54 in • p. 43

Wood Duck
size 18 in • p. 44

Gadwall
size 20 in • p. 45

Eurasian Wigeon
size 19 in • p. 46

American Wigeon
size 20 in • p. 47

American Black Duck
size 22 in • p. 48

Mallard
size 24 in • p. 49

Blue-winged Teal
size 15 in • p. 50

Northern Shoveler
size 19 in • p. 51

Northern Pintail
size 23 in • p. 52

Green-winged Teal
size 14 in • p. 53

Canvasback
size 20 in • p. 54

Redhead
size 20 in • p. 55

Ring-necked Duck
size 16 in • p. 56

Greater Scaup
size 18 in • p. 57

Lesser Scaup
size 17 in • p. 58

Harlequin Duck
size 17 in • p. 59

Surf Scoter
size 18 in • p. 60

White-winged Scoter
size 21 in • p. 61

Black Scoter
size 19 in • p. 62

Long-tailed Duck
size 19 in • p. 63

Bufflehead
size 14 in • p. 64

Common Goldeneye
size 18 in • p. 65

WATERFOWL

Hooded Merganser
size 17 in • p. 66

Common Merganser
size 24 in • p. 67

Red-breasted Merganser
size 22 in • p. 68

Ruddy Duck
size 15 in • p. 69

GROUSE & ALLIES

Ring-necked Pheasant
size 30 in • p. 70

Ruffed Grouse
size 17 in • p. 71

Wild Turkey
size 36 in • p. 72

Northern Bobwhite
size 10 in • p. 73

DIVING BIRDS

Red-throated Loon
size 25 in • p. 74

Common Loon
size 31 in • p. 75

Pied-billed Grebe
size 13 in • p. 76

Horned Grebe
size 14 in • p. 77

Red-necked Grebe
size 19 in • p. 78

Eared Grebe
size 13 in • p. 79

American White Pelican
size 63 in • p. 80

Double-crested Cormorant
size 29 in • p. 81

HERONLIKE BIRDS

American Bittern
size 25 in • p. 82

Least Bittern
size 13 in • p. 83

Great Blue Heron
size 51 in • p. 84

Great Egret
size 39 in • p. 85

Snowy Egret
size 24 in • p. 86

Little Blue Heron
size 24 in • p. 87

Cattle Egret
size 20 in • p. 88

Green Heron
size 18 in • p. 89

Black-crowned Night-Heron
size 24 in • p. 90

HERONLIKE BIRDS

Yellow-crowned Night-Heron
size 24 in • p. 91

Black Vulture
size 25 in • p. 92

Turkey Vulture
size 29 in • p. 93

BIRDS OF PREY

Osprey
size 24 in • p. 94

Bald Eagle
size 36 in • p. 95

Northern Harrier
size 20 in • p. 96

Sharp-shinned Hawk
size 12 in • p. 97

Cooper's Hawk
size 17 in • p. 98

Northern Goshawk
size 23 in • p. 99

Red-shouldered Hawk
size 19 in • p. 100

Broad-winged Hawk
size 17 in • p. 101

Red-tailed Hawk
size 22 in • p. 102

Rough-legged Hawk
size 21 in • p. 103

Golden Eagle
size 35 in • p. 104

American Kestrel
size 8 in • p. 105

Merlin
size 11 in • p. 106

Peregrine Falcon
size 17 in • p. 107

RAILS, COOTS & CRANES

King Rail
size 15 in • p. 108

Virginia Rail
size 10 in • p. 109

Sora
size 9 in • p. 110

Common Moorhen
size 13 in • p. 111

American Coot
size 15 in • p. 112

Sandhill Crane
size 45 in • p. 113

SHOREBIRDS

Black-bellied Plover
size 12 in • p. 114

American Golden-Plover
size 10 in • p. 115

Semipalmated Plover
size 7 in • p. 116

Killdeer
size 10 in • p. 117

American Avocet
size 17 in • p. 118

Greater Yellowlegs
size 14 in • p. 119

Lesser Yellowlegs
size 10 in • p. 120

Solitary Sandpiper
size 8 in • p. 121

Willet
size 15 in • p. 122

Spotted Sandpiper
size 7 in • p. 123

Upland Sandpiper
size 12 in • p. 124

Whimbrel
size 17 in • p. 125

Hudsonian Godwit
size 15 in • p. 126

Marbled Godwit
size 18 in • p. 127

Ruddy Turnstone
size 9 in • p. 128

Red Knot
size 10 in • p. 129

Sanderling
size 8 in • p. 130

Semipalmated Sandpiper
size 6 in • p. 131

Western Sandpiper
size 6 in • p. 132

Least Sandpiper
size 5 in • p. 133

White-rumped Sandpiper
size 7 in • p. 134

Baird's Sandpiper
size 7 in • p. 135

Pectoral Sandpiper
size 8 in • p. 136

Purple Sandpiper
size 9 in • p. 137

Dunlin
size 8 in • p. 138

Stilt Sandpiper
size 9 in • p. 139

Buff-breasted Sandpiper
size 8 in • p. 140

Short-billed Dowitcher
size 11 in • p. 141

Long-billed Dowitcher
size 12 in • p. 142

Wilson's Snipe
size 11 in • p. 143

American Woodcock
size 11 in • p. 144

Wilson's Phalarope
size 9 in • p. 145

Red-necked Phalarope
size 7 in • p. 146

Red Phalarope
size 8 in • p. 147

Pomarine Jaeger
size 21 in • p. 148

Laughing Gull
size 16 in • p. 149

Franklin's Gull
size 14 in • p. 150

Little Gull
size 10 in • p. 151

Bonaparte's Gull
size 13 in • p. 152

Ring-billed Gull
size 19 in • p. 153

Herring Gull
size 24 in • p. 154

Thayer's Gull
size 23 in • p. 155

Iceland Gull
size 22 in • p. 156

Lesser Black-backed Gull
size 20 in • p. 157

Glaucous Gull
size 27 in • p. 158

Great Black-backed Gull
size 30 in • p. 159

Sabine's Gull
size 13 in • p. 160

Black-legged Kittiwake
size 17 in • p. 161

Caspian Tern
size 21 in • p. 162

Common Tern
size 14 in • p. 163

Forster's Tern
size 15 in • p. 164

Black Tern
size 9 in • p. 165

REFERENCE GUIDE

DOVES & CUCKOOS

Rock Pigeon
size 12 in • p. 166

Mourning Dove
size 12 in • p. 167

Black-billed Cuckoo
size 12 in • p. 168

Yellow-billed Cuckoo
size 12 in • p. 169

OWLS

Barn owl
size 15 in • p. 170

Eastern Screech-Owl
size 8 in • p. 171

Great Horned Owl
size 22 in • p. 172

Snowy Owl
size 24 in • p. 173

Barred Owl
size 20 in • p. 174

Long-eared Owl
size 15 in • p. 175

Short-eared Owl
size 15 in • p. 176

Northern Saw-whet Owl
size 8 in • p. 177

NIGHTJARS, SWIFTS & HUMMINGBIRDS

Common Nighthawk
size 9 in • p. 178

Chuck-will's-widow
size 12 in • p. 179

Whip-poor-will
size 9 in • p. 180

Chimney Swift
size 5 in • p. 181

Ruby-throated Hummingbird
size 4 in • p. 182

Belted Kingfisher
size 12 in • p. 183

WOODPECKERS

Red-headed Woodpecker
size 9 in • p. 184

Red-bellied Woodpecker
size 10 in • p. 185

Yellow-bellied Sapsucker
size 8 in • p. 186

Downy Woodpecker
size 6 in • p. 187

Hairy Woodpecker
size 9 in • p. 188

Northern Flicker
size 13 in • p. 189

Pileated Woodpecker
size 17 in • p. 190

FLYCATCHERS

Olive-sided Flycatcher
size 7 in • p. 192

Eastern Wood-Pewee
size 6 in • p. 193

Yellow-bellied Flycatcher
size 6 in • p. 194

Acadian Flycatcher
size 6 in • p. 195

Alder Flycatcher
size 6 in • p. 196

Willow Flycatcher
size 6 in • p. 197

Least Flycatcher
size 5 in • p. 198

Eastern Phoebe
size 7 in • p. 199

Great Crested Flycatcher
size 8 in • p. 200

Eastern Kingbird
size 8 in • p. 201

SHRIKES & VIREOS

Northern Shrike
size 10 in • p. 202

White-eyed Vireo
size 5 in • p. 203

Bell's Vireo
size 5 in • p. 204

Yellow-throated Vireo
size 5 in • p. 205

Blue-headed Vireo
size 5 in • p. 206

Warbling Vireo
size 5 in • p. 207

Philadelphia Vireo
size 5 in • p. 208

Red-eyed Vireo
size 6 in • p. 209

JAYS & CROWS

Blue Jay
size 12 in • p. 210

American Crow
size 19 in • p. 211

LARKS & SWALLOWS

Horned Lark
size 7 in • p. 212

Purple Martin
size 7 in • p. 213

Tree Swallow
size 5 in • p. 214

Northern Rough-winged Swallow
size 5 in • p. 215

REFERENCE GUIDE

LARKS & SWALLOWS

Bank Swallow
size 5 in • p. 216

Cliff Swallow
size 5 in • p. 217

Barn Swallow
size 7 in • p. 218

CHICKADEES, NUTHATCHES & WRENS

Carolina Chickadee
size 5 in • p. 219

Black-capped Chickadee
size 5 in • p. 220

Tufted Titmouse
size 6 in • p. 221

Red-breasted Nuthatch
size 4 in • p. 222

White-breasted Nuthatch
size 5 in • p. 223

Brown Creeper
size 5 in • p. 224

Carolina Wren
size 5 in • p. 225

House Wren
size 5 in • p. 226

Winter Wren
size 4 in • p. 227

Sedge Wren
size 4 in • p. 228

Marsh Wren
size 5 in • p. 229

KINGLETS, BLUEBIRDS & THRUSHES

Golden-crowned Kinglet
size 4 in • p. 230

Ruby-crowned Kinglet
size 4 in • p. 231

Blue-gray Gnatcatcher
size 4 in • p. 232

Eastern Bluebird
size 7 in • p. 233

Veery
size 7 in • p. 234

Gray-cheeked Thrush
size 7 in • p. 235

Swainson's Thrush
size 7 in • p. 236

Hermit Thrush
size 7 in • p. 237

Wood Thrush
size 8 in • p. 238

American Robin
size 10 in • p. 239

Gray Catbird
size 9 in • p. 240

Northern Mockingbird
size 10 in • p. 241

Brown Thrasher
size 11 in • p. 242

European Startling
size 8 in • p. 243

American Pipit
size 6 in • p. 244

Cedar Waxwing
size 7 in • p. 245

Blue-winged Warbler
size 5 in • p. 246

Golden-winged Warbler
size 5 in • p. 247

Tennessee Warbler
size 5 in • p. 248

Orange-crowned Warbler
size 5 in • p. 249

Nashville Warbler
size 5 in • p. 250

Northern Parula
size 5 in • p. 251

Yellow Warbler
size 5 in • p. 252

Chestnut-sided Warbler
size 5 in • p. 253

Magnolia Warbler
size 5 in • p. 254

Cape May Warbler
size 5 in • p. 255

Black-throated Blue Warbler
size 5 in • p. 256

Yellow-rumped Warbler
size 5 in • p. 257

Black-throated Green Warbler
size 5 in • p. 258

Blackburnian Warbler
size 5 in • p. 259

Yellow-throated Warbler
size 5 in • p. 260

Pine Warbler
size 5 in • p. 261

Prairie Warbler
size 5 in • p. 262

Palm Warbler
size 5 in • p. 263

Bay-breasted Warbler
size 5 in • p. 264

Blackpoll Warbler
size 5 in • p. 265

WOOD-WARBLERS & TANAGERS

Cerulean Warbler
size 5 in • p. 266

Black-and-white Warbler
size 5 in • p. 267

American Redstart
size 5 in • p. 268

Prothonotary Warbler
size 5 in • p. 269

Worm-eating Warbler
size 5 in • p. 270

Ovenbird
size 6 in • p. 271

Northern Waterthrush
size 6 in • p. 272

Louisiana Waterthrush
size 6 in • p. 273

Kentucky Warbler
size 5 in • p. 274

Connecticut Warbler
size 5 in • p. 275

Mourning Warbler
size 5 in • p. 276

Common Yellowthroat
size 5 in • p. 277

Hooded Warbler
size 5 in • p. 278

Wilson's Warbler
size 5 in • p. 279

Canada Warbler
size 5 in • p. 280

Yellow-breasted Chat
size 7 in • p. 281

Summer Tanager
size 7 in • p. 282

Scarlet Tanager
size 7 in • p. 283

SPARROWS, GROSBEAKS & BUNTINGS

Eastern Towhee
size 8 in • p. 284

American Tree Sparrow
size 6 in • p. 285

Chipping Sparrow
size 5 in • p. 286

Clay-colored Sparrow
size 5 in • p. 287

Field Sparrow
size 5 in • p. 288

Vesper Sparrow
size 6 in • p. 289

Lark Sparrow
size 6 in • p. 290

Savannah Sparrow
size 5 in • p. 291

Grasshopper Sparrow
size 5 in • p. 292

Henslow's Sparrow
size 5 in • p. 293

Le Conte's Sparrow
size 5 in • p. 294

Nelson's Sharp-tailed Sparrow
size 5 in • p. 295

Fox Sparrow
size 6 in • p. 296

Song Sparrow
size 6 in • p. 297

Lincoln's Sparrow
size 6 in • p. 298

Swamp Sparrow
size 6 in • p. 299

White-throated Sparrow
size 7 in • p. 300

White-crowned Sparrow
size 6 in • p. 301

Dark-eyed Junco
size 6 in • p. 302

Lapland Longspur
size 6 in • p. 303

Snow Bunting
size 7 in • p. 304

Northern Cardinal
size 8 in • p. 305

Rose-breasted Grosbeak
size 8 in • p. 306

Blue Grosbeak
size 7 in • p. 307

Indigo Bunting
size 5 in • p. 308

Dickcissel
size 6 in • p. 309

BLACKBIRDS & ALLIES

Bobolink
size 7 in • p. 310

Red-winged Blackbird
size 8 in • p. 311

Eastern Meadowlark
size 9 in • p. 312

Western Meadowlark
size 9 in • p. 313

Yellow-headed Blackbird
size 9 in • p. 314

Rusty Blackbird
size 9 in • p. 315

Brewer's Blackbird
size 9 in • p. 316

Common Grackle
size 12 in • p. 317

Brown-headed Cowbird
size 7 in • p. 318

Orchard Oriole
size 7 in • p. 319

Baltimore Oriole
size 7 in • p. 320

FINCHLIKE BIRDS

Purple Finch
size 5 in • p. 321

House Finch
size 5 in • p. 322

Red Crossbill
size 6 in • p. 323

White-winged Crossbill
size 6 in • p. 324

Common Redpoll
size 5 in • p. 325

Pine Siskin
size 5 in • p. 326

American Goldfinch
size 5 in • p. 327

Evening Grosbeak
size 8 in • p. 328

House Sparrow
size 6 in • p. 329

INTRODUCTION

BIRDING IN OHIO

In recent decades, birding has evolved from an eccentric pursuit practiced by relatively few individuals to a continent-wide activity that involves millions. In 2001, a U.S. Fish and Wildlife Service study estimated that 46 million Americans birded in one form or another, and that collectively they spent nearly 85 billion dollars annually on their hobby! This same study calculated that 20 percent of Ohioans, or about 1.9 million people, watch birds. Birding draws money into the Buckeye state, too; an economic study of the Ottawa National Wildlife Refuge estimated that the 130,000 annual visitors—mostly birders—contribute 5.6 million dollars to the local economy. And that represents just a little slice of Ohio birding!

Ohio birders are fortunate in that our state encompasses a great diversity of habitat. We have prairie influences from the west, unglaciated Appalachian hill country in the south and, of course, one of the great inland seas—Lake Erie—to the north. As a result, Ohio's total bird list currently has 412 species, surpassing all surrounding states except Michigan, which barely eclipses us with 419 species. Although Ohio has changed greatly since pre-settlement days when the state was 95 percent forested, many interesting birding habitats remain, and an abundance of great birding opportunities await the Ohio birdwatcher. Furthermore, in recent decades new birds have been added to the state list at the rate of about 1.5 species per year, so that by 2033, Ohio could have 457 species!

In addition to simply watching birds, there are ample opportunities for birders to become involved in ornithology, at both amateur and professional levels. Christmas bird counts, breeding bird surveys, winter bird surveys, nest box programs, workshops, migration monitoring and rare species censuses are just some of the ways to put your birding interests to work.

American Black Duck

BEGINNING TO LEARN THE BIRDS

While sometimes challenging, birding (or birdwatching) is always rewarding. Learning to recognize all of the birds that occur in Ohio is a long process, but the majority of the regularly occurring species can be learned fairly quickly. This guide should be a great help in mastering the local bird life of Ohio, and with Ohio under your belt, branching out into birding in other areas of North America should be a bit less daunting.

In addition to guides like this one, other more experienced birders are great sources of information. As birders tend to gather at favorite birding locales, finding like-minded individuals shouldn't be too difficult. Also, events such as those sponsored by the Ohio Ornithological Society bring birders together and are great places to meet. Don't be shy about asking questions of others with more experience; most birders love to share their knowledge.

TECHNIQUES OF BIRDING

Learning about the relationships between birds and habitat is a key to maximizing bird finding. Although there's always a bit of serendipity involved in stumbling across interesting birds, the best birders are the ones who know what types of birds are found in certain habitats. It's no coincidence that many birders are also excellent amateur botanists—conifer stands are good places to look for Red-breasted Nuthatches and roosting owls, open meadows often harbor Eastern Meadowlarks and cottonwoods support Warbling Vireos. Even artificial habitats like power line towers are great places to look for perched raptors.

Understanding migration is also important to becoming a good birder. Spring and fall are the most obvious times of movement, but at least some species are involved in some type of migration in every month of the year. Spring migration, with its waves of returning neotropical migrants, is certainly the most popular time of year for birders. Male songbirds are in full breeding plumage and also are actively singing. A visit to the Magee Marsh Wildlife Area on International Migratory Bird Day (the second Saturday in May) will provide ample evidence of the popularity of spring migration among birders.

Eastern Meadowlark

Optics

In the old days, before the development of good optical equipment, ornithologists shot birds. While a good shot certainly allowed for a good look, such techniques took their toll, and fortunately we've progressed well beyond that point. Today's binoculars and spotting scopes can provide stellar views of distant birds, making identification much easier and providing the chance to really appreciate the details of a bird.

Binoculars

There are two types of binoculars: porro prism and roof prism. Each has its advantages, but without going into great detail, roof prisms probably make the better "bins" for most birders. There is plenty of information available to help you select your binoculars, but don't ignore the opinions of other birders. Check out their gear and ask which type and brand they like and why. In general, though, roof-prism binoculars with 7x35 or 8x42 magnification and a wide, comfortable neck strap are best. A good pair can be had for as little as $300. Expensive models usually have better optics and generally stand up better to abuse.

The optical power of binoculars is described with a two-number code. For example, a compact pair might be "8x21," while a larger pair might have "7x40" stamped on it. In each case, the first number states the magnification, while the second number records the diameter, in millimeters, of the front lenses. Seven-power binoculars are easier to hold and to find birds with; 10-power binoculars give a shakier but more magnified view. Larger lenses gather more light, so a 40mm or 50mm lens will perform better at dusk than a 20mm or 30mm lens of the same magnification.

Spotting Scopes

A tool that most advanced birders have in their chest is the spotting scope, which enables study of more distant birds than do binoculars—ducks in the middle of a lake, for example. Great scopes can be purchased for about the same price as an inexpensive set of binoculars. For most birders, 20x magnification is probably best. Of course, a good tripod will also be necessary. Once you've become accustomed to "scoping" birds, you'll be hooked!

Cameras

It's difficult to practice serious photography and birding simultaneously; photography simply requires too much time. But, having a camera along can be a tremendous asset if you stumble across something incredible that no one would believe, such as an Ivory-billed Woodpecker. Sometimes, a common species will present itself in an irresistible way, and it's wonderful to be able to capture the moment for posterity. The increasing sophistication of digital cameras is creating a whole new venue of photography, called "digiscoping." This simply involves holding the camera to the lens of a telescope—although the technique is much more refined among aficionados—and snapping away. Amazingly good pictures can be taken in this way.

For serious photography, a single-lens reflex camera, either 35mm film or high-resolution digital, with a long telephoto lens is essential. As with spotting scopes, a good tripod is important.

Birding by Ear

Almost all birders start by learning birds visually, then often progress into learning songs and calls. Knowing vocalizations is a great way to enhance the birding experience, for a number of reasons. First, you'll find more birds—way more birds! Almost all species call or sing regularly, and a bird otherwise undetectable in the bush will instantly be revealed when it sounds off. In fact, in leafed-out, summer Ohio woodlands, a birder who knows sounds might find 10 times as many birds as the birder who is oblivious to calls. Second, songs are sometimes tremendously helpful in making an identification. Drab, similarly colored flycatchers are a good example. Is that an Eastern Wood-Pewee or an Olive-sided Flycatcher way up on that branch? When it whistles its plaintive *pee-ah-wee*, the identification will be clinched. Finally, birds create some of the most beautiful sounds on earth. It's empowering to know what species is making what sound, and in addition to impressing your companions, this adds tremendously to the birding experience. There are many good tapes and CDs available now that are great aids in learning bird songs.

Eastern Wood-Pewee

BIRDING BY HABITAT

Ohio has a wealth of habitats, and the state has been divided into five physiographic regions, each with its own characteristics. Learning to recognize which species prefer what habitat types will help tremendously in understanding where to find birds and will often help with identification. While not every bird will always be found where it's "supposed" to be, more often than not birds stick to certain habitats.

Some birds, such as loons and many diving ducks, frequent open waters of lakes—a habitat rather general in description. Others, such as breeding Blue-headed Vireos and Black-throated Green Warblers, occur only in hemlock gorges, a habitat that is very narrowly defined and rare in Ohio. Your understanding of the relationship between birds and habitat can become quite refined as you become

Horned Grebe

more experienced. For instance, you may notice that wintering Hermit Thrushes are often associated with brushy thickets containing sumacs, as the fruit of these small trees is a dietary staple of the thrushes. Even in migration, many birds remain faithful to certain habitats; for example, Winter Wrens tend to lurk about root masses, downed trees and brush piles, just as they do on the breeding grounds.

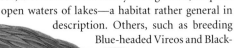

Making an effort to learn at least the more common plants will help a lot, not only in assisting your learning of habitats, but also in describing to fellow birders where a bird is. Rather than saying "Yellow-throated Warbler at 9 o'clock in the big tree," it's often more helpful to be able to identify the "big tree" as a sycamore. After a while, certain plants will send up "red flags"; for example, ornamental birches loaded with catkins in winter are always worth a double look, as redpolls are often attracted to them.

Yellow-throated Warbler

BIRD LISTING

Keeping lists is immensely popular among birders and can take many forms. Most bird watchers keep a life list, many maintain a state list and backyard lists are common. The more fanatical might make county lists or lists for every state they've visited. So popular is listing that the American Birding Association publishes a regular bulletin featuring the accomplishments of its members. Careful listing of birds seen on trips—particularly if numbers are recorded—serves a purpose, too. It provides a record that can be submitted to editors of birding journals that track birds, and also creates a memento of your birding experiences.

BIRDING ACTIVITIES
Birding Groups

A great way to increase the pleasure of birding is to join with like-minded people. Not only is it fun to have companionship, fraternizing with more experienced birders is an excellent way to accelerate learning. Becoming involved with Christmas bird counts, birding festivals, breeding or winter bird surveys and local Audubon societies, as well as attending natural history meetings, all provide ways to advance your skills and increase your involvement in the birding world. Following are some contacts that can help you get involved.

ORGANIZATIONS

American Birding Association
P.O. Box 6599
Colorado Springs, CO 80934-6599
800-850-2473
www.americanbirding.org

Audubon Ohio
692 N. High Street,-Suite 208-
Columbus, OH 43215-
614-224-3303-
www.audubon.org/chapter/oh/oh

Black Swamp Bird Observatory
P.O. Box 228
Oak Harbor, OH 43449
419-898-4070
www.bsbobird.org

Cincinnati Bird Club
www.cincinnatibirds.com/birdclub

Kelleys Island Audubon Club
P.O. Box 42
Kelleys Island, OH 43438
www.kelleysislandnature.com

Kirtland Bird Club
www.kirtlandbirdclub.org

Ohio Bluebird Society
PMB 111
343 W. Milltown Rd.
Wooster, OH 44691
www.obsbluebirds.com

Ohio Environmental Council
1207 Grandview Ave., Suite 201
Columbus, OH 43212-3449
614-487-7506
www.theoec.org

Ohio Ornithological Society
P.O. Box 14051
Columbus, OH 43214
www.ohiobirds.org

Sierra Club, Ohio Chapter
36 West Gay Street, Suite 314
Columbus OH 43215-2811
614-461-0734
www.ohio.sierraclub.org

The Nature Conservancy, Ohio Chapter
6375 Riverside Drive, Suite 50
Dublin, OH 43017
614-717-2770
nature.org/ohio

Toledo Naturalists' Association
www.toledonaturalist.org

Cerulean Warbler

BIRD CONSERVATION

Although Ohio has the seventh-largest population of all the states and is known for industry and agriculture, it is also a leader in conservation efforts. Multitudes of private and public organizations and agencies have protected nearly one million acres of Ohio lands. The Ohio Department of Natural Resources alone owns over 470,000 acres, which includes 127 state nature preserves, 96 wildlife areas and 74 state parks. Many birders belong to conservation groups, and joining or supporting any of the groups listed above is a good way to help further habitat and bird protection in our state. The Ohio Ornithological Society, in particular, is a statewide network of birders geared toward promoting education, raising awareness of habitat protection and providing resources to those interested in birdwatching.

BIRD FEEDING

If it's sometimes difficult to get out birding in far-flung haunts, why not bring the birds to you? A great way to do this is with backyard bird feeding. Depending on nearby habitat, a great variety of birds can readily be lured into suburban yards—some homeowners have accumulated yard lists in excess of 100 species! Some wonderful sources of information on bird feeding and backyard habitat improvement can be found at local stores that specialize in products for feeding birds and attracting birds to your yard.

A common myth has it that birds become dependent on feeders and will perish if we quit feeding. This is absolutely false; while birds are opportunists and will readily avail themselves of our handouts, they are quite mobile and quickly find natural food sources if the feeder runs dry. Some people also believe that feeding interferes with migration, stimulating birds—particularly hummingbirds—to remain too far north and thus perish with the onset of severe winter weather. This is also false; migratory instincts are quite strong and override the allure of artificial feeding. In fact, hummingbird feeders should remain up until at least late October, as this is when vagrant western hummers such as the Rufous Hummingbird tend to appear. These wayward westerners almost always show up at feeders, are extremely hardy and can remain until surprisingly late in the year, often well into winter before departing for warmer climates.

Rufous Hummingbird

NEST BOXES

Another popular way to attract birds is to set out nest boxes. Many interesting species will often nest in suburban backyards if an appropriate box is erected, including House Wrens, Carolina Chickadees, Black-capped Chickadees and Tufted Titmice. In more rural, open locations, Eastern Bluebirds, Tree Swallows and Purple Martins might be enticed with suitable housing. If big enough boxes are placed and you're really lucky, even Eastern Screech-Owls might make a home! While 27 species of Ohio's regularly breeding birds depend upon tree cavities for nest sites, only a relative handful routinely adopt nest boxes. Nevertheless, boxes have become quite important in bolstering populations of some birds, such as Eastern Bluebirds, Purple Martins and even Prothonotary Warblers. The widespread establishment of introduced European Starlings and House Sparrows, which also use cavities and can aggressively out-muscle native birds, has had a serious impact on cavity nesters. By maintaining suitable nest boxes, we can have a very positive effect on the populations of these native, cavity-nesting birds.

Black-capped Chickadee

Cleaning Feeders and Nest Boxes

Nest boxes and feeding stations must be kept clean to prevent birds from becoming ill or spreading disease. Old nesting material may harbor parasites and their eggs. Once the birds have left for the season, remove the old nesting material and wash and scrub the nest box with detergent or a 10 percent bleach solution (1 part bleach to 9 parts water). You can also scald the nest box with boiling water. Rinse it well and let it dry thoroughly before you remount it.

Feeding stations should be cleaned monthly. Feeders can become moldy and any seed, fruit or suet that is moldy or spoiled must be removed. Unclean bird feeders can also be contaminated with salmonellosis and possibly other avian diseases. Clean and disinfect feeding stations with a 10 percent bleach solution, scrubbing thoroughly. Rinse the feeder well and allow it to dry completely before refilling it. Discarded seed and feces on the ground under the feeding station should also be removed.

We advise that you wear rubber gloves and a mask when cleaning nest boxes or feeders.

Purple Martin

TOP BIRDING SITES IN OHIO

There are hundreds of excellent birding locales in Ohio, from top spots such as the Magee Marsh bird trail to local parks. There simply isn't space in this book to include all of them. The following sites are noted for producing good birding, at least seasonally, and have been selected to represent a broad range of habitats, good geographic representation and accessibility. In sites that are primarily seasonal, the best birding months are noted in parentheses. Boldface type indicates sites that are described in the Top 20 Ohio Birding Sites section. For the top 20 sites, interesting species that can be expected, as well as past rarities, are included. Directions to all of these sites can be found at the Ohio Ornithological Society website (www.ohiobirds.org).

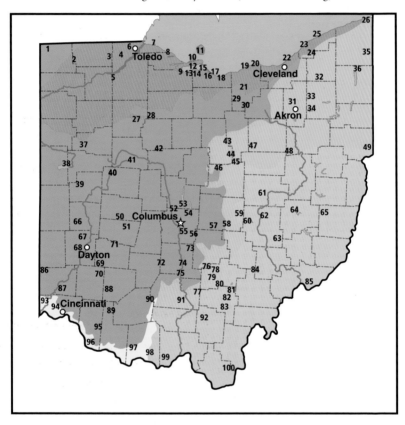

Physiographic Regions of Ohio

- Till Plains
- Huron-Erie Plains
- Bluegrass Section
- Glaciated Allegheny Plateaus
- Allegheny Plateaus

Abbreviations

MP . . . . Metropark
NRA . . . . . . National Recreation Area
NWR . . . . . National Wildlife Refuge
SF . . . . State Forest
SP . . . . State Park
SNP . . . . . . State Nature Preserve
WA . . . . Wildlife Area

OHIO'S TOP 100 BIRDING SITES

1. Lake La Su An WA
2. Goll Woods SNP
3. Maumee SF
4. **Oak Openings MP**
5. Maumee River Rapids
6. Woodlawn Cemetery
7. Maumee Bay SP / Mallard Club Marsh WA
8. **Magee Marsh WA / Metzger Marsh WA / Ottawa NWR**
9. Pickerel Creek WA
10. East Harbor SP
11. **Kelleys Island**
12. Medusa Marsh / Sandusky Bay
13. Resthaven WA
14. Castalia Pond (October-April)
15. Pipe Creek WA
16. **Sheldon Marsh SNP**
17. Huron Municipal Pier
18. Old Woman Creek SNP
19. **Lorain Harbor** (November-March)
20. Avon Lake Power Plant (November-March)
21. Oberlin Reservoir (October-April)
22. Cleveland Lakefront SP (November-March)
23. Eastlake Power Plant (November-March)
24. Holden Arboretum
25. **Fairport Harbor / Headlands Dunes SNP**
26. **Conneaut Harbor**
27. Findlay Reservoir
28. Springville Marsh SNP
29. Wellington Reservoir
30. Findley SP
31. Cuyahoga Valley NRA
32. La Due Reservoir
33. Lake Rockwell
34. Mogadore Reservoir
35. Pymatuning SP
36. Mosquito Creek WA
37. Bresler Reservoir / Ferguson Reservoir / Kendrick Woods SNP
38. **Grand Lake St. Marys**
39. Lake Loramie SP
40. Indian Lake SP
41. Lawrence Woods SNP
42. **Big Island WA / Killdeer Plains WA**
43. Charles Mill Lake
44. Pleasant Hill Reservoir
45. **Mohican Memorial SF**
46. Knox Lake
47. **Funk Bottoms WA / Killbuck Marsh WA**
48. Stark Wilderness Center
49. Beaver Creek SP
50. Cedar Bog State Memorial
51. **Buck Creek SP / C.J. Brown Reservoir**
52. Highbanks MP
53. Alum Creek SP
54. **Hoover Reservoir**
55. **Green Lawn Cemetery / Greenlawn Avenue Dam**
56. Pickerington Ponds MP
57. Hebron Fish Hatchery
58. Dawes Arboretum
59. Blackhand Gorge SNP
60. Dillon SP
61. Woodbury WA
62. Tri-Valley WA
63. **The Wilds**
64. Salt Fork SP
65. Egypt Valley WA
66. Brukner Nature Center
67. Aullwood Audubon Center
68. Great Miami River / Englewood Reserve / Germantown Reserve
69. Spring Valley WA
70. Caesar Creek SP
71. Clifton Gorge SNP / John Bryan SP / Glen Helen
72. Deer Creek SP / WA
73. Slate Run MP
74. Calamus Swamp
75. Charlie's Pond
76. **Clear Creek MP**
77. Tar Hollow SF
78. Cantwell Cliffs SP
79. Conkle's Hollow SNP
80. Cedar Falls SP
81. Lake Hope SP
82. Zaleski SF
83. MeadWestvaco Experimental Forest
84. Burr Oak SP
85. **Newell's Run / Ohio River / Willow Island Lock and Dam**
86. Hueston Woods SP
87. Gilmore Ponds Preserve
88. Cowan Lake SP
89. Indian Creek WA
90. Paint Creek SP / WA
91. Scioto Trail SF
92. Lake Katherine SNP
93. **Miami-Whitewater Wetlands**
94. Spring Grove Cemetery
95. East Fork SP
96. Meldahl Locks and Dam (November-March)
97. Chaparral Prairie SNP
98. Edge of Appalachia Preserve
99. **Shawnee SF**
100. **Crown City WA**

TOP 20 OHIO BIRDING SITES

Big Island–Killdeer Plains Wildlife Areas

Occupying the former expanse of wet prairie known as the Sandusky Plains (now largely destroyed), these two sites collectively total nearly 14,000 acres. Birding is good year-round and can be fantastic during spring and fall migration. These areas are renowned for wintering raptors and owls, including the Northern Harrier, Rough-legged Hawk, Northern Saw-whet Owl, Short-eared Owl and Long-eared Owl as well as the occasional Northern Goshawk. Bald Eagles nest at both wildlife areas. Breeding birds include rare prairie species such as the Upland Sandpiper, Dickcissel and sometimes the Western Meadowlark. Large numbers and a great diversity of waterfowl can be seen on the wetlands, and if shorebird habitat is available, waders can be abundant. Many rarities have been seen, including the Long-billed Curlew, Ruff, White Ibis and Northern Wheatear. About 275 species have been seen in this area.

Short-eared Owl

Buck Creek State Park

This park is located near Springfield, which was named for the abundance of artesian springs in the region that form ecologically interesting wetlands known as fens. Some of these, such as Cedar Bog State Memorial, still exist in the area and harbor many species of rare flora and fauna. Buck Creek State Park comprises just over 4000 acres, with the centerpiece the 2120-acre C.J. Brown Reservoir. Birding is best during migration, though this may be the most reliable Ohio locale for the breeding Bell's Vireo. Lower water levels in the lake can create an abundance of shorebird habitat at the north end, and the beach near the dam is also good for shorebirds and gulls. Waterbird migration is excellent and includes large numbers of Common Loons and sometimes a Red-throated Loon. Rarities spotted include the Least Tern, Piping Plover and American White Pelican.

Clear Creek Metropark

Steep, cool slopes carpeted with hemlock characterize this nearly 5000-acre forested valley, owned and managed by Franklin County Metroparks. The botanical diversity is astounding—over 800 species of plants have been identified, including a number of rarities. The hemlock communities support concentrations of breeding birds normally found far to the north of Ohio, such as the Hermit Thrush, Blue-headed Vireo and Canada, Magnolia and Black-throated Green warblers. Birding is most interesting in late spring and early summer, when the nesters are at the peak of activity. About 120 nesting species have been recorded, including 22 species of warblers. Noteworthy nocturnal species include the Whip-poor-will, and occasionally Chuck-will's-widow.

Whip-poor-will

Conneaut Harbor

Located in the extreme northeastern corner of the state, Conneaut Harbor was created by a series of rock jetties that form a protected embayment. Mudflats lure a great variety and number of shorebirds, although they often don't linger long as they migrate along Lake Erie's shoreline. At least 33 species have been recorded, and uncommon birds such as the Whimbrel, American Avocet and Red Phalarope are regular. Snowy Owls are often found perched on breakwaters in winter, and migratory gulls can be numerous, including rarities such as Black-headed, California and Little gulls. Occasionally a jaeger will make an appearance, and this is one of the best places to find Brant. Almost anything can drop in here, and Conneaut is a great place to just sit and watch the action.

Crown City Wildlife Area

This 11,171-acre site lies just north of the Ohio River and straddles the Gallia-Lawrence county line. Donated to the Ohio Department of Natural Resources by the Richard King Mellon Foundation in 1997, the Crown City Wildlife Area occupies reclaimed strip mine lands and is a mosaic of open grassland, woodlands and brushy thickets with some small ponds. Many areas resemble the African savanna and are quite unlike other Ohio habitats. This is an excellent place for Blue Grosbeaks—as many as 20 pairs can be present and at times one is never out of earshot of a singing male. Dickcissels can also be frequent, as are Henslow's Sparrows, Grasshopper Sparrows and Northern Bobwhite. Black locust thickets harbor Orchard Orioles, White-eyed Vireos, Yellow-breasted Chats and Prairie Warblers. Winter birding can also be good, as Northern Harriers, Rough-legged Hawks and Short-eared Owls use the area. Peregrine Falcons sometimes appear—probably the pair that nests on a nearby bridge across the Ohio River.

Blue Grosbeak

Fairport Harbor and Headlands Dunes State Nature Preserve

This complex of beaches, original dune grass, stone jetties and the open water of Lake Erie is one of the most productive birding locales in the state. The total species list is in excess of 300 and includes many rarities. King Eiders, Common Eiders and Mew Gulls have been spotted, and this may be the best place to find Harlequin Ducks and Purple Sandpipers. Scruffy woods along the south end of the beach can harbor tremendous fallouts of migrant passerines, including all of the regularly occurring warblers. There can be good raptor migrations along the lakeshore in spring and fall, and many diving ducks stop in the harbor, including all three scoter species. A visit to Headlands Dunes also affords the chance to see one of the best remaining natural dune communities along Lake Erie, with a dominance of American beach grass, coastal little bluestem and sand dropseed peppered with rarities such as inland beach pea and seaside spurge.

Funk Bottoms–Killbuck Marsh Wildlife Areas

One of the largest wetland complexes remaining in the state, these two wildlife areas collectively cover over 6900 acres of the Killbuck Valley, in the area where Ashland, Holmes and Wayne counties meet. The area is fantastic for shorebirds and waterfowl in migration, including big flocks of Tundra Swans. If conditions are suitable in early spring, more than 20 species of waterfowl can be present, with flocks numbering in the thousands. Rarities have included the Greater White-fronted Goose and Eurasian Wigeon. Funk Bottoms is one of the best interior shorebird spots in Ohio and can attract tremendous concentrations of waders—at least 33 species have been recorded. Sandhill Cranes are regular migrants and have nested in the area in recent years.

Tundra Swan

Grand Lake St. Marys

When it was constructed as a feeder lake for the Ohio & Erie Canal in the 1830s and 1840s, Grand Lake St. Marys was the largest artificial lake in the world, covering 17,500 acres. Although the nature of the lake has changed considerably since its early days and many of the former habitats no longer exist, Grand Lake still serves as a beacon for migrating birds passing through heavily agricultural western Ohio. In addition to the open waters of the lake, the 1408-acre Mercer Wildlife Area is located on the shoreline and attracts large numbers of migrant waterfowl, particularly geese. Greater White-fronted Geese and Ross's Geese have been spotted here, and a large heron rookery is present, with an active Bald Eagle nest nearby. The St. Marys Fish Hatchery at the eastern end of the lake covers more than 160 acres and attracts waterfowl, shorebirds and gulls. Over 300 species have been tallied around Grand Lake, including unusual species such as the Lesser Black-backed Gull, Red-necked Grebe, Whimbrel and Sabine's Gull. Although not well covered by birders anymore, trips to Grand Lake can be productive and may produce some great birding.

Green Lawn Cemetery–Greenlawn Avenue Dam

Founded in 1849 and covering 360 acres, Green Lawn Cemetery is Ohio's second-largest cemetery, and is lushly carpeted with large trees, forming an oasis for migrants in an otherwise urban setting. Over 230 species have been seen here and fallouts of spring migrants can be breathtaking. Many rarities have been found over the years, including Kirtland's Warbler, Swainson's Warbler, Harris's Sparrow and both crossbill species. A side benefit is the arboretum-like aspect of the cemetery, which has almost every native Ohio tree growing within its confines. The nearby Greenlawn Avenue Dam, on the Scioto River near downtown Columbus, forms a large pool as well as rapids below the dam. This site attracts numerous waterbirds in migration, including at least 23 species of waterfowl. Yellow-crowned Night-Herons and Prothonotary Warblers have nested in the riparian vegetation below the dam. Many interesting birds have appeared here, including the Great Black-backed Gull, Black-legged Kittiwake, Red-necked Grebe and White-winged Scoter.

Yellow-crowned Night-Heron

Hoover Reservoir

Completed in 1955, Hoover Reservoir impounds 3300 acres and has 45 miles of shoreline, much of which is owned by the City of Columbus. The lake has a well-deserved reputation for attracting rarities, including the Long-tailed Jaeger, Western Grebe and Least Tern, in addition to good numbers of migrant loons, ducks, gulls and terns. Most years, lower water levels in autumn produce extensive mudflats at the lake's upper end, and shorebirding is often better here than at any other central Ohio site. At least 31 species of shorebirds have been documented and the mudflats are good for central Ohio rarities such as the Red Knot, American Avocet, Willet, Buff-breasted Sandpiper and both godwit species. Migrant Merlins and Peregrine Falcons occasionally strafe shorebirds. Swamp woods buffering the reservoir support large numbers of breeding Prothonotary Warblers, and terrestrial migrants can be numerous in shoreline woods.

Kelleys Island

A trip to Kelleys Island is an adventure, because the best way to get there is via ferry from Marblehead on the mainland. The boat ride gives you the opportunity to view waterbirds on the open waters of Lake Erie, adding another dimension to the trip. One of 21 islands in the shallow western basin of Lake Erie, Kelleys Island covers nearly 3000 acres with 18 miles of shoreline and is the largest U.S. island in the region. Nearly one-third of the island has been protected, and many of the most significant habitats are now under the ownership of the Department of Natural Resources or the Cleveland Museum of Natural History. Spring and fall passerine migrations can be astounding! One of the best places to observe migrant land birds is North Pond State Nature Preserve. The total list for the island is well over 250 species, and includes great finds such as the Kirtland's Warbler, Harlequin Duck, Golden Eagle, Purple Sandpiper and all three scoter species. Offshore flocks of Buffleheads can number nearly a thousand birds in November and December—one of the densest concentrations on the Great Lakes.

Lorain Harbor

Very few places can boast the number of rarities that have been found at Lorain Harbor over the years. The California Gull, Northern Gannet, Barrow's Goldeneye, Black-headed Gull and others, as well as Ohio's first and only records of Heermann's Gull and the Tufted Duck, have been found here. The harbor was at its peak of birding glory back in the days when the municipal power plant released hot water, keeping the harbor open even during the coldest weather. As this site is situated on Lake Erie, hordes of ducks and gulls were attracted to the shoals of dying gizzard shad that concentrated in the open water. Today, the plant no longer releases warm water, but Lorain birding can be quite good nevertheless. Large flocks of Bonaparte's Gulls appear in November and December, and oftentimes are accompanied by Little Gulls and occasionally a Black-headed Gull. A variety of waterfowl drop into the sheltered harbor and spectacular flocks of Red-breasted Mergansers

Red-breasted Merganser

29

gather offshore in late fall and early winter, perhaps numbering into the tens of thousands. The dredge-spoil impoundment at the harbor's east side has wetland vegetation and scruffy thickets that attract marsh birds and migrant passerines. An occasional Snowy Owl is found along the breakwaters in winter.

Magee Marsh–Metzger Marsh Wildlife Areas and Ottawa National Wildlife Refuge

This wetland complex encompasses over 8000 acres along the western Lake Erie shoreline and is the number one birding destination in Ohio. Well over 100,000 visitors come to the area annually, many of them seeking the spectacular birding opportunities found here. The bird trail at Magee Marsh is legendary and is known to birders nationwide. Spring migration can be unbelievable, with warblers, tanagers, orioles and grosbeaks seeming to festoon every tree. Even ordinary birds become extraordinary—Blue Jays, for example, can form enormous migratory flocks in May. Well over 300 species have been seen in this region, including many rarities. Bald Eagles are ubiquitous and unusual marsh dwellers such as the Yellow-headed Blackbird are regularly found. During waterfowl migrations, the marshes teem with ducks and geese, including large flocks of Tundra Swans. Depending on water levels, a great variety of shorebirds use the marshes, and it's always wise to watch overhead, as tremendous numbers of raptors move through the area when conditions are favorable.

Bald Eagle

Miami-Whitewater Wetlands

Often considered the premier wetland birding locale in southwestern Ohio, the 4279-acre Miami Whitewater Forest is a Hamilton County park with 130 acres of restored wetlands. Rare breeders have included both species of bitterns, Marsh Wrens, Sedge Wrens and Ruddy Ducks. The diversity and number of migrant waterbirds is outstanding, and there is always a chance to find rarities. The Eared Grebe, Greater White-fronted Goose, Little Blue Heron, Glossy Ibis, King Rail, Purple Gallinule and Red Phalarope have all been detected. Grasslands buffer the wetlands and the park district is gradually planting these grasslands to native prairie vegetation. These fields can be good for Northern Harriers, Rough-legged Hawks and Short-eared Owls during winter, and sometimes Dickcissels and Upland Sandpipers are discovered in migration.

Mohican-Memorial State Forest

Located within an easy drive of the major central Ohio cities, this 4500-acre state forest is well known for its scenic attributes, which include the pristine Clear Fork of the Mohican River. This stream carves through rugged topography and steep slopes carpeted with hemlock trees that support rare boreal breeders such as Canada, Magnolia and Blackburnian warblers, Winter Wrens and Hermit Thrushes. Over 100 species of birds have been recorded nesting here, making Mohican one of the most diverse Ohio woodlands for breeding avifauna. Over half of Ohio's 61 species of neotropical birds nest here. The nearby Pleasant Hill Reservoir can be good for migrant loons, waterfowl, gulls, Bald Eagles and Ospreys.

Newell's Run–Ohio River–Willow Island Lock and Dam

Just upstream from Marietta, Newell's Run flows into the Ohio River, where it forms a large, shallow estuary-like bay at the confluence. As the Ohio River is a major flyway for waterbirds, many migrants are enticed to stop and rest in this placid backwater. Good numbers of most of the ducks and geese that pass through Ohio are recorded here, as are many species of shorebirds. Rarities that have been spotted include the Red-necked Grebe, Western Grebe and Tricolored Heron. The deeper waters of the adjacent Ohio River attract diving ducks in migration, including Common Goldeneyes, Lesser Scaup and all three merganser species. Bald Eagles are often seen in the area. Willow Island Lock and Dam is one of a series of dams created to pool the river for shipping purposes, and the turbulent waters downstream from the dam often attract large numbers of gulls. Unfortunately, this area is not well known to many Ohio birders other than the locals, and does not receive the attention that it deserves.

Oak Openings Metropark

The Oak Openings ecosystem, which covers 130 square miles, is situated on relict beach ridges of the preglacial ancestor of Lake Erie and is characterized by dry sandy ridges, wet sedge meadows and open oak savannas. The 3600-acre Oak Openings Metropark, owned by Metroparks of the Toledo Area, is the largest holding, but The Nature Conservancy and Ohio Department of Natural Resources collectively own over 4000 acres. Without a doubt, the Oak Openings is Ohio's most unusual habitat, harboring far more rare plants than any other region in the state, and many rare animals. This is the only regular Ohio breeding locale for the Lark Sparrow, and many birds that are quite hard to find in northwestern Ohio occur, such as the Whip-poor-will, Pine Warbler and Summer Tanager. Golden-crowned Kinglets have nested in the red and white pine plantations, and Blue Grosbeaks have summered in the area. Migratory raptor movements over the Oak Openings can be sensational, as the birds circle around Lake Erie and pass over the area in big numbers. Probably the best chance to see a migratory Golden Eagle is here, as several are normally detected each fall.

Lark Sparrow

Shawnee State Forest

At more than 60,000 acres, Shawnee is the largest state forest in Ohio. Situated in the rugged hills of Scioto and Adams counties along the Ohio River, the area is reminiscent of the Appalachian Mountains and is sometimes called "Ohio's Little Smokies." The forest is incredibly diverse botanically, harboring around 1000 species of vascular plants, including many very rare species. Perhaps the greatest concentrations of breeding interior forest birds are found here, and over 100 species of birds have been documented as nesting. Flashy jewels such as the Hooded Warbler and Scarlet Tanager are commonplace, and more localized and secretive birds such as the Worm-eating Warbler can be found in abundance. A network of little-traveled roads allows easy exploration of the forest; driving these early in the morning often yields surprises such as a gray fox or Ruffed Grouse along the roadway. In May, fantastic wildflower displays complement the birding.

Sheldon Marsh State Nature Preserve

This site contains a diversity of habitats, including an extensive barrier beach, mature woodlands, open fields, large marshes and mudflats. Although Sheldon Marsh is one of the best birding sites along Lake Erie, the more widely known Magee Marsh to the west often overshadows it. Nevertheless, over the years birders have compiled an extensive species list, numbering nearly 300 species, from the 465-acre Sheldon Marsh. Spring and fall movements of land birds can be stellar, and unusual species such as the Golden-winged Warbler are frequently detected. The grapevine tangles along the trails and the few conifers near the entrance are always worth scrutiny, as Northern Saw-whet Owls sometimes stop over and spend time in the dense cover. Although much of the barrier beach along Lake Erie is closed to foot traffic, good views can still be had from the eastern end, and a variety of gulls and terns often roost on the sand. Bald Eagles are almost always in evidence and many species of ducks gather offshore in migration. The large mudflats in the marsh behind the beach have attracted at least 32 species of shorebirds, including the American Avocet, Red Knot and Red-necked Phalarope.

Northern Saw-whet Owl

The Wilds

The Wilds is so unlike any other Ohio habitat that, upon entering, you might think you have somehow been transported to the savannas of Africa. This impression would only be reinforced by the presence of free-ranging rhinos, giraffes and fringe-eared oryx, among other very exotic beasts. Established in 1986, The Wilds is a conservation and research facility dedicated to preserving imperiled animals. The strip-mined lands have been reclaimed into gently rolling hills carpeted with grasslands and dotted with small ponds. Grassland species such as Grasshopper, Henslow's and Savanna sparrows are plentiful. Dickcissels occur occasionally, and Eastern Meadowlarks and Bobolinks serenade birders throughout the area. Winter raptor watching can be unbelievable—one year, more than 60 Short-eared Owls were present! Rough-legged Hawks, both light and dark phases, and Northern Harriers are always about in winter. In recent years, two or three Golden Eagles have overwintered and can often be found fairly easily, making The Wilds the most reliable place in Ohio to find this difficult-to-locate species. In February 2004, a Prairie Falcon was discovered and widely seen—only the second Ohio record!

Northern Harrier

ABOUT THE SPECIES ACCOUNTS

This book gives detailed accounts of the 295 species of birds that are considered regular by the Ohio Bird Records Committee; these species can be expected annually. An additional 117 species have been recorded in Ohio, giving the state a list of 412 species to date. Of the rarities, 51 of the species that appear more regularly are briefly discussed in an illustrated appendix. Most of these can be expected every one to five years, and there are multiple records of all of them. The remaining 66 species are included in the checklist at the back of the book. These are not treated in the book, as most of these birds are very rare vagrants that have been spotted only once or a few times in the state, and can't predictably be expected to occur in Ohio.

Yellow-headed Blackbird

The order of the birds and their common and scientific names follow the American Ornithologists' Union's *Check-list of North American Birds* (7th edition) and the *Forty-fourth Supplement 2003*. Readers familiar with other bird guides will note that the species accounts now start with geese, swans and ducks, rather than with loons. This rather radical shift in phylogenetic order results from recent studies that have determined that waterfowl actually represent the least-evolved group of birds, followed by grouse, turkeys and other species in the family *Phasianidae*. As with any good bird guide, bird families are placed in phylogenetic order, starting with the most primitive groups and ending with the most modern.

One of the innovations of this book is that space is allocated in each species account to describe interesting traits of character, life history and other facts. In order to provide more information about each species, fewer illustrations have been used, so only the most typical plumages that a birder could expect to see in Ohio are depicted. Many species have other plumages that depend on whether the bird is in breeding (alternate) or nonbreeding (basic) plumage, or is male or female, juvenile or immature. To try to depict all plumages for some species would require too much space and would completely shift the focus of this book.

ID: While illustrations are very helpful in visualizing what a bird looks like, written descriptions also serve a purpose in describing a bird's appearance. In conjunction with the illustrations, these descriptors should give a good impression of the bird in question. Where appropriate, the description is subdivided to highlight the differences between males and females, breeding and nonbreeding plumages, and immature and adult birds. The use of highly technical terms has been minimized; rather, words such as "eyebrow," "chin," and "jaw line" have been used, as most readers easily understand them. Some of the most common features of birds are pointed out in the Glossary illustration (p. 345).

Size: The size measurement—average length from bill to tip of tail—gives a feel for the size of the bird as it is seen in nature. Larger birds often have measurements given as a range, as there is variation among individuals, particularly between males and females. Note that birds with long tails often have large measurements that don't necessarily reflect "body" size. In addition, wingspan (wing tip to wing tip measurement) is given for all birds.

Habitat: In many cases, a general habitat description serves to illustrate where a species is typically found. In other cases, species tend to occupy a specific habitat niche; in these cases the habitat will be described with precision. Of course, birds have wings and may turn up in odd places not associated with typical habitats, particularly in migration. However, most species require a specific suite of environmental conditions that offer appropriate vegetation cover, food, water and, in the case of breeders, nesting cover—this is where these species will normally be found.

Nesting: In all cases where the species breeds in Ohio, specifics of nesting ecology are discussed. Each account includes descriptions of nest structure and location, clutch size, incubation period, and parental duties. Remember that birding ethics prohibit the disturbance of active nests. If you disturb a nest, you may drive off the parents during a critical period or expose the defenseless young to predators. The nesting behavior of birds that do not nest in Ohio is not described.

Feeding: Birds spend a great deal of time foraging for food. Knowledge of what a bird eats and where it forages can be very useful in locating that species in the field; by following this information, your chances of locating a particular species will improve. Also, diet can provide information that may help to support an identification.

Voice: The very best birders are those who know vocalizations, and with good reason. By knowing songs and calls, you'll find many more birds than the birder who is oblivious to bird sounds. Many species are real skulkers, remaining hidden in thick cover, but by knowing their calls, you'll be able to locate them. Mnemonics—memory aids—can help birders to remember songs, and we've often employed such aids in this book. Attempting to convert a bird song to human words doesn't always work, but in many cases it provides an easy and understandable way to remember the song.

Red-tailed Hawk

Similar Species: To the extent practicable, easily confused species are discussed. Fortunately, in many cases there are only a few species that might readily be confused with the bird in question, but in some cases it gets trickier. For instance, space limitations preclude a thorough analysis of similar gulls in all plumages—to do so might require several pages for certain species! Most birds won't be nearly as problematic as gulls, though, and consulting this section should prove useful in confirming a bird's identity.

Selected Sites: Some sites are better than others for finding a particular species, and this section notes areas that tend to be good locales for the bird in question. Some species, such as the Mourning Dove, are wide-ranging and ubiquitous, so singling out specific sites in which to look for them is hardly necessary. Other birds tend to favor very specific places, even though the range map may depict a broad distribution. All of the sites mentioned are easily accessible, and for the most part access is not restricted.

Seasonal Abundance Graphs: A general comment, such as "common," "uncommon" or "rare" often suffices to describe the relative abundance of a species. However, status can vary considerably throughout the course of a year, and the bar graph included for each bird should be very helpful in determining the abundance for each species at different times of year. These graphs are broken into weekly intervals, and thus are able to accurately depict the arrival and departure periods for migrants, as well as their typical status when present. The graphs depict three abundance levels as follows:

Common: to be expected, sometimes in large numbers.

Uncommon: observed infrequently, usually in small numbers.

Rare: normally occurs annually, but with only a few records on average.

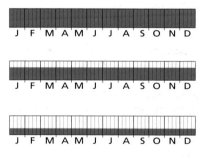

We wish to thank the Ohio Bird Records Committee for their kind permission to use the graphs from their "Checklist of the Birds of Ohio."

Range Maps: The range map for each species represents the typical distribution in an average year. Most birds will confine their annual movements to this area, although each year some individuals will wander beyond the depicted boundaries. Differences in abundance, from common to uncommon, within the mapped area are not generally shown—some regions may support significantly higher populations owing to better habitat, while in others the species might be uncommon. We have used crosshatching to indicate areas where a species is regular, but rare to casual in occurrence. Areas of very rare or limited distributions, such as rare breeders that may only nest in a particular region on an irregular basis, may not be shown. These maps, when used in tandem with the bar graphs of seasonal abundance, should give an accurate portrayal of each species' status within the state.

Unlike many other field guides, we have attempted to show migratory pathways—areas of Ohio where birds tend to concentrate on their way to or from breeding or wintering grounds. Sometimes these corridors are not well defined, and the mapped areas tend to show the regions in which the majority of documented sightings occur. Exceptional flights may bring many observations of a particular species away from mapped areas.

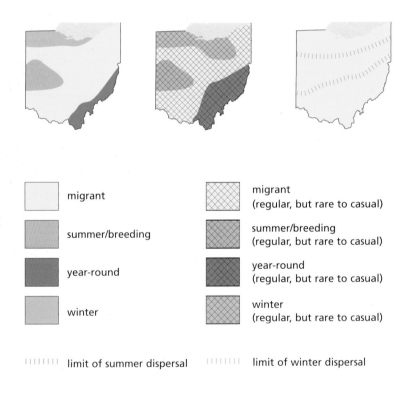

migrant

summer/breeding

year-round

winter

migrant
(regular, but rare to casual)

summer/breeding
(regular, but rare to casual)

year-round
(regular, but rare to casual)

winter
(regular, but rare to casual)

limit of summer dispersal   limit of winter dispersal

NONPASSERINES

Waterfowl

Grouse & Allies

Diving Birds

Heronlike Birds

Birds of Prey

Rails, Coots & Cranes

Shorebirds

Gulls & Allies

Doves & Cuckoos

Owls

Nightjars, Swifts & Hummingbirds

Woodpeckers

Nonpasserine birds represent 18 of the 19 orders of birds found in Ohio, about 57 percent of the species in our region. They are grouped together and called "nonpasserines" because, with few exceptions, they are easily distinguished from the "passerines," or "perching birds," which make up the 19th order. Being from 18 different orders, however, means that nonpasserines vary considerably in their appearance and habits—they include everything from the 5-foot-tall Great Blue Heron to the 3¾-inch-long Ruby-throated Hummingbird.

Generally speaking, nonpasserines do not "sing." Instead, their vocalizations are referred to as "calls." There are also other morphological differences. For example, the muscles and tendons in the legs of passerines are adapted to grip a perch, and the toes of passerines are never webbed. Many nonpasserines are large, so they are among our most noticeable birds. Waterfowl, raptors, gulls, shorebirds and woodpeckers are easily identified, at least to group, by most people. Some of the smaller nonpasserines, such as doves, swifts and hummingbirds, are frequently thought of as passerines by novice birders, and can cause those beginners some identification problems. With a little practice, however, they will become recognizable as nonpasserines. By learning to separate the nonpasserines from the passerines at a glance, birders effectively reduce by half the number of possible species for an unidentified bird.

GREATER WHITE-FRONTED GOOSE

Anser albifrons

Greater White-fronted Geese breed on the arctic tundra and winter in the southern U.S. and Mexico. Most Ohio sightings are of spring migrants, usually single birds or very small groups, though fall and winter sightings have been increasing. The largest flock in Ohio, reported on March 16, 1985, consisted of 97 birds. • White-fronts often travel among flocks of Canada Geese. The bright orange feet of the slightly smaller White-fronted Geese shine like beacons as the birds stand on fields and frozen wetlands. • With a nearly circumpolar distribution, the Greater White-fronted Goose has the widest range of any species in its genus. It is the only New World representative of the five species of gray geese found in the Old World. • Like most geese, White-fronts are long-lived birds that mate for life, with both parents caring for the young. • There have been several records of the Greenland subspecies *flavirostris* in Ohio; it is perhaps best recognized by its orange bill.

ID: brown overall; black speckling on belly; pinkish bill; white around bill and on forehead; white hindquarters; black band on upper tail; orange feet. *Immature:* pale belly without speckles; little or no white on face.
Size: *L* 27–33 in; *W* 4½–5 ft.
Habitat: croplands, fields, open areas and shallow marshes in migration.
Nesting: does not nest in Ohio.
Feeding: dabbles in water and gleans the ground for grass shoots, sprouting and waste grain and occasionally aquatic invertebrates.
Voice: high-pitched "laugh."
Similar Species: *Canada Goose* (p. 40): white "chin strap"; black neck; lacks speckling on belly. *Snow Goose* (p. 39): blue morph has white head and upper neck and all-dark breast and belly.
Best Sites: Killdeer Plains WA; Big Island WA; western L. Erie marshes. May appear anywhere there are large concentrations of Canada Geese.

J F M A M J J A S O N D

SNOW GOOSE
Chen caerulescens

Ohio lies between major flyways used by the Snow Goose, and thus normally doesn't get enormous concentrations of this species. Typically singles or small groups are seen here, so it's hard to imagine the flock of 150,000 seen in Mahoning County on October 25, 1952. In recent years, Snow Goose populations have increased dramatically in North America, as they have taken advantage of human-induced changes in the landscape and in the food supply. • Snow Geese grub for their food, often targeting the underground parts of plants. Their strong, serrated bills are well designed for pulling up the roots of marsh plants and for gripping slippery grasses. • Unlike Canada Geese, which fly in V-formations, migrating Snow Geese usually form oscillating, wavy lines. • As with Sandhill Cranes, Snow Goose plumage is often stained rusty red from iron in the water. This species has two color morphs, a white and a blue, that until 1983 were considered different species. The scientific name *caerulescens* means "bluish" in Latin and describes the blue morph.

blue morph

ID: white overall; black wing tips; pink feet and bill; dark "grinning patch" on bill; plumage is occasionally stained rusty red. *Blue morph:* white head and upper neck; dark, blue gray body. *Immature:* gray or dusty white plumage; dark bill and feet.
Size: *L* 28–33 in; *W* 4½–5 ft.
Habitat: shallow wetlands, lakes and fields.
Nesting: does not nest in Ohio.

Feeding: grazes on waste grain and new sprouts; also eats aquatic vegetation, grasses, sedges and roots.
Voice: loud, constant, nasal *houk-houk* in flight.
Similar Species: *Ross's Goose* (p. 330): smaller; shorter neck; stubbier bill; lacks dark "grinning patch." *Tundra* (p. 43), *Mute* (p. 42) and *Trumpeter swans:* much larger; white wing tips.
Best Sites: western L. Erie marshes; Ottawa NWR; larger interior wetlands and reservoirs.

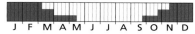

J F M A M J J A S O N D

CANADA GOOSE

Branta canadensis

The Canada Goose is now possibly Ohio's most widely recognized bird—it's hard to believe that this species was once restricted to migrants passing through the state. By the 1940s, overhunting and other factors had reduced numbers to perilously low levels, and various game agencies embarked on a recovery program. Unfortunately, this effort included introducing geese to areas in which they had not formerly bred, such as Ohio. These geese, like many waterfowl, domesticate easily, and now breeding birds can be found almost anywhere. Most resident Canada Geese are largely nonmigratory and display little or no fear of people. They can also be quite a nuisance as they despoil lawns, golf courses and parks with their prodigious guano. • Ohio still gets large, spectacular migrant flocks, which are much wilder and can number in the thousands. The collective honking of these huge gaggles, which resonates for long distances and is truly a sound of the wilderness, heralds their presence. Rarely, one of the wild flocks will have a diminutive "Richardson's Goose," which is mallard-sized but otherwise identical to our "Giant" subspecies.

ID: long, black neck; white "chin strap"; white undertail coverts; light brown underparts; dark brown upperparts; short, black tail; black bill and legs.
Size: *L* 25–45 in; *W* 3½–5 ft.
Habitat: lakeshores, riverbanks, ponds, farmlands and city parks.
Nesting: on an island or shoreline; usually on the ground, a muskrat lodge or "gander lander" nesting platform; may use a heron rookery; female builds a nest of plant material lined with down and incubates 3–8 white eggs for 25–28 days while the male stands guard.

Feeding: grazes on new sprouts, aquatic vegetation, grass and roots; tips up for aquatic roots and tubers.
Voice: loud, familiar *ah-honk*, often answered by other Canada Geese.
Similar Species: *Greater White-fronted Goose* (p. 38): brown neck and head; orange legs; white around base of bill; dark speckling on belly; lacks white "chin strap." *Brant* (p. 41): white "necklace"; black upper breast; lacks white "chin strap."
Best Sites: almost anywhere; spectacular movements of "wild" migrants can be seen in western L. Erie marshes.

J F M A M J J A S O N D

BRANT

Branta bernicla

This little goose is primarily coastal, with small numbers passing over Lake Erie en route to or from their high-arctic breeding grounds. There are two principal populations: the "Black" Brant of the Pacific coast, and the "Atlantic" Brant, which winters on the Atlantic coast. • Formerly, the Brant depended heavily on eelgrass, a plant that was decimated by disease in the 1930s, thus triggering massive declines in Brant populations. Fortunately, eelgrass eventually rebounded and Brant also began to utilize alternative foods, a strategy that kept the population from disappearing altogether; their numbers are now stable. • In Ohio, Brant can be found along Lake Erie but are very rare inland. • Both "Brant" and *Branta* are derived from the Anglo-Saxon word for "burned" or "charred," a reference to this bird's dark plumage. The species name *bernicla* means "barnacle." These geese are so closely allied with the sea that legend has it that Brant are hatched from the shells of these crustaceans.

ID: black neck, head and upper breast; dark upperparts; white "necklace"; white hindquarters; black feet; pale brown sides and flanks; white belly. *Juvenile:* "necklace" is less conspicuous.
Size: *L* 24–28 in; *W* 3½ ft.
Habitat: L. Erie shoreline; rarely inland reservoirs or ponds.
Nesting: does not nest in Ohio.
Feeding: grazes on aquatic and wetland vegetation; elsewhere, eats eelgrass almost exclusively.

Voice: deep, prolonged *c-r-r-r-uk*, with hissing.
Similar Species: *Canada Goose* (p. 40): white "chin strap"; brown upperparts and upper breast. *Greater White-fronted Goose* (p. 38): brown neck and head; orange legs; white around base of bill; dark speckling on belly.
Best Sites: L. Erie shoreline dredge-spoil impoundments such as at Huron Municipal Pier and Conneaut Harbor; lawns near the beach at Maumee Bay SP.

J F M A M J J A S O N D

MUTE SWAN

Cygnus olor

This large Eurasian native was introduced to eastern North America in the mid-1800s to adorn estates and city parks. Over the years, the Mute Swan has adapted well to North American wetlands. The first "wild" Ohio nesting took place in 1987—the species is now locally common along Lake Erie and is increasing in numbers. • Although this species is not usually migratory, more northerly nesters have established short migratory routes to milder wintering areas. • As Mute Swans have increased in number, there has been more concern expressed about their impact. Like many nonnative species, Mute Swans are often fierce competitors for nesting areas and food sources. They can be very aggressive toward geese and ducks, sometimes displacing native species. Mute Swans also adversely affect aquatic vegetation by overgrazing. • A reliable long-distance characteristic for distinguishing the Mute Swan from other swans is the way it holds its neck in a graceful curve with its orange bill hanging down.

ID: all-white plumage; orange bill with downturned tip and bulbous, black knob at base; neck usually held in an S-shape; wings often held in an arch over back while swimming. *Immature:* plumage may be white to grayish brown.

Size: *L* 5 ft; *W* 7½–8 ft.

Habitat: freshwater marshes, lakes and ponds.

Nesting: on the ground along a shoreline; female builds a mound of vegetation; male may help gather material; female incubates 5–10 pale green eggs for about 36 days; pair tends the young.

Feeding: tips up or dips its head below the water's surface for aquatic plants; grazes on land.

Voice: generally silent; may hiss or issue hoarse barking notes; loud wingbeats can be heard for long distances.

Similar Species: *Tundra Swan* (p. 43) and *Trumpeter Swan:* lack orange bill with black knob at base; hold necks straight; shorter tails.

Best Sites: Sandusky Bay; Medusa Marsh; East Harbor SP.

J F M A M J J A S O N D

TUNDRA SWAN

Cygnus columbianus

The Tundra Swan is Ohio's only native swan, and the only swan to be encountered in large flocks. A true harbinger of spring, the Tundra Swan passes through Ohio in late February and March on its way to its nesting grounds in the High Arctic. Flocks can sometimes number in the thousands, but groups of a dozen or two are more typical. This species is also a common fall migrant, with peak numbers occurring in November and December along Lake Erie. • Formerly known as "Whistling Swan," this massive bird can weigh up to 16 pounds. • All swans require a considerable distance to "run" across the water to gain flight, which makes Niagara Falls—a favorite stopover—particularly hazardous. In some years, the rapid current of the Niagara River has swept many birds over the falls to their deaths before they could become airborne. • In the early 19th century, members of the Lewis and Clark expedition collected the first specimen of this bird near the Columbia River, thus its scientific name *columbianus*.

ID: white plumage; large, black bill; black feet; often shows yellow lores; neck is held straight up; neck and head show rounded, slightly curving profile.
Immature: gray brown plumage; gray bill.
Size: L 4–5 ft; W 6½ ft.
Habitat: shallow areas of lakes and wetlands, agricultural fields and flooded pastures.
Nesting: does not nest in Ohio.
Feeding: tips up, dabbles and surface gleans for aquatic vegetation and aquatic invertebrates; grazes for tubers, roots and waste grain.
Voice: high-pitched, quivering *oo-oo-whoo*, constantly repeated by migrating flocks.
Similar Species: *Trumpeter Swan:* larger; loud, bugling voice; lacks yellow lores; neck and head have more angular profile. *Mute Swan* (p. 42): orange bill with black knob on upper base; neck is usually held in an S-shape; downpointed bill; often holds wings in an arch over back while swimming.
Best Sites: western L. Erie marshes; Killdeer Plains WA; migrant flocks occur almost anywhere along L. Erie shoreline.

J	F	M	A	M	J	J	A	S	O	N	D

WOOD DUCK

Aix sponsa

Six species of North American ducks regularly nest in cavities, and the Wood Duck is the most successful of the lot, often using natural cavities such as Pileated Woodpecker nest holes. • Wood Ducks are quite common in Ohio today, but in the early 1900s populations had plummeted so badly that some ornithologists were predicting that the species might be extinct before 1930. Loss of woodland and wetland habitats played a major role in this decline, but the popular practice of erecting nest boxes for the Wood Duck has since aided population growth in many areas. The beaver has also helped boost Wood Duck numbers. As that large rodent has become more prevalent, so have the Wood Duck-friendly wetlands that it creates. • Baby "Woodies" have a quick and rude introduction to life's realities. Within 24 hours of hatching, they must leap to the ground from their nest, which may be 75 feet high or more in a tree. Then, it can be quite a trek to the nearest water, which is why the female and young are sometimes seen in woods far from any water body.

ID: *Male:* glossy, green head with some white streaks; crest is slicked back from crown; white "chin" and throat; purple chestnut breast spotted with white; black-and-white shoulder slash; golden sides; dark back and hindquarters. *Female:* distinctive, white, teardrop-shaped eye patch; white streaking on mottled brown breast; brownish gray upperparts; white belly.
Size: *L* 15–20 in; *W* 30 in.
Habitat: beaver ponds, swamps, streams, marshes and lakeshores with wooded edges.
Nesting: in a hollow or tree cavity (may be 30 ft or more above the ground); also in an artificial nest box; usually near water; lines cavity with down; female incubates 9–14 white to buff eggs for 25–35 days.
Feeding: gleans the water's surface and tips up for aquatic vegetation, especially duckweed, aquatic sedges and grasses; eats more fruits and nuts than other ducks.
Voice: *Male:* ascending *ter-wee-wee*. *Female:* squeaky *woo-e-e-k*.
Similar Species: males are unique and unmistakable. *Hooded Merganser* (p. 66): long tails similar in flight, but fly with head down.
Best Sites: Blackhand Gorge SNP; Killbuck Marsh WA. Canoeing streams is a sure way to see Wood Ducks, but they can be found on most lakes and marshes, at least in migration.

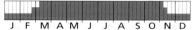

J F M A M J J A S O N D

GADWALL

Anas strepera

Male Gadwalls lack the striking plumage of most other male ducks, but they nevertheless have a dignified appearance and a subtle beauty. Once you learn their field marks, a black rump and white wing patches, Gadwalls are surprisingly easy to identify. • Ducks in the genus *Anas,* the dabbling ducks, are most often observed tipping up their hindquarters and submerging their heads to feed, but Gadwalls dive more often than others of this group. These ducks feed equally during the day and night, a strategy that reduces the risk of predation because the birds avoid spending long periods of time sleeping or feeding. • Gadwall numbers have greatly increased in our area since the 1950s, and this duck has expanded its range throughout North America. The majority of Gadwalls winter on the Gulf Coast of the U.S. and Mexico, though increasing numbers overwinter on inland lakes across the country, including the Great Lakes region. In Ohio, Gadwalls are common migrants throughout the state and very rare nesters in the western Lake Erie marshes.

ID: white speculum; white belly. *Male:* mostly gray; black hindquarters; dark bill. *Female:* mottled brown overall; brown bill with orange sides.

Size: L 18–22 in; W 33 in.

Habitat: shallow wetlands, lake borders and beaver ponds.

Nesting: in tall vegetation, sometimes far from water; nest is well concealed in a scraped-out hollow, often with grass arching overhead; nest is made of grass and other dry vegetation and lined with down; female incubates 8–11 white eggs for 24–27 days.

Feeding: dabbles and tips up for water plants; grazes on grass and waste grain during migration; also eats aquatic invertebrates, tadpoles and small fish; one of the few dabblers to routinely dive for food.

Voice: *Male:* simple, singular quack; often whistles harshly. *Female:* high *kaak kaaak kak-kak-kak,* given in series and oscillating in volume.

Similar Species: *American Wigeon* (p. 47): green speculum; male has white forehead and green swipe trailing from each eye; female lacks black hindquarters. *Mallard* (p. 49), *Northern Pintail* (p. 52) and *other dabbling ducks* (pp. 44–53): generally lack white speculum, black hindquarters of male Gadwall and orange-sided beak of female.

Best Sites: Castalia Pond, Pickerington Ponds MP, or most marshlands in migration; small numbers overwinter at warm water outlets of L. Erie power plants.

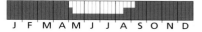

J F M A M J J A S O N D

EURASIAN WIGEON

Anas penelope

Finding a Eurasian Wigeon among packs of migrant dabbling ducks in a marsh is always a treat—only a few are found each year in Ohio. The males are conspicuous and easily identified, and even the novice birder will have no problem recognizing a drake. Females, however, probably go largely undetected, as they are quite similar to female American Wigeons. They are best separated by the underwing color—gray in Eurasian, white in American—but this is difficult to see in the field. • The Eurasian Wigeons seen in Ohio are vagrants from across the Atlantic, and are most often seen in association with their American cousins. Conversely, American Wigeons are also wanderers—they have been seen among flocks of Eurasian Wigeons in Europe and eastern Russia. • Although Eurasian Wigeons are not recorded breeders in North America, the increased number of sightings each spring has convinced some people that there is a small breeding population somewhere in Canada, possibly in Ontario.

ID: black-tipped, gray blue bill. *Male:* rufous head; cream forehead; rosy breast; gray sides; black hindquarters; dark feet. *Female:* rufous hints on predominantly brown head and breast. *In flight:* large, white wing patch; dusky gray "wing pits."
Size: *L* 17–21 in; *W* 32 in.
Habitat: shallow wetlands, lake edges and ponds.
Nesting: does not nest in Ohio.

Feeding: primarily vegetarian; dabbles and grazes for grass, leaves and stems; occasionally pirates food from American Coots.
Voice: high-pitched, 2-note whistle.
Similar Species: *American Wigeon* (p. 47): white "wing pits"; male lacks reddish-brown head; female may have browner head.
Best Sites: L. Rockwell; Medusa Marsh; can appear rarely anywhere in the state; found within large concentrations of American Wigeons.

J F M A M J J A S O N D

AMERICAN WIGEON

Anas americana

Hunters have coined colorful nicknames for many waterfowl, and the American Wigeon's is "Baldpate," in reference to its white crown and forehead. • American Wigeons give a characteristic, three-syllable whistle that is easily heard among the cacophony of sounds in a busy marsh. Indeed, "wigeon" is French, meaning "whistling duck." • Wigeons seem to have an insatiable appetite for the roots and leaves of deepwater submergent plants, which is a drawback for dabbling ducks that cannot dive very well. To satisfy their cravings, American Wigeons have become avian pirates, often snatching succulent vegetation from just-surfaced divers such as Lesser Scaup, Redheads and Canvasbacks. • Unlike most ducks, the American Wigeon is a good walker and is commonly observed grazing on shore. • Although a common migrant statewide in Ohio, the American Wigeon is also a rare nester in western Lake Erie marshes.

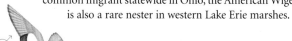

ID: large, white wing patch; cinnamon breast and sides; white belly; black-tipped, gray blue bill; green speculum; white "wing pits." *Male:* white forehead; green swipe extends back from eye. *Female:* grayish head; brown underparts.
Size: *L* 18–22½ in; *W* 32 in.
Habitat: shallow wetlands, lake edges and ponds.
Nesting: always on dry ground, often far from water; nest is well concealed in tall vegetation and is built with grass, leaves and down; female incubates 8–11 white eggs for 23–25 days.
Feeding: dabbles and tips up for aquatic leaves and the stems of pondweeds; also grazes and uproots young shoots in fields; may eat some invertebrates; occasionally pirates food from other birds.
Voice: *Male:* nasal, frequently repeated whistle: *whee WHEE wheew. Female:* soft, seldom-heard quack.
Similar Species: *Gadwall* (p. 45): white speculum; lacks large, white forewing patch; male lacks green eye swipe; female has orange swipes on bill. *Eurasian Wigeon* (p. 46): gray "wing pits"; male has rufous head, cream forehead and rosy breast; lacks green eye swipe; female usually has browner head.
Best Sites: Castalia Pond; Funk Bottoms WA; Ottawa NWR.

J F M A M J J A S O N D

AMERICAN BLACK DUCK

Anas rubripes

Once a very common duck, the American Black Duck population was estimated at 800,000 birds in the 1950s, but had declined to 300,000 by the 1990s. Unusual in the modern era of increasingly science-based wildlife management, overhunting was a primary reason for the decline. A lawsuit by the Humane Society of the U.S. in 1983 forced severe changes to the legal harvest, and American Black Duck numbers have since slowly risen. • Mallards frequently hybridize with American Black Ducks, and this is the most common wild hybrid waterfowl seen in Ohio. It was thought that "genetic swamping" owing to interbreeding between these two species was contributing to declines in American Black Duck numbers, but this doesn't appear to be a major factor. • Male and female American Black Ducks are remarkably similar in appearance, which is unusual for waterfowl. • This duck is a rare nester in the northern third of Ohio. • The scientific name *rubripes* means "red foot" in Latin.

ID: dark, brown black body; light brown head and neck; bright orange feet; violet speculum. *Male:* yellow olive bill. *Female:* dull green bill mottled with gray or black. *In flight:* whitish underwings; dark body.
Size: *L* 20–24½ in; *W* 35 in.
Habitat: lakes, wetlands, rivers and agricultural areas.
Nesting: usually on the ground among clumps of dense vegetation near water; female fills a shallow depression with plant material and lines it with down; female incubates 7–11 white to greenish buff eggs for 26–28 days; second clutches are common, usually to replace lost broods.
Feeding: tips up and dabbles in shallows for seeds and roots of pondweeds; also eats aquatic invertebrates, larval amphibians and fish eggs.
Voice: *Male:* a croak. *Female:* a loud quack.
Similar Species: *Mallard* (p. 49): white belly; blue speculum bordered with black and white bars; female is lighter overall and has white tail. *Gadwall* (p. 45): black hindquarters; white speculum.
Best Sites: *In migration:* western L. Erie marshes; common statewide. *Winter:* Scioto R. south of Columbus.

J F M A M J J A S O N D

MALLARD

Anas platyrhynchos

This abundant and adaptable species ranges throughout the North American continent, and is probably the most recognizable duck in the world. • With the exception of the Muscovy Duck *(Cairina moschata)*, all domestic ducks share their genetic heritage with Mallards, and all manner of feral, Mallard-derived crosses with other ducks are encountered. • Reflecting its high fecundity and versatility, Mallards are the most heavily hunted of North American waterfowl, yet their overall population remains high. Mallards are far more abundant in Ohio now than in the early 20th century or even prior to European settlement. Early Ohio ornithologist Lynds Jones, writing in 1903, said of the Mallard that it was "absent in many localities and mostly seen in small flocks." This is hard to believe today, when Mallards occupy every type of wetland and water body—sometimes even visiting swimming pools!

ID: dark blue speculum bordered with white; orange feet. *Male:* glossy, green head; yellow bill; chestnut brown breast; white "necklace"; gray body plumage; black tail feathers curl upward. *Female:* mottled brown overall; orange bill spattered with black.
Size: *L* 20–27½ in; *W* 35 in.
Habitat: lakes, wetlands, rivers, city parks, agricultural areas and sewage lagoons.
Nesting: in tall vegetation or under a bush, often near water; nest of grass and other plant material is lined with down; female incubates 7–10 light green to white eggs for 26–30 days.

Feeding: tips up and dabbles in shallows for the seeds of sedges, willows and pondweeds; also eats aquatic invertebrates, larval amphibians and fish eggs.
Voice: *Male:* deep, quiet quacks. *Female:* loud quacks; very vocal.
Similar Species: *Northern Shoveler* (p. 51): much larger bill; male has white breast. *American Black Duck* (p. 48): darker than female Mallard; purple speculum lacks white borders.
Best Sites: any wetland, pond or lake; large seasonal concentrations at Castalia Pond and Blendon Woods MP.

J	F	M	A	M	J	J	A	S	O	N	D

BLUE-WINGED TEAL

Anas discors

Blue-winged Teals are very common migrants, but they can sometimes be overlooked because they prefer to lurk in wetland vegetation rather than frequent open water. They are one of the last ducks to return in spring and the last to establish pair bonds, often waiting until they are on their breeding grounds. They nest sparingly throughout areas of Ohio that were glaciated, more commonly to the north. • Most Blue-winged Teals are probably produced in the great "duck factories" of the famed Prairie Pothole region of the Great Plains. This area is liberally dotted with small wetlands, and drought years negatively affect reproductive success. Blue-winged Teal numbers dropped to near record lows in 1990 after several successive dry years, but have rebounded nicely since then. • Despite the similarity of their names, the Green-winged Teal is not the Blue-winged Teal's closest relative. The Blue-winged Teal is more closely related to the Cinnamon Teal and the Northern Shoveler.

ID: *Male:* blue gray head; white crescent on face; black-spotted breast and sides. *Female:* mottled brown overall. *In flight:* blue forewing patch; green speculum.

Size: *L* 14–16 in; *W* 23 in.

Habitat: shallow lake edges and wetlands; prefers areas of short but dense emergent vegetation.

Nesting: in grass along shorelines and in meadows; nest is built with grass and considerable amounts of down; female incubates 8–13 white eggs (may be tinged with olive) for 23–27 days.

Feeding: gleans the water's surface for sedge and grass seeds, pondweeds, duckweeds and aquatic invertebrates.

Voice: *Male:* soft *keck-keck-keck. Female:* soft quacks.

Similar Species: *Green-winged Teal* (p. 53): female has smaller bill, black-and-green speculum and lacks blue forewing patch. *Northern Shoveler* (p. 51): much larger bill with paler base; male has green head and lacks spotting on body.

Best Sites: Big Island WA; Gilmore Ponds Preserve; Springville Marsh SNP; virtually any marshy area.

| J | F | M | A | M | J | J | A | S | O | N | D |

NORTHERN SHOVELER

Anas clypeata

At first glance, the male Northern Shoveler resembles a male Mallard with an extremely large bill. A closer look, however, will reveal other differences—the Northern Shoveler has a white breast and chestnut flanks, while the Mallard has a chestnut breast and white flanks. • A large, spoonlike bill allows this handsome duck to strain small invertebrates from the water and from the mud on the bottoms of ponds. The shoveler's specialized feeding strategy means that it is rarely seen tipping up, but is more often found in the shallows of ponds and marshes where the mucky bottom is easier to reach. • Northern Shovelers are global players in the waterfowl world, breeding not only in North America, but also Britain, central Europe, Asia and Siberia. They are not players regarding fidelity—pairs remain attached for longer than any other species of duck. Males are also extremely territorial, and will not tolerate the intrusion of other shovelers. • Northern Shovelers are common migrants in Ohio.

ID: large, spatulate bill; blue forewing patch; green speculum. *Male:* green head; white breast; chestnut brown flanks. *Female:* mottled brown overall; orange-tinged bill.
Size: *L* 18–20 in; *W* 34 in.
Habitat: shallow marshes and lakes with muddy bottoms and emergent vegetation, usually in open and semi-open areas.
Nesting: in a shallow hollow on dry ground, usually within 160 ft of water; female builds the nest with dry grass and down and incubates 10–12 pale, greenish buff eggs for 21–28 days.

Feeding: dabbles in shallow and often muddy water; strains out plant and animal matter, especially aquatic crustaceans, insect larvae and seeds; rarely tips up.
Voice: generally quiet; occasionally a raspy chuckle or quack; most often heard during spring courtship.
Similar Species: *Mallard* (p. 49): blue speculum bordered by white; lacks pale blue forewing; male has chestnut brown breast and white flanks. *Blue-winged Teal* (p. 50): much smaller bill; smaller overall; male has spotted breast and sides.
Best Sites: Charles Mill L.; Dillon Reservoir; regularly winters at Castalia Pond; rare nester in western L. Erie marshes.

J F M A M J J A S O N D

NORTHERN PINTAIL

Anas acuta

Northern Pintails are hardy ducks, pushing northward with the spring thaw and commencing nesting soon after ice-out, even in many northern locales. • Flooded agricultural fields tend to attract the largest pintail flocks. These flocks often have many more males than females, probably the result of summer predation at the fairly open, ground nesting sites. • In Ohio, Northern Pintails are fairly common migrants, primarily in glaciated regions of the state, and very rare nesters, mostly in western Lake Erie marshes. • According to early naturalists, pintails were easily the most abundant ducks passing through Ohio up to the early 1900s, greatly surpassing Mallards in number. Large-scale habitat changes resulted in a halving of the overall population—dropping from 6 million to less than 3 million—from 1970 into the early 1990s. Fortunately, recent aggressive conservation measures aimed at improving habitat in core nesting regions, coupled with harvest restrictions, seem to have caused an upsurge in numbers.

ID: long, slender neck; dark, glossy bill. *Male:* chocolate brown head; long, tapering tail feathers; white on breast extends up sides of neck; dusty gray body plumage; black-and-white hindquarters. *Female:* mottled, light brown overall. *In flight:* slender body; brownish speculum with white trailing edge to wing.

Size: *L* 21–25 in; 34 in.

Habitat: shallow wetlands, fields and lake edges.

Nesting: in a small depression in vegetation; nest of grass, leaves and moss is lined with down; female incubates 6–12 greenish buff eggs for 22–25 days.

Feeding: tips up and dabbles in shallows for the seeds of sedges, willows and pondweeds; also eats aquatic invertebrates and larval amphibians; eats waste grain in agricultural areas during migration; diet is more varied than that of other dabbling ducks.

Voice: *Male:* soft, whistling call. *Female:* rough quack.

Similar Species: male is distinctive. *Mallard* (p.-49) and *Gadwall* (p. 45): females are chunkier, usually with dark or 2-tone bills, and lack tapering tail and long, slender neck. *Blue-winged Teal* (p. 50): female is smaller; green speculum; blue forewing patch.

Best Sites: Funk Bottoms WA; Killdeer Plains WA; western L. Erie marshes.

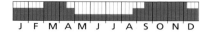

J F M A M J J A S O N D

GREEN-WINGED TEAL

Anas crecca

Weighing less than a pound, the Green-winged Teal is the smallest North American dabbling duck. This bird's small size, coupled with rocketlike acceleration and erratic, zigzagging flight, mean that this teal is unmatched by any other duck in aerial maneuverability. • There are three recognized Green-winged Teal subspecies—birders should be on the watch for a rare sighting of the Eurasian type, known as "Common Teal" (*A. c. crecca*). The prominent horizontal white stripe on the sides of this subspecies, rather than the vertical stripe of the Green-winged Teal subspecies commonly found in North America, makes for easy identification of the males. There are a few records of the Eurasian form in Ohio, and discovery of one should excite those avid listers who anticipate that this form may eventually be recognized as a separate species. • The Green-winged Teal is a common migrant throughout Ohio and is also a rare nester in the western Lake Erie marshes.

ID: small bill; green-and-black speculum. *Male:* chestnut brown head; green swipe extends back from eye; white shoulder slash; creamy breast spotted with black; pale gray sides. *Female:* mottled brown overall; light belly.
Size: *L* 12–16 in; *W* 23 in.
Habitat: shallow lakes, wetlands, beaver ponds and rivers.
Nesting: well concealed in tall vegetation; nest is built of grass and leaves and lined with down; female incubates 6–14 cream to pale buff eggs for 20–24 days.

Feeding: dabbles in shallows, particularly on mudflats, for aquatic invertebrates, larval amphibians, marsh plant seeds and pondweeds.
Voice: *Male:* crisp whistle. *Female:* soft quack.
Similar Species: *American Wigeon* (p. 47): male lacks white shoulder slash and chestnut brown head. *Blue-winged Teal* (p. 50): female has blue forewing patch.
Best Sites: Calamus Swamp; Funk Bottoms WA; Slate Run MP; migrants on most wetlands statewide.

J F M A M J J A S O N D

CANVASBACK

Aythya valisineria

Canvasbacks present one of the more distinctive profiles of our diving ducks. A long sloping bill segues into a flat forehead, allowing either sex to be identified at long range by silhouette. The drakes—arguably our most handsome diving ducks—are conspicuous at long range owing to their large size and bright white bodies. • Canvasbacks are devoted deep divers and seldom stray into areas of wetlands that are too shallow to allow foraging dives, so birders often need binoculars to properly admire the male's red eyes and rich, mahogany head. • The interrelationship between plants and animals is especially apparent with the Canvasback. The duck's scientific name *valisineria* refers to water celery *(Vallisneria americana)*, an underwater plant that is a dietary staple. Water celery growth has recently increased in Lake Erie, and migrant and wintering Canvasbacks have also become more frequent in the last decade.

ID: head slopes upward from bill to forehead. *Male:* bright white back; chestnut brown head; black breast and hindquarters; red eyes. *Female:* duller brown-and-gray plumage.

Size: *L* 19–22 in; *W* 29 in.

Habitat: marshes, ponds, shallow lakes and other wetlands; large lakes in migration.

Nesting: does not nest in Ohio.

Feeding: dives to depths of up to 30 ft (average is 10–15 ft); feeds on roots, tubers, the basal stems of plants (including pondweeds and water celery) and bulrush seeds; occasionally eats aquatic invertebrates.

Voice: generally quiet. *Male:* occasional coos and "growls" during courtship. *Female:* low, soft, purring quack or *kuck;* also "growls."

Similar Species: *Redhead* (p. 55): rounded rather than sloped forehead; male has gray back and bluish bill.

Best Sites: Maumee Bay SP and vicinity; Sandusky Bay; Avon Lake Power Plant.

J F M A M J J A S O N D

REDHEAD

Aythya americana

Most North American ducks can be put into three general categories: "dabblers," "sea ducks" and "bay ducks." Dabblers, which include Mallards, teals and Northern Pintails, normally don't dive, but tip up and submerge just the upper body. Sea ducks, which include scoters and eiders, are rare in Ohio, having primarily maritime wintering and migration patterns. Bay ducks, such as Redheads, scaup and Canvasbacks, are common migrants in Ohio, and to a lesser extent, winter residents. They get their catchall name of "bay ducks" because of an affinity for larger inland water bodies such as reservoirs and large rivers and, on the coast, bays. • Historically, Redheads were probably the most numerous of North American diving ducks, but unfortunately the overall population has declined markedly. One factor in this decline has been the introduction of the common carp. These large, bottom-feeding fish have destroyed vegetation upon which Redheads were heavily dependent in formerly prime Atlantic coast wintering habitats. • In Ohio, Redheads may be locally common in winter along Lake Erie and are rare nesters in the western Lake Erie marshes.

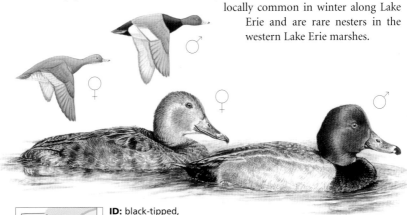

ID: black-tipped, blue gray bill. *Male:* rounded, red head; black breast and hindquarters; gray back and sides. *Female:* dark brown overall; light "chin" and "cheek" patches.
Size: *L* 18–22 in; *W* 29 in.
Habitat: large, shallow, well-vegetated wetlands, ponds and lakes; sometimes large rivers.
Nesting: usually in shallow water, sometimes on dry ground; deep basket nest of reeds and grass is lined with down and suspended over water at the base of emergent vegetation; female incubates 9–14 greenish eggs for 23–29 days; female may lay eggs in other ducks' nests.
Feeding: dives to depths of 10 ft; primarily eats aquatic vegetation, especially pondweeds, duckweeds and the leaves and stems of plants; occasionally eats aquatic invertebrates.
Voice: generally quiet. *Male:* catlike meow in courtship. *Female:* rolling *kurr-kurr-kurr; squak* when alarmed.
Similar Species: *Canvasback* (p. 54): clean white back; bill extends onto forehead. *Ring-necked Duck* (p. 56): female has more prominent, white eye ring, white ring on bill and peaked head. *Lesser Scaup* (p. 58) and *Greater Scaup* (p. 57): prominent white wing bar; male has dark head and whiter sides; female has more white at base of bill.
Best Sites: Mogadore Reservoir; Big Island WA; Oberlin Reservoir. Migrants can occur in any inland marsh or reservoir.

J F M A M J J A S O N D

RING-NECKED DUCK

Aythya collaris

Ring-necked Ducks are much more like dabbling ducks than any other diver. They spring readily into the air from the water's surface, without the running takeoff of other diving ducks, tend to frequent shallow waters and are generally less winter hardy than the other divers. • This species is likely a primary agent of dispersal for various pondweeds, coontails and other aquatic plants that invade newly created ponds. The plant seeds are transported in the stomachs of migrant Ring-necked Ducks, and deposited into far-flung water bodies via the birds' excretory system. • A more fitting name for this species would have been "Ring-billed Duck." The faint cinnamon neckband, to which the scientific name *collaris* refers, is virtually impossible to see in the field. However, the prominent white ring on the drake's bill jumps out like a beacon and is a good field mark. • In Ohio, Ring-necks prefer inland ponds, borrow pits and reservoirs, where they feast on aquatic plants. Occasional summering birds have been noted.

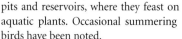

Feeding: dives underwater for aquatic vegetation, including seeds, tubers and pondweed leaves; also eats aquatic invertebrates and mollusks.
Voice: seldom heard. *Male:* low-pitched, hissing whistle. *Female:* growling *churr.*
Similar Species: *Lesser Scaup* (p. 58) and *Greater Scaup* (p. 57): lack white ring near tip of bill; male lacks black back; female has broad, clearly defined white border around base of bill and lacks eye ring. *Redhead* (p. 55): rounded rather than peaked head; less white on front of face; female has less prominent eye ring.
Best Sites: Mogadore Reservoir; Killbuck Marsh WA; Knox L.; borrow pits along freeways in migration.

ID: *Male:* angular, dark purple head; black breast, back and hindquarters; white shoulder slash; gray sides; blue gray bill with black and white bands at tip; thin, white border around base of bill. *Female:* dark brown overall; white eye ring; dark bill with black and white bands at tip; pale crescent on front of face.
Size: *L* 14–18 in; *W* 25 in.
Habitat: reservoirs, shallow wooded ponds, swamps, marshes and sloughs with emergent vegetation.
Nesting: does not nest in Ohio.

J F M A M J J A S O N D

GREATER SCAUP

Aythya marila

It wasn't until relatively modern times that ornithologists learned to reliably distinguish Greater Scaup from Lesser Scaup, and some of them still have trouble. In Ohio, Greater Scaup—the most northerly breeders of the genus *Aythya*—are almost exclusively winter visitors and early spring migrants, and are casual at best away from Lake Erie. Reports of large inland flocks are usually misidentified Lesser Scaup. • The frequently cited identifying characteristic of head coloration—purple gloss in Lessers, green in Greaters—is quite difficult to accurately ascertain and is not normally of much use in the field. • Cold winters that cause Lake Erie to ice over produce the best numbers of Greater Scaup. During such times, these hardy ducks will form large rafts in open water, sometimes numbering into the thousands. • Scaup have been dubbed "Bluebills," an apropos moniker as males have conspicuous, chalky blue bills.

ID: rounded head; golden eyes. *Male:* iridescent, dark green head; black breast; white belly and flanks; light gray back; dark hindquarters; black-tipped, blue bill. *Female:* brown overall; well-defined white patch at base of bill. *In flight:* white stripe through wing extends well into primary feathers.
Size: L 16–19 in; W 28 in.
Habitat: primarily open waters of L. Erie; inland, large reservoirs and big rivers.
Nesting: does not nest in Ohio.
Feeding: dives underwater, to greater depths than other *Aythya* ducks, for aquatic invertebrates and vegetation; favors freshwater mollusks in winter.

Voice: generally quiet in migration; alarm call is a deep *scaup*. *Male:* may issue a 3-note whistle and a soft *wah-hooo*. *Female:* may give a subtle "growl."
Similar Species: *Lesser Scaup* (p. 58): slightly smaller; shorter white wing stripe in flight is confined to secondaries; slightly smaller bill; male has peaked, purplish head; female usually has peaked head. *Ring-necked Duck* (p. 56): black back; white shoulder slash; white ring around base of bill. *Redhead* (p. 55): male has red head and darker sides; female has less white at base of bill.
Best Sites: Cleveland Lakefront SP; Conneaut Harbor; Eastlake and Avon Lake power plants.

J F M A M J J A S O N D

57

LESSER SCAUP

Aythya affinis

Unlike the hardy northern Greater Scaup, the Lesser Scaup ranges farther south than any other *Aythya* duck, even regularly visiting the Caribbean. Lesser Scaup are generally plentiful migrants throughout interior Ohio, frequenting borrow pits, reservoirs, wetlands and large rivers. The largest numbers occur in the Maumee Bay area of western Lake Erie, where single-day counts of up to 80,000 birds have been made in March and early April. • As with a number of other northern ducks, Ohio is on the extreme southern edge of the Lesser Scaup's breeding range. The last confirmed nesting was in 1937, although summering birds are occasionally seen, and they may still rarely breed. • It can be quite difficult to reliably distinguish the two scaup species. Perhaps the best way is to note the extent of white on the wings. In Lesser Scaup, this white is largely confined to the secondary flight feathers, whereas in the Greater Scaup it extends well into the primary flight feathers.

ID: yellow eyes. *Male:* dark, purplish head is peaked; black breast and hindquarters; dusty white sides; grayish back; black-tipped, blue gray bill. *Female:* dark brown overall; well-defined white patch at base of bill; white wing stripe does not extend far onto primaries as in Greater Scaup.

Size: *L* 15–18 in; *W* 25 in.
Habitat: lakes, ponds, marshes, borrow pits, large rivers.
Nesting: does not nest in Ohio.
Feeding: dives underwater for aquatic invertebrates, mostly mollusks, crustaceans and insect larvae; occasionally takes aquatic vegetation.
Voice: alarm call is a deep *scaup*. *Male:* soft *whee-oooh* in courtship. *Female:* purring *kwah*.
Similar Species: *Greater Scaup* (p. 57): slightly larger bill; longer white wing flash; rounded head; male has greenish head. *Ring-necked Duck* (p. 56): male has white shoulder slash and black back; female has white-ringed bill. *Redhead* (p. 55): male has red head and darker sides; female has less white at base of bill.
Best Sites: Blendon Woods MP; East Harbor SP; Maumee Bay SP; virtually any reservoir, lake or sizable marsh in migration.

J F M A M J J A S O N D

HARLEQUIN DUCK

Histrionicus histrionicus

The small, surf-loving Harlequin Duck is a rare migrant and winter visitor to our region, where it occurs along the shoreline of Lake Erie. Early spring Lake Erie birds are likely ones that overwintered. • Eastern Harlequin populations have dwindled to only about 1500 birds, making it perhaps the rarest breeding duck in eastern North America. In spite of the small overall population, this species has slowly become more regular in Ohio, now averaging about three birds annually. It went unrecorded in the state until 1949. • Harlequin Ducks are unique among our waterfowl in that their breeding habitat is turbulent, fast-flowing rivers and streams. • This duck is named after a character from traditional Italian comedy and pantomime, the Harlequin, who wore a diamond-patterned costume and performed "histrionics," or tricks. • Perhaps the best chance of seeing a Harlequin Duck in Ohio is to watch Lake Erie from a good vantage point on a cold, blustery day in November, preferably with north winds, and hope for a flyby.

ID: small, rounded duck; blocky head; short bill; raises and lowers tail while swimming. *Male:* blue gray body; chestnut brown sides; white spots and stripes outlined in black on head, neck and flanks. *Female:* dusky brown overall; light underparts; 2–3 light patches on head.
Size: *L* 14–19 in; *W* 26 in.
Habitat: large, freshwater lakes; appears rarely on large rivers such as the Great Miami R.
Nesting: does not nest in Ohio.

Feeding: dabbles and dives for aquatic invertebrates, mostly crustaceans and mollusks in freshwater lakes.
Voice: generally silent outside the breeding season.
Similar Species: male is distinctive. *Bufflehead* (p. 64): smaller; female lacks white between eye and bill. *Surf Scoter* (p. 60): female has bulbous bill. *White-winged Scoter* (p. 61): female has white wing patch and bulbous bill.
Best Sites: Huron Municipal Pier; Avon Lake Power Plant; Mentor Headlands.

J F M A M J J A S O N D

SURF SCOTER
Melanitta perspicillata

Scoters are the most frequently seen "sea ducks" that pass through Ohio, and the Surf Scoter is the second most commonly encountered of the three species, after the Black Scoter. • The Surf Scoter remains one of the least understood and studied of North American ducks; for instance, only a few nests have been described. In the last few decades, though, our knowledge of this species has changed markedly. • Prior to the 1980s, Surf Scoters were considered rare visitors in Ohio with just a few reported annually. In the last 20 years, large fall movements numbering 100 or more birds have been detected along Lake Erie on some days. This increase in sightings may in part reflect more sophistication on the part of birders, who have learned that watching offshore waters of Lake Erie during blustery November or December weather often produces large numbers of sightings of migrating waterbirds. • The scientific name *Melanitta* means "black duck"; *perspicillata* is Latin for "spectacular," which refers to this bird's colorful, bulbous bill.

ID: large, stocky duck; large bill; sloping forehead. *Male:* black overall; white on forehead and back of neck; orange bill and legs; black spot, outlined in white, at base of bill. *Female:* brown overall; dark gray bill; 2 whitish patches on sides of head.
Size: *L* 16–20 in; *W* 30 in.
Habitat: large lakes; occasionally big rivers.
Nesting: does not nest in Ohio.
Feeding: dives to depths of 30 ft; eats mostly mollusks; also takes aquatic insect larvae, crustaceans and some aquatic vegetation.

Voice: generally quiet; infrequently utters low, harsh croaks. *Male:* occasionally gives a low, clear whistle. *Female:* guttural *krraak krraak.*
Similar Species: *White-winged Scoter* (p. 61): white wing patches; male lacks white on forehead and nape. *Black Scoter* (p. 62): male is all black; female has well-defined, pale "cheek."
Best Sites: Huron Municipal Pier; Cleveland Lakefront SP; Avon Lake Power Plant; occasional on larger inland reservoirs and rivers.

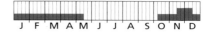

J F M A M J J A S O N D

WHITE-WINGED SCOTER

Melanitta fusca

Distinctively plumaged adult male White-winged Scoters are easy to identify; unfortunately, virtually all Ohio scoters are females or subadults. Distant, resting flocks can be tricky to distinguish, but sooner or later a resting White-winged Scoter will move or stretch, revealing the diagnostic white wing speculum. • Once considered Ohio's most common scoter, the White-winged Scoter has now become the scarcest. This may reflect an overall population decline, but may also be the result of a shift in the other two species' migratory patterns. • White-winged Scoters breed throughout wide areas of America, Europe and Asia. Some authorities believe the European subspecies is distinct and should be classified as a separate species. • The name "scoter" may be derived from the way this bird scoots across the water's surface, as if to take flight. Scooting can be a means of traveling quickly from one foraging site to another. • Scoters have small wings relative to the weight of their bodies, so they require long stretches of water for takeoff.

ID: stocky; large, bulbous bill; sloping forehead; base of bill is fully feathered. *Male:* black overall; white patch below eye. *Female:* brown overall; gray brown bill; 2 whitish patches on sides of head. *In flight:* white wing patches.
Size: *L* 18–24 in; *W* 34 in.
Habitat: large lakes and rivers.
Nesting: does not nest in Ohio.
Feeding: deep, underwater dives last up to 1 minute; eats mostly mollusks; may also take crustaceans, aquatic insects and some small fish.
Voice: courting pair produces harsh, guttural noises, between a *crook* and a *quack.*
Similar Species: *Surf Scoter* (p. 60): lacks white wing patches; male has white forehead and nape. *Black Scoter* (p. 62): lacks white patches on wings and around eyes.
Best Sites: Huron Municipal Pier; Avon Lake Power Plant; Cleveland Lakefront SP; occasional on inland reservoirs and large rivers.

J F M A M J J A S O N D

BLACK SCOTER

Melanitta nigra

Occasional large flights of Black Scoters occur in Ohio, such as one on November 11, 1985, when 600 were seen near Vermilion. • Adult male Black Scoters are unmistakable, possessing a luminescent yellow bill protuberance, but Ohio birds are mostly females or subadults, and are sometimes mistaken for the much smaller Ruddy Duck in nonbreeding plumage. • The Black Scoter is the most frequently occurring Ohio scoter, and as with the other two Ohio scoters, the vast majority are seen in the central basin of Lake Erie, from Huron eastward. Peak numbers occur in fall, with spring birds being relatively rare. • While floating on the water's surface, Black Scoters tend to hold their heads high, unlike other scoters, which tend to look downward. • Black Scoters are the most genetically isolated of the three species, and provide an evolutionary link to the genus *Clangula*—the Long-tailed Duck. Sea ducks rarely hybridize, so imagine the confusion of the person who found the only known Black Scoter hybrid—with a Eurasian Wigeon!

ID: *Male:* black overall; large orange knob on bill. *Female:* light "cheek"; dark cap; brown overall; dark gray bill.
Size: *L* 17–20 in; *W* 28 in.

Habitat: large lakes and rivers.
Nesting: does not nest in Ohio.
Feeding: dives underwater; eats mostly mollusks and aquatic insect larvae; occasionally eats aquatic vegetation and small fish.
Voice: generally quiet; occasionally an unusual *cour-loo;* wings whistle in flight.

Similar Species: *White-winged Scoter* (p. 61): white wing patches; male has white slash below eye. *Surf Scoter* (p. 60): male has white on head; female has 2 whitish patches on sides of head. *Ruddy Duck* (p. 69): nonbreeding male is much smaller with fatter bill and whiter cheeks.
Best Sites: Huron Municipal Pier; Avon Lake Power Plant; Cleveland Lakefront SP; occasional on inland reservoirs.

J F M A M J J A S O N D

LONG-TAILED DUCK

Clangula hyemalis

The Long-tailed Duck has long been known as the "Oldsquaw," which was an apt name because groups keep up a continual stream of melodious calls— a large flock can be heard up to a mile away! • Unfortunately, the peak migration route of this fascinating sea duck just misses Ohio; nearby Lake Michigan gets concentrations of up to 40,000 birds. In Ohio, Long-tailed Ducks are normally encountered as single birds along Lake Erie and on large inland reservoirs. They are rare here; five to ten reports annually is about average. Occasional larger flights result in scattered sightings of small flocks. • These sea ducks are truly adapted to deep water—they tend to remain in deeper waters well away from shore and have been caught in fishing nets at depths of up to 200 feet! • Long-tailed Ducks are unique among waterfowl in that they have three distinct plumages, while other ducks have only two. Most Ohio birds are females, rather than the spectacularly plumaged "long-tailed" males.

♂

nonbreeding

♀

♂

nonbreeding

♀

ID: *Nonbreeding male:* pale head with dark patch; pale neck and belly; dark breast; long, white patches on back; pink bill with dark base; long, dark central tail feathers. *Nonbreeding female:* similar, but generally lighter, especially on head; lacks long tail feathers.

Size: *L* 17–20 in; *W* 28 in.

Habitat: large, deep lakes and, sometimes, big rivers.

Nesting: does not nest in Ohio.

Feeding: dives for mollusks, crustaceans and aquatic insects; occasionally eats roots and young shoots; may also take some small fish.

Voice: courtship call, *owl-owl-owlet*, is rarely heard outside the breeding range.

Similar Species: a very distinctive species in any plumage; unlikely to be confused with any other duck.

Best Sites: Eastlake Power Plant; Maumee River Rapids; La Due Reservoir; may appear anywhere along L. Erie and rarely inland on large water bodies.

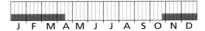

J F M A M J J A S O N D

BUFFLEHEAD

Bucephala albeola

Our smallest duck, the tiny Bufflehead is quite hardy and routinely overwinters on the frigid waters of Lake Erie, where it mingles—and competes quite well—with other larger species of ducks. • The Bufflehead is one of six North American species of cavity-nesting ducks. Its small size allows it to routinely exploit Northern Flicker cavities as nest sites. • This duck is one of the few North American waterfowl that have increased in numbers in recent decades. As of 1992, the overall population was estimated at about 1.4 million. Ohio harbors a significant concentration, with the recent documentation of several thousand in fall and early winter in the Lake Erie waters off Kelleys Island. • The Bufflehead is actually a small goldeneye, as similarities in its profile, behavior and scientific name will attest. Both the common name, Bufflehead, and the scientific name *Bucephala* (which means "ox-headed" in Greek), refer to this bird's large head.

ID: very small, rounded duck; short gray bill; short neck; white speculum in flight. *Male:* white wedge on back of head; head is otherwise iridescent dark green or purple, usually appearing black; dark back; white neck and underparts. *Female:* dark brown head; white, oval ear patch; light brown sides.
Size: *L* 13–15 in; *W* 21 in.
Habitat: open water of lakes, large ponds and rivers.
Nesting: does not nest in Ohio.

Feeding: dives for aquatic invertebrates; favors mollusks, particularly snails, and crustaceans in winter; also eats some small fish and pondweeds.
Voice: *Male:* "growling" call. *Female:* harsh quack.
Similar Species: *Hooded Merganser* (p. 66): white crest outlined in black. *Harlequin Duck* (p. 59): female has several light spots on head. *Common Goldeneye* (p. 65): males are larger and have white patch between eye and bill. *Other diving ducks* (pp. 54–69): females are much larger.
Best Sites: Kelleys I.; Mogadore Reservoir; Caesar Creek Reservoir. *In migration:* fairly common statewide. *Winter:* open water.

J F M A M J J A S O N D

COMMON GOLDENEYE

Bucephala clangula

Few birds are as hardy as Common Goldeneyes—in Ohio, they generally aren't seen in good numbers unless Lake Erie freezes. Then, they queue up in open water leads among the ice. • Common Goldeneyes don't dally when it comes to courting season, cold or not, and we're fortunate to be able to observe their courtship antics, which commence in February. Goldeneyes form small packs called "display groups," in which one or two females are treated to several males vying for their attention with one of the most spectacular avian displays of any North American bird. The male alternately thrusts his neck backward until his bill is vertical, then launches forward while kicking his brilliant orange feet forward. This is combined with various other head movements and even short flights. • Fortunately, populations of this beautiful and interesting duck have increased in Ohio in modern times. • Goldeneyes are often called "Whistlers" because of the loud and distinctive noise made by their wings in flight.

ID: steep forehead with peaked crown; black wings with large, white patches; golden eyes. *Male:* dark, iridescent, green head; round, white "cheek" patch; dark bill; dark back; white sides and belly. *Female:* chocolate brown head; lighter breast and belly; gray brown body plumage; dark bill, tipped with yellow in spring and summer.
Size: *L* 16–20 in; *W* 26 in.
Habitat: open water of lakes and large rivers.
Nesting: does not nest in Ohio.

Feeding: dives for crustaceans, mollusks and aquatic insect larvae; may also eat tubers, leeches, frogs and small fish.
Voice: *Male:* courtship calls are a nasal *peent* and a hoarse *kraaagh*. *Female:* harsh croak.
Similar Species: *Barrow's Goldeneye* (p. 331): very rare in Ohio; male has large, white, crescent-shaped "cheek" patch and purplish head; female has more orange on bill and more steeply sloped forehead.
Best Sites: Maumee Bay SP; Cleveland Lakefront SP; Mosquito Creek Reservoir.

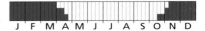

J F M A M J J A S O N D

65

HOODED MERGANSER

Lophodytes cucullatus

This is our smallest merganser, and the only one that is restricted to North America. A common spring and fall migrant in Ohio, the Hooded Merganser is routinely encountered in small flocks on all types of water bodies throughout the state, and is even locally common in winter. It also nests sparingly here, primarily in the northern third of the state, but this may be increasing. It nests in cavities and seems to be increasingly using Wood Duck nest boxes, which are a common fixture in many wetlands. • Like Wood Ducks, Hooded Mergansers frequent streams and rivers, and even resemble that species in flight, as both have long, projecting tails that lend a distinctive silhouette and make in-flight identification from a distance easy. Hooded Mergansers' lowered heads help distinguish these birds from distant, in-flight Wood Ducks with their raised heads.

ID: slim body; crested head; thin, dark, pointed bill. *Male:* black head and back; bold, white crest outlined in black; white breast with 2 black slashes; rusty sides. *Female:* dusky brown body; shaggy, reddish brown crest. *In flight:* small, white wing patches.
Size: *L* 16–18 in; *W* 24 in.
Habitat: ponds, wetlands, lakes and rivers.
Nesting: usually in a tree cavity 15–40 ft above the ground; may also use a nest box; cavity is lined with leaves, grass and down; female incubates 10–12 spherical, white eggs for 29–33 days; some females may lay their eggs in other birds' nests, including nests of other species.

Feeding: very diverse diet; dives for small fish, caddisfly and dragonfly larvae, snails, amphibians and crayfish.
Voice: low grunts and croaks. *Male:* frog-like *crrrrooo* in courtship display. *Female:* generally quiet; occasionally a harsh *gak* or a croaking *croo-croo-crook*.
Similar Species: *Red-breasted Merganser* (p. 68) and *Common Merganser* (p. 67): females are larger with much longer, orange bills and gray backs. *Other small diving ducks* (pp. 54–69): females lack crest.
Best Sites: Greenlawn Avenue Dam (Columbus); Hoover Reservoir; Deer Creek Reservoir.

J F M A M J J A S O N D

COMMON MERGANSER

Mergus merganser

The Common Merganser is the most common North American merganser, and the second most frequently occurring merganser in Ohio, after the Red-breasted Merganser. Although, historically, the Common Merganser was a common migrant throughout much of Ohio—with reports of 2000 on Buckeye Lake on March 20, 1924, for example—the species is now generally encountered in large numbers only along Lake Erie. There, it peaks in winter around Maumee Bay and the Cleveland area, where high counts of about 10,000 birds have been recorded. • The Common Merganser is the largest duck found in Ohio, and the gleaming white plumage of the male stands out at a great distance. • This merganser nests along forested rivers and lakes that are generally "wild" and pollution-free. Therefore, it was exciting that this species was recently discovered nesting in Ohio, along the beautiful and remote Little Beaver Creek in Columbiana County.

ID: large, elongated body. *Male:* glossy, green head without crest; blood red bill and feet; white body plumage; black stripe on back; dark eyes. *Female:* rusty neck and crested head; clean white "chin" and breast; orange bill; gray body; orangy eyes. *In flight:* shallow wingbeats; body is compressed and arrowlike.

Size: *L* 22–27 in; *W* 34 in.

Habitat: large rivers and deep lakes.

Nesting: often in a tree cavity 15–20 ft above ground; occasionally on the ground, under a bush or log, on a cliff ledge or in a large nest box; usually not far from water; female incubates 8–11 pale buff eggs for 30–35 days.

Feeding: dives to depths of 30 ft for small fish, usually whitefish, trout, suckers, perch and minnows; young eat surface insects and aquatic invertebrates before switching to small fish.

Voice: *Male:* harsh *uig-a*, like a guitar twang. *Female:* harsh *karr karr*.

Similar Species: *Red-breasted Merganser* (p. 68): male has shaggy, green crest and spotted, red breast; female lacks cleanly defined white throat.

Best Sites: Maumee Bay SP region; Cleveland Lakefront SP; Findlay Reservoir. *In migration:* any large, inland reservoir or river.

J F M A M J J A S O N D

RED-BREASTED MERGANSER

Mergus serrator

This species and other mergansers are often referred to as "Sawbills" because of the serrated mandibles of their bills. This specialized adaptation allows these fish-eaters to more easily grasp their slippery prey, which is caught in underwater dives. • One of the most fantastic avian events in Ohio, and one that didn't come to light until the 1960s, is the incredible November migration of Red-breasted Mergansers along Lake Erie. Occurring primarily from the Lake Erie islands to Cleveland, immense groups stage, their numbers staggering the imagination with flocks so dense that they resemble low-lying storm clouds on the horizon. One-day estimates near 250,000 have been made, and as many as 100,000 have been viewed in a 10-minute period. This is particularly significant when one takes into account a 1996 study that estimated the entire North American population at 250,000 birds!

ID: large, elongated body; red eyes; thin, serrated, orange bill; shaggy, slicked-back head crest. *Male:* green head; light rusty breast spotted with black; white "collar"; gray sides; black-and-white shoulders. *Female:* gray brown overall; reddish head; white "chin," foreneck and breast. *In flight:* male has large, white wing patch crossed by 2 narrow, black bars; female has 1 dark bar separating white speculum from white upperwing patch.
Size: *L* 19–26 in; *W* 30 in.
Habitat: lakes and large rivers, especially those with rocky shorelines and islands.

Nesting: does not nest in Ohio.
Feeding: dives underwater for small fish; also eats aquatic invertebrates, fish eggs and crustaceans.
Voice: generally quiet. *Male:* catlike *yeow* during courtship and feeding. *Female:* harsh *kho-kha.*
Similar Species: *Common Merganser* (p. 67): male has clean, white breast, blood red bill and lacks head crest; female has rusty foreneck with white "chin" and breast.
Best Sites: Huron Municipal Pier; Cleveland Lakefront SP; Alum Creek Reservoir. Migrants occur on lakes, rivers and large wetlands statewide.

J F M A M J J A S O N D

RUDDY DUCK

Oxyura jamaicensis

Taking flight is a big ordeal for the "chunky" Ruddy Duck—it requires an extensive run across the water's surface into the wind, then, once airborne, tends to fly low and erratically. • These interesting ducks are rather like grebes in their ability to slowly sink from sight, their preference for diving rather than flying to avoid predators and their virtual inability to walk on land. • This small duck normally holds its fan-shaped tail cocked upward like a wren—hence the colloquial name "Stiff-tail." • Migration occurs largely at night, and by and large, Ruddy Ducks are nocturnal in many of their habits. • Female Ruddies commonly lay an average of eight eggs at a time—a remarkable feat, considering that their eggs are bigger than a Mallard's, and that a Ruddy Duck is significantly smaller than a Mallard. Females take part in a strange practice, often dumping eggs in a communal "dummy" nest, which may finally accumulate as many as 60 eggs that will receive no motherly care. • In Ohio's western Lake Erie marshes, the Ruddy Duck is a rare nester but shows large fall concentrations.

breeding

nonbreeding

ID: large bill and head; short neck; long, stiff tail feathers (often cocked upward). *Breeding male:* white "cheek"; chestnut red body; blue bill; black tail and crown. *Female:* brown overall; dark "cheek" stripe; darker crown and back. *Nonbreeding male:* similar to female but with white "cheek."
Size: *L* 15–16 in; *W* 18½ in.
Habitat: *Breeding:* shallow marshes with dense emergent vegetation and muddy bottoms. *In migration* and *winter:* lakes with open, shallow water.
Nesting: in cattails, bulrushes or other emergent vegetation; female suspends a woven platform nest over water; may use an abandoned duck or coot nest, muskrat

lodge or exposed log; female incubates 5–10 rough, whitish eggs for 23–26 days; an occasional brood parasite.
Feeding: dives to the bottom of wetlands for the seeds of pondweeds, sedges and bulrushes and for the leafy parts of aquatic plants; also eats a few aquatic invertebrates.
Voice: *Male:* courtship display is *chuck-chuck-chuck-chur-r-r-r. Female:* generally silent.
Similar Species: fairly distinctive. *Black Scoter* (p. 62): female and subadult are much larger with thinner bills and duskier cheeks; similar to nonbreeding Ruddy Duck.
Best Sites: Findlay Reservoir; La Due Reservoir; Magee Marsh WA. Migrants occur statewide.

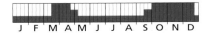

J F M A M J J A S O N D

RING-NECKED PHEASANT

Phasianus colchicus

The spectacular Ring-necked Pheasant was introduced to Ohio from its native Eurasia in 1896 to provide a replacement for native grouselike birds, which were at all-time lows. Introductions continue to this day, although not at the level of the 1920s and 1930s, when the state game agency seemed determined to make this bird a permanent part of our landscape—in those days they released up to 25,000 annually. • Unfortunately, unlike its native brethren such as the Northern Bobwhite and the Ruffed Grouse, the Ring-necked Pheasant does not have feathered legs and feet to insulate it through the winter months, and it cannot survive on native plants alone. The availability of grain and corn crops, as well as sheltered hedgerows and woodlots, have allowed this pheasant to survive in our area. • Populations fluctuate widely in local areas depending on weather conditions, predation and farming practices, but Ring-necked Pheasants have not done well in recent years. Their continued existence is largely a result of regular releases of fresh stock.

ID: large; long, barred tail; unfeathered legs. *Male:* green head; white "collar"; bronze underparts; naked, red face patch. *Female:* mottled brown overall; light underparts.

Size: *Male: L* 30–36 in; *W* 31 in. *Female: L* 20–26 in; *W* 28 in.

Habitat: grasslands, grassy ditches, hayfields and grassy or weedy fields, fencelines and crop margins.

Nesting: on the ground, among grass or sparse vegetation or next to a log or other natural debris; in a slight depression lined with grass and leaves; female incubates 10–12 olive buff eggs for 23–28 days; male takes no part in parental duties.

Feeding: *Summer:* gleans the ground and vegetation for weed seeds, grains and insects. *Winter:* eats mostly seeds, corn kernels and buds.

Voice: *Male:* loud, raspy, roosterlike crowing *ka-squawk;* whirring of wings, usually just before sunrise.

Similar Species: distinctive; no similar species shares its habitat.

Best Sites: areas where regular stocking occurs, such as Killdeer Plains WA; also Maumee Bay SP.

J F M A M J J A S O N D

RUFFED GROUSE

Bonasa umbellus

Starting in mid- to late March, one of nature's most unusual sounds can be heard in eastern Ohio woodlands—the drumming of the Ruffed Grouse. Displaying males locate a log, stump or other elevated platform, and using it as a perch, quickly rotate their wings, creating an air vacuum that produces a noise akin to a distant sonic boom. Starting slowly, these "booms" rapidly accelerate into a series lasting perhaps 10 seconds. So low in frequency is this sound, it's almost as if the observer feels, rather than hears, the drumming. • Ruffed Grouse are generally not birds of mature woodlands, but instead favor successional growth such as tangled clear-cuts. • Grouse are normally quite secretive, but a good way to find them is to look for dense tangles of brush, green briers or honeysuckle, and then to slowly and erratically walk among the brush. This seems to make grouse nervous and they will often flush, sometimes almost at your feet. Occasionally, males become extremely territorial and will attack people, and even cars, that dare enter their domain. • The Ruffed Grouse is strictly confined to eastern Ohio, with the greatest numbers found in logged-over areas of the southeastern counties. Historically, this bird was found statewide.

♂

gray morph

ID: small head crest; mottled, gray brown overall; black feathers on sides of lower neck (visible when fluffed out in courtship); gray- or reddish-barred tail has broad, dark, subterminal band and white tip. *Female:* incomplete subterminal tail band.
Size: *L* 15–19 in; *W* 22 in.
Habitat: deciduous and mixed forests and riparian woodlands; favors young, second-growth stands caused by clear-cutting.
Nesting: in a shallow depression among leaf litter; often beside boulders, under logs or at the base of a tree; female incubates 9–12 buff eggs for 23–25 days.

Feeding: gleans from the ground and vegetation; omnivorous diet includes seeds, buds, flowers, berries, catkins, leaves, insects, spiders and snails; may take small frogs.
Voice: *Male:* uses wings to produce a hollow, drumming courtship sound of accelerating, deep booms, like a lawnmower starting up and stalling. *Female:* clucks and "hisses" around her chicks.
Similar Species: no similar species shares its habitat.
Best Sites: extensive overgrown clear-cuts and successional forests in Vinton Co.; Tar Hollow SF; Shawnee SF; MeadWestvaco Experimental Forest.

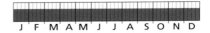

J F M A M J J A S O N D

WILD TURKEY

Meleagris gallopavo

The Wild Turkey is the largest North American member of the family Galliformes, or grouselike birds. If Benjamin Franklin had had his way, this bird, rather than the Bald Eagle, would be our national symbol. • Prior to European settlement, turkeys were abundant throughout the great forests that cloaked virtually all of Ohio. Their numbers quickly declined with the arrival of settlers, who recognized turkeys as avian delicacies. As hunting pressures increased and forests were felled, Wild Turkeys continued to decline until, by 1900, none were left in the state. Beginning in 1956 in southeastern Ohio, and continuing yearly thereafter, Wild Turkeys were reintroduced into the forests. Today, they are once again common throughout the hill country, and smaller numbers can be found in woodlands throughout much of the rest of Ohio. • Turkeys often form large flocks of up to 50 or 60 birds, and are fond of foraging in meadows and cornfields near woods. Perhaps the easiest way to locate Wild Turkeys is by driving roads in likely areas just after dawn, and carefully watching distant fields near woodland borders for feeding flocks.

ID: naked, red blue head; dark, glossy, iridescent body plumage; barred, copper-colored tail; largely unfeathered legs. *Male:* long central breast tassel; colorful head and body; red wattles. *Female:* smaller; blue gray head; less iridescent body.
Size: *Male: L* 3–3½ ft; *W* 5½ ft. *Female: L* 3 ft; *W* 4 ft.
Habitat: deciduous, mixed and riparian woodlands; occasionally eats waste grain and corn in late fall and winter.
Nesting: in woodlands or at field edges; in a depression on the ground under thick cover; nest is lined with grass and leaves;

female incubates 10–12 brown-speckled, pale buff eggs for up to 28 days.
Feeding: usually in fields near protective woods; forages on the ground for seeds, fruits, bulbs and sedges; also eats insects, especially beetles and grasshoppers; may take small amphibians.
Voice: wide array of sounds; courting male gobbles loudly; alarm call is a loud *pert;* gathering call is a *cluck;* contact call is a loud *keouk-keouk-keouk.*
Similar Species: unmistakable; may be confused with escaped domestic turkeys.
Best Sites: The Wilds; Hocking Hills region; Crown City WA.

J F M A M J J A S O N D

NORTHERN BOBWHITE

Colinus virginianus

The unmistakable, cheery whistle of the male Northern Bobwhite was a common sound throughout much of Ohio until the mid-1970s. The Northern Bobwhite's history has been one of boom and bust, and prior to settlement, bobwhites were probably not found here, as they require open landscapes of meadows, fencerows and pastures. Forest clearing created excellent bobwhite habitat, in the form of small, bird-friendly farms—at their peak in the mid-1930s, it was estimated that 20 pairs of Northern Bobwhites occupied every square mile. The gradual shift from small farms to large, neatly manicured mega-farms contributed greatly to the reduction of the Northern Bobwhite population. • As with other southern birds that rapidly expanded north as a result of human-induced changes to the landscape, bobwhites are very susceptible to brutal winters—the savage winters of 1976–78 caused a 90 percent population reduction.

ID: mottled brown, buff and black upperparts; white crescents and spots edged in black on upper breast and chestnut brown sides; short tail. *Male:* white throat; broad white "eyebrow." *Female:* buff throat and "eyebrow."
Size: *L* 10 in; *W* 13 in.
Habitat: farmlands, open woodlands, woodland edges, grassy fencelines, and brushy, open country.
Nesting: in a shallow depression on the ground; often concealed by surrounding vegetation or a woven, partial dome; nest is lined with grass and leaves; pair incubates 12–16 white to pale buff eggs for 22–24 days.

Feeding: seasonally available seeds, berries, leaves, roots, nuts, insects and other invertebrates.
Voice: whistled *hoy* is given year-round. *Male:* a whistled, rising *bob-white* given in spring and summer.
Similar Species: *Ruffed Grouse* (p. 71): lacks conspicuous throat patch and broad "eyebrow"; long, fan-shaped tail has broad, dark subterminal band; black patches on sides of neck; normally found in different habitat.
Best Sites: parts of Adams, Brown, Lawrence and Gallia counties; Woodbury WA (augmented by introductions); Crown City WA; Tranquility WA.

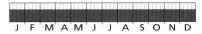

J F M A M J J A S O N D

73

RED-THROATED LOON

Gavia stellata

Red-throated Loons are unique among loons in that they require almost no room to take flight—other loon species must run for considerable distances across the water before becoming airborne. In flight, which is often how they are seen on Lake Erie, they differ from other loons by their drooping neck and deeper wingbeats. • The Red-throated Loon is most often detected in fall on Lake Erie and on large inland reservoirs, with November being the peak month. This bird is often found in association with large flocks of migrant Common Loons, and it stands out from that species by its paler overall look, smaller, slimmer shape and thinner bill held at an upward angle. In Ohio, this species is much scarcer than the Common Loon. • The Red-throated Loon is the smallest loon, and is quite striking in its breeding colors. Unfortunately, it is seldom seen in that plumage in Ohio. • The scientific name *stellata* refers to the starlike, white speckles on this bird's back in its nonbreeding plumage.

nonbreeding

nonbreeding

ID: slim bill is held upward. *Breeding:* red throat; gray face and neck; black and white stripes from nape to back of head; plain, brownish back. *Nonbreeding:* back is speckled with white; white face; dark gray on crown and back of head; often thin, brownish "chin strap." *In flight:* hunched back; legs trail behind tail; rapid wingbeats.
Size: *L* 23–27 in; *W* 3½ ft.
Habitat: L. Erie and large inland lakes.
Nesting: does not nest in Ohio.

Feeding: dives deeply and captures small fish; occasionally eats aquatic insects and amphibians as well as aquatic vegetation in early spring.
Voice: Mallard-like *kwuk-kwuk-kwuk-kwuk* in flight.
Similar Species: *Common Loon* (p. 75): larger; heavier bill; lacks white speckling on back in nonbreeding plumage.
Best Sites: L. Erie and large reservoirs in November; Alum Creek, Hoover and C.J. Brown reservoirs.

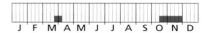

J	F	M	A	M	J	J	A	S	O	N	D

COMMON LOON

Gavia immer

The quavering wail of the Common Loon is one of the classic sounds of northern lakes during breeding season; unfortunately this haunting call is rarely heard in Ohio. • These divers are well adapted to their aquatic lifestyle: most birds have hollow bones, but loons have nearly solid bones that make them less buoyant. As a result, loons float low on the water, resembling partially submerged submarines. Also, these birds' feet are placed well back on their bodies for underwater propulsion. There are records of loons being snared in fishing nets at depths of up to 200 feet. • These excellent underwater hunters feast on a variety of small fish. When diving for food, they may emerge far away from the point of entry, leaving the observer to wonder where they went. • Peak numbers occur in November, when over 400 have been seen together on large reservoirs, and one-day counts of more than 900 have been made along Lake Erie. They are very rare in winter.

nonbreeding

nonbreeding

ID: *Breeding:* green black head; stout, thick, black bill; white "necklace"; white breast and underparts; black-and-white "checkerboard" upperparts; red eyes. *Nonbreeding:* much duller plumage; sandy brown back; light underparts. *In flight:* wings beat constantly; hunchbacked appearance; legs trail behind tail.
Size: *L* 28–35 in; *W* 4–5 ft.
Habitat: deep lakes with unfrozen water.
Nesting: does not nest in Ohio.
Feeding: pursues small fish underwater; occasionally eats large, aquatic invertebrates and larval and adult amphibians.

Voice: alarm call is a quavering tremolo, often called "loon laughter"; contact call is a long but simple wailing note. *Male:* territorial call is an undulating, complex yodel.
Similar Species: *Red-throated Loon* (p. 74): smaller; slender bill; red throat in breeding plumage; sharply defined white face and white-spotted back in nonbreeding plumage.
Best Sites: *In migration:* L. Erie and inland lakes; Alum Creek Reservoir.

J F M A M J J A S O N D

PIED-BILLED GREBE

Podilymbus podiceps

P ied-billed Grebes nest in Ohio's larger wetlands, and are most often detected by their calls, which sound like loud, maniacal laughter. • This species has declined, along with most other wetland birds, because of habitat loss. The Pied-billed has the widest distribution of any North American grebe, and is most closely related to the extinct Giant Pied-billed Grebe *(Podilymbus gigas)* of Lake Atitlan, Guatemala. • Masterful swimmers, these grebes can sink stealthily from sight as if pulled from below, or can swim with only their eyes above the water's surface. Hence, the colloquial name "Helldiver." • Pied-billed Grebes build their floating nests among sparse vegetation, a strategy that allows them to see predators approaching. When frightened by an intruder, they cover their eggs and slide underwater, leaving a nest that looks like nothing more than a mat of debris. • This is still a common species in migration, and may be expected on nearly any water body, even the bigger rivers, though it is rarely seen in flight as most migration is nocturnal. • As with other grebes, the boldly striped young often ride on the adult's back.

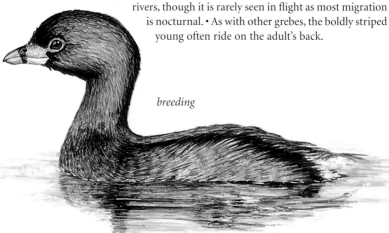

breeding

ID: *Breeding:* all-brown body; black ring on pale bill; laterally compressed "chicken bill"; black throat; very short tail; white undertail coverts; pale belly; pale eye ring. *Nonbreeding:* yellow eye ring; yellow bill lacks black ring; white "chin" and throat; brownish crown.
Size: *L* 12–15 in; *W* 16 in.
Habitat: wetlands with emergent vegetation. *In migration:* lakes, ponds and rivers.
Nesting: in sheltered ponds and marshes; floating platform nest of wet and decaying plants is anchored to or placed among emergent vegetation; pair incubates 4–5 white to buff eggs and raises the striped young together.

Feeding: makes shallow dives and gleans the surface for aquatic invertebrates, small fish, adult and larval amphibians and occasionally aquatic plants.
Voice: loud, whooping call begins quickly, then slows down: *kuk-kuk-kuk cow cow cow cowp cowp cowp.*
Similar Species: *Eared Grebe* (p. 79) and *Horned Grebe* (p. 77): much thinner bills; black-and-white heads and dusky white flanks contrast with dark backs in non-breeding plumage.
Best Sites: *Summer:* Magee Marsh WA; Big Island WA. *In migration:* any body of water.

J F M A M J J A S O N D

HORNED GREBE

Podiceps auritus

A Horned Grebe in breeding plumage is a striking sight, with its golden head tufts set off by its rufous and black body feathers. Breeding plumage is primarily acquired after the birds arrive on their breeding grounds, so most birds in Ohio are seen in the drab, black-and-white nonbreeding plumage. • This species' numbers, as it passes through Ohio, vary from year to year. This bird is rare in winter and summering individuals are almost unknown. • Unlike the fully webbed front toes of many swimming birds, grebe toes are individually webbed or lobed—the three forward-facing toes have individual flanges that are not connected to the other toes. • Although we do not see the display in Ohio, Horned Grebes, like many other grebes, perform spectacular and elaborate pair-bonding ceremonies—the avian equivalent of synchronized swimming. • Both the Horned Grebe's common name and its scientific name, *auritus* (eared), refer to the golden feather tufts, or "horns," that these grebes acquire in breeding plumage.

nonbreeding

ID: *Breeding*: rufous neck and flanks; black head; golden "ear" tufts ("horns"); black back; white underparts; red eyes; flat crown. *Nonbreeding*: lacks "ear" tufts; black upperparts; white "cheek," foreneck and underparts. *In flight*: wings beat constantly; hunchbacked appearance; legs trail behind tail.
Size: *L* 12–15 in; *W* 18 in.
Habitat: wetlands, lakes and ponds.
Nesting: does not nest in Ohio.
Feeding: makes shallow dives and gleans the water's surface for aquatic insects, crustaceans, mollusks, small fish, and adult and larval amphibians.

Voice: loud series of croaks and shrieking notes and a sharp *keark keark* during courtship; usually quiet outside the breeding season.
Similar Species: *Eared Grebe* (p. 79): black neck in breeding plumage; black "cheek" and darker neck in nonbreeding plumage, with slighter, upturned, dark-tipped bill. *Pied-billed Grebe* (p. 76): thicker, stubbier bill; mostly brown body. *Red-necked Grebe* (p. 78): much larger; far bigger bill; dark eyes; lacks "ear" tufts; white "cheek" in breeding plumage.
Best Sites: best on L. Erie; any body of water, especially larger lakes; often frequents marshes in spring migration.

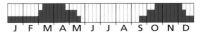

J F M A M J J A S O N D

RED-NECKED GREBE

Podiceps grisegena

The discovery of a Red-necked Grebe is always exciting, as this species, along with the Eared Grebe, is Ohio's rarest regularly occurring grebe. Large water bodies are the best places to look—most Red-necks are encountered on Lake Erie or on big reservoirs. • The Red-necked Grebe is a rare spring and fall migrant in Ohio, and some may winter on open waters of Lake Erie. Occasionally though, larger than normal flights pass through, as happened in 1994, when more than 60 birds were reported across the state. • Red-necked Grebes have a propensity for mistaking iced-over roadways and fields for water, and will sometimes "ground" themselves in large numbers during ice storms. • Red-necked Grebes may be confused with loons because of their large size, although these grebes are fairly unmistakable in breeding plumage. Unfortunately, this plumage is not often seen in Ohio.

nonbreeding

ID: *Breeding:* rusty neck; whitish "cheek"; black crown; straight, heavy bill is dark above and yellow underneath; black upperparts; light underparts; dark eyes. *Nonbreeding:* grayish white foreneck, "chin" and "cheek."
Size: *L* 17–22 in; *W* 24 in.
Habitat: open, deep lakes and sometimes large wetlands.
Nesting: does not nest in Ohio.
Feeding: dives and gleans the water's surface for small fish, aquatic invertebrates and amphibians.

Voice: often-repeated, laughlike, excited *ah-ooo ah-ooo ah-ooo ah-ah-ah-ah-ah.*
Similar Species: *Horned Grebe* (p. 77): dark "cheek" and golden "horns" in breeding plumage; red eyes, all-dark bill and bright white "cheek" in nonbreeding plumage. *Eared Grebe* (p. 79): black neck in breeding plumage; black "cheek" in nonbreeding plumage. *Western Grebe* (p. 331): red eyes; black-and-white neck in breeding plumage. *Red-throated Loon* (p. 74) and *Common Loon* (p. 75): larger; paler throat; lack yellowish bill.
Best Sites: L. Erie; also any of the larger reservoirs.

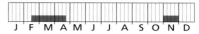

J F M A M J J A S O N D

EARED GREBE

Podiceps nigricollis

The first Eared Grebe wasn't recorded in Ohio until 1941, possibly because it was overlooked owing to confusion with nonbreeding Horned Grebes, which are similar in appearance. A few Eared Grebes are now seen here annually. This species also inhabits parts of Europe, Asia, Central Africa and South America, and it is the most abundant grebe not only in North America, but also the world. • Eared Grebes undergo cyclical periods of atrophy and hypertrophy throughout the year, meaning that their internal organs and pectoral muscles shrink or swell, depending on whether or not the birds need to migrate. This strategy leaves Eared Grebes flightless for 9 to 10 months per year—a longer period than any other flying bird in the world. • Like other grebes, the Eared Grebe eats feathers. The feathers pack the digestive tract, and it is thought that they protect the stomach lining and intestines from sharp fish bones or parasites, or perhaps slow the passage of food, allowing more time for complete digestion.

nonbreeding

ID: *Breeding:* black neck, "cheek," forehead and back; red flanks; fanned-out, golden "ear" tufts; white underparts; thin, straight bill; red eyes; slightly raised crown. *Nonbreeding:* dark "cheek" and upperparts; light underparts; dusky upper foreneck and flanks. *In flight:* wings beat constantly; hunchbacked appearance; legs trail behind tail.
Size: *L* 11½–14 in; *W* 16 in.
Habitat: wetlands; larger lakes including L. Erie; rarely small ponds.
Nesting: does not nest in Ohio.
Feeding: makes shallow dives and gleans the water's surface for aquatic insects,

crustaceans, mollusks, small fish and larval and adult amphibians.
Voice: mellow *poo-eee-chk* during courtship; usually quiet outside the breeding season.
Similar Species: *Horned Grebe* (p. 77): thicker, straight bill; rufous neck in breeding plumage; white "cheek" in nonbreeding plumage. *Pied-billed Grebe* (p. 76): thicker, stubbier bill; mostly brown body. *Red-necked Grebe* (p. 78): larger overall; longer bill; rusty neck and whitish "cheek" in breeding plumage; grayish white "cheek" in nonbreeding plumage.
Best Sites: C.J. Brown Reservoir; possible on L. Erie and any interior reservoir.

J F M A M J J A S O N D

AMERICAN WHITE PELICAN

Pelecanus erythrorhynchos

Few birds are as easily recognized as the White Pelican, and though still uncommon and irregular, an overall expansion of the population has resulted in increased sightings of these large birds in Ohio. Normally they appear as single birds, but occasionally small groups of up to six are seen. • Groups of foraging pelicans deliberately herd fish into schools, then dip their bills and scoop up the prey. In a single scoop, a pelican can hold over 3 gallons of water and fish, which is about two to three times as much as its stomach can hold. White Pelicans eat about 4 pounds of fish per day, but because they prefer nongame fish they do not pose a threat to the potential catches of fishermen. • Pelicans are inclined to soar, and sometimes will catch thermals and rise so high that they disappear from sight. All other large, white birds with black wing tips fly with their necks extended; the American White Pelican is the only one to fly with its neck pulled back toward its wings.

nonbreeding

ID: very large, stocky, white bird; long, orange bill and throat pouch; black primary and secondary wing feathers; short tail; naked, orange skin patch around eye. *Breeding:* small, keeled plate develops on upper mandible; pale yellow crest on back of head. *Nonbreeding* and *immature:* white plumage is tinged with brown.
Size: *L* 4½–6 ft; *W* 9 ft.
Habitat: large lakes or rivers.

Nesting: does not nest in Ohio.
Feeding: surface dips for small fish and amphibians; small groups of pelicans often feed cooperatively by herding fish into large concentrations.
Voice: generally quiet; adults rarely issue piglike grunts.
Similar Species: no other large, white bird has a long beak with a pouch.
Best Sites: L. Erie; Hoover and C.J. Brown reservoirs; any inland reservoir; sometimes large rivers.

J F M A M J J A S O N D

DOUBLE-CRESTED CORMORANT

Phalacrocorax auritus

No other Ohio waterbird has undergone the tremendous population explosion of the Double-crested Cormorant. Unregulated use of pesticides such as DDT decimated its numbers from the 1950s to the 1970s. These chemicals were detrimental to the birds' reproduction, and the cormorant's comeback is indicative of the disappearance of these toxins from the environment. Now, aggregations of 10,000 or more birds gather in western Lake Erie and Double-crested Cormorants may be seen on any inland water body. In fact, there are so many in the Great Lakes that control measures have been proposed, as some think that these birds are harmful to sport fish populations. Cormorants are strictly fish eaters but research suggests they primarily consume nongame species. • Increased breeding colonies have had harmful effects on other bird species. Cormorants will nest in heronries and can displace herons, as has been the case at West Sister Island in Lake Erie.

breeding

juvenile

ID: all-black body; long, crooked neck; thin bill, hooked at tip; blue eyes. *Breeding:* throat pouch becomes intense orange yellow; fine, black plumes trail from "eyebrows." *Immature:* brown upperparts; buff throat and breast; yellowish throat patch. *In flight:* rapid wingbeats; kinked neck.
Size: *L* 26–32 in; *W* 4½ ft.
Habitat: large lakes and rivers.
Nesting: colonial; on a low-lying island, often with pelicans, terns and gulls, or high in a tree; nest platform is made of sticks, aquatic vegetation and guano; pair

incubates 3–6 bluish white eggs for 25–33 days; young are fed by regurgitation.
Feeding: long underwater dives to depths of 30 ft or more; eats small fish or rarely, amphibians and invertebrates; brings prey to the surface to swallow.
Voice: generally quiet; may issue piglike grunts or croaks, especially near nesting colonies.
Similar Species: *Common Loon* (p. 75): shorter neck; black bill lacks hooked tip; spotted back in breeding plumage; white underparts in nonbreeding plumage.
Best Sites: any large lake or reservoir; most common along L. Erie, although rare there in winter.

J F M A M J J A S O N D

AMERICAN BITTERN

Botaurus lentiginosus

The American Bittern produces one of the most bizarre and unmistakable sounds emanating from marshes. Often given at night, its "song" is suggestive of a stake-driver operated underwater—a loud *oonka-choonk,* given repeatedly—and justifies the nickname "Thunder-Pumper." • Although hearing an American Bittern isn't difficult, seeing one is. The bittern relies on its cryptic plumage for camouflage; it blends in among the cattails so well that you might walk within five feet of one and miss it. When alarmed, the bird freezes, compressing its body feathers and pointing its bill skyward. It may even sway slightly like a reed in the breeze. • Unfortunately, American Bitterns are symbolic of wetland losses in Ohio, and are now listed as endangered. Ohio is estimated to have lost over 90 percent of its original wetlands, and fascinating denizens of our marshes, like the American Bittern, have declined accordingly.

nonbreeding

ID: brown upperparts; brown streaking from "chin" through breast; straight, stout bill; yellow legs and feet; black outer wings; black streaks from bill down neck to shoulder; short tail.
Size: *L* 23–27 in; *W* 3½ ft.
Habitat: marshes, wetlands and lake edges with tall, dense grasses, sedges, bulrushes and cattails.
Nesting: singly; above the waterline in dense vegetation; nest platform is made of grass, sedges and dead reeds; nest often has separate entrance and exit paths; female incubates 3–5 pale olive or buff eggs for 24–28 days.

Feeding: patient stand-and-wait predator; strikes at small fish, crayfish, amphibians, reptiles, mammals and insects.
Voice: deep, slow, resonant, repetitive *pomp-er-lunk* or *oonka-choonk,* often heard in the evening or at night.
Similar Species: *Black-crowned Night-Heron* (p. 90), *Yellow-crowned Night-Heron* (p. 91), *Least Bittern* (p. 83) and *Green Heron* (p. 89): immatures lack dark streak from bill to shoulder; immature nightherons have white-flecked upperparts.
Best Sites: generally occurs only in the largest wetlands, such as those buffering western L. Erie, Big Island WA and Springville Marsh SNP; migrants may appear in odd places.

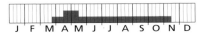

J F M A M J J A S O N D

LEAST BITTERN

Ixobrychus exilis

The Least Bittern is the smallest North American heron and one of the most secretive. It inhabits freshwater marshes where tall, impenetrable stands of cattails conceal most of its movements. • Like the larger American Bittern, habitat loss has led to the Least Bittern's current listing as threatened in Ohio. A relatively new threat to its dense, cattail-marsh habitat is the invasion of nonnative plants such as purple loosestrife and *Phragmites*, which can displace the cattails. • Where suitable habitat is present, Least Bitterns can occur in surprising densities, sometimes as many as 15 nests per 2^1/$_2$ acres. • At productive feeding sites, these birds may build small platforms from which they snare small fish and dragonflies. • One of the rarest color forms of any North American bird was the "Cory's Least Bittern," rufous chestnut in color, with one Ohio record in 1907. Rediscovery of this type would be a spectacular ornithological event!

♂

nonbreeding

ID: rich buff flanks and sides; streaking on foreneck; white underparts; mostly pale bill; short tail; dark primary and secondary feathers. *Male:* black crown and back. *Female* and *immature:* chestnut brown head and back; immature has darker streaking on breast and back. *In flight:* large, buffy shoulder patches.
Size: *L* 11–15 in; *W* 17 in.
Habitat: large, freshwater marshes with cattails and other dense, emergent vegetation.
Nesting: mostly the male constructs a platform of dry plant stalks in marsh vegetation; nest site is well concealed within dense vegetation; pair incubates 4–5 pale green or blue eggs for 17–20 days; pair feeds the young by regurgitation.

Feeding: stabs prey with its bill; eats mostly small fish; also takes large insects, tadpoles, frogs, small snakes, leeches and crayfish; may build a hunting platform.
Voice: *Male:* guttural *uh-uh-uh-oo-oo-oo-ooah*. *Female:* a ticking sound. Both issue a *tut-tut* call or a *koh* alarm call.
Similar Species: *American Bittern* (p. 82): larger; bold, brown streaking on underparts; black streak from bill to shoulder. *Black-crowned Night-Heron* (p. 90) and *Yellow-crowned Night-Heron* (p. 91): immatures have dark brown upperparts with white flecking. *Green Heron* (p. 89): immature has dark brown upperparts.
Best Sites: large marshes along western L. Erie, such as Mallard Marsh WA and Magee Marsh WA.

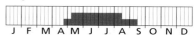

| J | F | M | A | M | J | J | A | S | O | N | D |

GREAT BLUE HERON
Ardea herodias

This majestic bird is our largest heron, as well as the most common and conspicuous. Many nonbirders mistakenly refer to these birds as "cranes," but herons fly with their necks folded into their bodies, and Sandhill Cranes fly with their necks outstretched. • The Great Blue Heron has fared well in association with humans, and there are nesting colonies in most Ohio counties. This stately wader can be seen in nearly any type of water body, and unlike most other herons, is quite hardy and may be seen in winter wherever open water persists. • Great Blues are quite varied in their culinary selection—no small animal is safe near them, as they capture fish, snakes, frogs, mice and even small birds. • Great Blue Herons sometimes form large nesting colonies, occasionally with 300 or more nests, with the largest peaking at 2400 nests on West Sister Island.

breeding

ID: large, blue gray bird; long, curving neck; long, dark legs; straight, yellow bill; chestnut brown thighs. *Breeding:* richer colors; plumes streak from crown and throat. *In flight:* neck folds back over shoulders; legs trail behind body; slow, steady wingbeats.
Size: *L* 4–4½ ft; *W* 6 ft.
Habitat: forages along the edges of rivers, lakes and marshes; also seen in fields and wet meadows.
Nesting: colonial; usually in a tree; stick-and-twig platform is often added onto over years, and can be up to 4 ft in diameter; pair incubates 4–7 pale blue eggs for about 28 days.
Feeding: patient, stand-and-wait predator; strikes at small fish, amphibians, small mammals, aquatic invertebrates and reptiles; rarely scavenges.
Voice: usually quiet away from the nest; occasionally a deep, harsh *frahnk frahnk frahnk,* mostly during takeoff.
Similar Species: *Green Heron* (p. 89), *Black-crowned Night-Heron* (p. 90) and *Yellow-crowned Night-Heron* (p. 91): much smaller; shorter legs. *Egrets* (pp. 85–88): all are predominantly white. *Sandhill Crane* (p. 113): red "cap"; flies with neck outstretched. *Little Blue Heron* (p. 87): smaller; dark overall; purplish head; lacks yellow on bill. *Tricolored Heron* (p. 331): smaller; darker upperparts; white underparts.
Best Sites: widespread; abundant in western L. Erie marshes.

J F M A M J J A S O N D

GREAT EGRET

Ardea alba

Great Egrets lend a very Floridian feel to our wetlands, and we are fortunate to be able to enjoy them at all. Plume hunters nearly exterminated this species and many others in the early 1900s in their quest to provide feathers for the millinery trade. The National Audubon Society was formed to protect herons from this senseless plunder, and the Great Egret remains its symbol to this day. • Although egrets may be found scattered anywhere in Ohio during migration and late summer dispersals, they reach peak abundance in the large marshes of western Lake Erie, appearing there by the dozens. • Great Egrets were discovered nesting in Ohio in 1940, and the few colonies along Lake Erie have steadily increased in number since then. These birds are not nearly as hardy as Great Blue Herons, and there are very few winter records.

nonbreeding

breeding

ID: all-white plumage; black legs; yellow bill. *Breeding:* white plumes trail from throat and rump; green skin patch between eyes and base of bill. *In flight:* neck folds back over shoulders; legs extend backward.
Size: *L* 3–3½ ft; *W* 4 ft.
Habitat: marshes, open riverbanks, irrigation canals and lakeshores.
Nesting: colonial, often with Great Blue Herons and Double-crested Cormorants,

but may nest in isolated pairs; in a tree or tall shrub; pair builds a platform of sticks and incubates 3–5 pale blue green eggs for 23–26 days.
Feeding: patient stand-and-wait predator; occasionally stalks slowly, stabbing at frogs, lizards, snakes and small mammals.
Voice: rapid, low-pitched, loud *cuk-cuk-cuk*.
Similar Species: *Snowy Egret* (p. 86): smaller; black bill; yellow feet. *Cattle Egret* (p. 88): smaller; stockier; orange bill and legs.
Best Sites: Magee Marsh WA; Ottawa NWR.

J F M A M J J A S O N D

SNOWY EGRET

Egretta thula

Like the Great Egret, Snowy Egret populations were once decimated to provide adornments for women's hats. In 1886, an ounce of their plumes was worth $32—more than the cost of an ounce of gold! • This species is currently listed as endangered in the state because of its very rare breeding status. A few pairs were found in the huge West Sister Island heronry in 1983, and the species maintains a tenuous foothold there. • Snowy Egrets are dynamic feeders, engaging in all manner of activity to capture food. They can be quite active, dashing and hopping about, and stirring the substrate with their feet. An interesting form of feeding is "wing-shading," in which the egret creates pockets of shade by standing in the shallows with its wings spread. When a fish succumbs to the lure of the cooler, shaded spot, the egret promptly seizes and eats it.

breeding

nonbreeding

ID: white plumage; black bill and legs; bright yellow feet. *Breeding:* long plumes on throat and rump; erect crown; orange red lores. *Immature:* similar to adult, but with more yellow on legs. *In flight:* yellow feet are obvious.

Size: *L* 22–26 in; *W* 3½ ft.

Habitat: large, mixed-emergent marshes; occasionally along the margins of ponds, lakes and other wetlands; rarely in flooded fields.

Nesting: colonial, often among other herons; in a tree or tall shrub; pair builds a platform of sticks and incubates 3–5 pale blue green eggs for 20–24 days.

Feeding: stirs wetland muck with its feet; stands and waits with wings held open; occasionally hovers and stabs; eats small fish, amphibians and invertebrates.

Voice: low croaks; bouncy *wulla-wulla-wulla* on breeding grounds.

Similar Species: *Great Egret* (p. 85): larger; yellow bill; black feet. *Cattle Egret* (p. 88): orange yellow legs and bill. *Little Blue Heron* (p. 87): juvenile has pale greenish legs and 2-tone bill.

Best Sites: along the causeway at Magee Marsh WA or in adjacent Ottawa NWR.

| J | F | M | A | M | J | J | A | S | O | N | D |

LITTLE BLUE HERON

Egretta caerulea

Many southern wading birds, such as herons, storks and ibis, engage in a northward, late summer migration, termed a postbreeding dispersal, that sometimes brings large numbers of birds to Ohio and beyond. Little Blue Herons, the best example of this phenomenon in Ohio, have staged periodic large invasions—the biggest in 1930 when nearly 1200 were seen statewide. Only a few of these birds, though, are found in a typical year. • The Little Blue Heron is more dimorphic in plumage (showing different coloration) than our other herons. First-year birds are white and can be confused with egrets—they take two years to reach the distinct, completely dark plumage of adult birds. • Feeding behavior is often helpful in distinguishing herons. Larger herons seem graceful even while lunging for a fish, whereas the Little Blue Heron often seems tentative and stiff in its hunting maneuvers as it awkwardly jabs at prey. • Little Blues once bred rarely on West Sister Island, but are not currently known to nest in the state.

breeding

nonbreeding

ID: medium-sized heron; purplish blue overall. *Breeding:* shaggy, maroon-colored head and neck; black legs and feet. *Nonbreeding:* smooth, purple head and neck; dull green legs and feet. *Juvenile:* white, becoming mottled with gray as bird ages; dusky tips on primaries; greenish legs; thick bill.
Size: *L* 24 in; *W* 3½ ft.
Habitat: marshes, ponds, larger interior wetlands, lakes, streams and meadows.
Nesting: nests in a shrub or tree above water; pair builds large nest of sticks; pair

incubates 3–5 pale greenish blue eggs for 22–24 days.
Feeding: patient, stand-and-wait predator; may also wade slowly to stalk prey; eats mostly fish, crabs and crayfish; also takes grasshoppers and other insects, frogs, lizards, snakes and turtles.
Voice: generally silent.
Similar Species: *Egrets* (pp. 85–88): resemble juvenile Little Blues, but lack greenish legs and dark wing tips.
Best Sites: large marshes buffering western L. Erie; individuals may appear in larger interior wetlands, particularly in invasion years.

J F M A M J J A S O N D

CATTLE EGRET

Bubulcus ibis

These African natives have managed to colonize every continent on the globe except Antarctica. Cattle Egrets were first recorded in North America in the early 1950s; Ohio had its first report in 1958. These birds have been discovered nesting in very small numbers in a Lake Erie-area heronry, and are listed as endangered in Ohio. In migration, they may be seen anywhere in the state. • The feeding behavior of Cattle Egrets is quite different from our other herons. They shun wetlands, foraging instead in upland pastures, fields and even yards, often associating with livestock. They follow cattle, seeking the insects kicked up by their hooves, and sometimes even perch on the beasts' backs! • Cattle Egrets are quite gregarious and are almost always seen in small flocks. In recent years, their numbers appear to have declined.

breeding

nonbreeding

ID: mostly white; yellow orange bill and legs. *Breeding:* long plumes on throat and rump; buff orange throat, rump and crown; orange red legs and bill; purple lores. *Immature:* similar to adult, but with black feet and dark bill.
Size: *L* 19–21 in; *W* 3 ft.
Habitat: agricultural fields and marshes.
Nesting: colonial; often among other herons; in a tree or tall shrub; male supplies sticks for the female who builds a platform or shallow bowl; pair incubates 3–4 pale blue eggs for 21–26 days.
Feeding: picks grasshoppers, other insects, worms, small vertebrates and spiders from fields; often associated with livestock.
Voice: generally silent.
Similar Species: *Great Egret* (p. 85): larger; black legs and feet. *Snowy Egret* (p. 86): black legs; yellow feet; black bill.
Best Sites: pastures near the shore of western L. Erie, particularly fields with cattle.

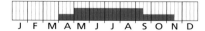

J F M A M J J A S O N D

GREEN HERON

Butorides virescens

These crow-sized herons tend to be secretive, and their favored haunts are small streams, rivers and ponds. They are fairly common throughout Ohio—probably all counties have breeding pairs, though there are no overwintering records. • The Green Heron is often first detected by its loud, penetrating *kyow* call. This is frequently given when the bird is flushed, and is often accompanied by an explosive release of feces. Many birds release feces when startled in order to drop weight and improve takeoff. • This species is not normally colonial, and typically builds solitary nests located fairly low in dense growth. These nests rival those of the Mourning Dove for flimsiness, being little more than a loose, unlined platform of sticks. • Green Herons use "tools" to improve fishing success. They are known to drop feathers or debris on the water's surface, apparently in an attempt to lure small fish.

ID: stocky; green black crown; chestnut brown face and neck; white foreneck and belly; blue gray back and wings mixed with iridescent green; short, green yellow legs; bill is dark above and greenish below; short tail. *Breeding male:* bright orange legs. *Immature:* heavy streaks along neck and underparts; dark brown upperparts.
Size: *L* 15–22 in; *W* 26 in.
Habitat: freshwater marshes, lakes, ponds and streams with dense shoreline or emergent vegetation.
Nesting: singly or in small, loose groups; male begins and female completes construction of a stick platform in a tree or shrub, close to water; pair incubates 3–5 pale blue green eggs for 19–21 days; young are fed by regurgitation.

Feeding: eats mostly small fish; also takes frogs, tadpoles, crayfish, aquatic and terrestrial insects, small rodents, snakes, snails and worms; stabs prey with its bill after slowly stalking or standing and waiting.
Voice: generally silent; alarm and flight call is a loud *kowp, kyow* or *skow;* aggression call is a harsh *raah.*
Similar Species: *Black-crowned Night-Heron* (p. 90): larger; white "cheek"; pale gray and white neck; 2 long, white plumes trail down from crown; immature has streaked face and white flecking on upperparts. *Least Bittern* (p. 83): buffy yellow shoulder patches, sides and flanks. *American Bittern* (p. 82): larger; more tan overall; black streak from bill to shoulder.
Best Sites: smaller watercourses and ponds.

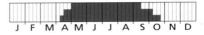

J F M A M J J A S O N D

BLACK-CROWNED NIGHT-HERON

Nycticorax nycticorax

This odd bird nearly rivals the Cattle Egret in its cosmopolitan distribution—Black-crowned Night-Herons occupy every continent except Australia and Antarctica. • These birds are given to nocturnal habits and possess large, light-sensitive eyes that give excellent night vision. Fortunately, they often emit a loud, readily identifiable *quark*, otherwise many would pass unnoticed in the dark. • Black-crowneds are rare breeders in Ohio, and are now listed as threatened, with only a few known colonies. The biggest colony, on West Sister Island in Lake Erie, has declined alarmingly from 1200 nests to less than 100 in the last decade. Competition from the Double-crested Cormorant, with its exploding numbers also nesting on West Sister, may be the culprit. • Black-crowned Night-Herons are surprisingly hardy, and small numbers regularly overwinter near the warm-water out-flows of Lake Erie power plants.

immature

breeding

ID: black "cap" and back; white "cheek," foreneck and under-parts; gray neck and wings; dull yellow legs; stout black bill; large, red eyes. *Breeding:* 2 white plumes trail down from crown. *Immature:* lightly streaked underparts; brown upper-parts with white flecking.
Size: *L* 23–26 in; *W* 3½ ft.
Habitat: shallow cattail and bulrush marshes, lakeshores and along slowly mov-ing rivers.
Nesting: colonial, often with Great Blue Herons and Great Egrets; in a tree or shrub; loose nest platform of twigs and sticks; male gathers the material and female builds the nest; pair incubates 3–4 pale green eggs for 21–26 days.
Feeding: often at dusk; patient, stand-and-wait predator; stabs for small fish,

amphibians, aquatic invertebrates, reptiles, young birds and small mammals.
Voice: deep, guttural *quark* or *wok,* often heard as the bird takes flight.
Similar Species: *Yellow-crowned Night-Heron* (p. 91): white plumes, crown and "cheek" patch on black head; gray back; immature has all-dark bill, longer neck and legs, and narrower breast streaks. *Green Heron* (p. 89): chestnut brown face and neck; blue gray back with green iri-descence; immature has heavily streaked underparts. *American Bittern* (p. 82): similar to immature Black-crowned Night-Heron, but bittern has black streak from bill to shoulder and is lighter tan overall.
Best Sites: Gilmore Ponds Preserve; west-ern L. Erie marshes.

J F M A M J J A S O N D

YELLOW-CROWNED NIGHT-HERON

Nyctanassa violacea

The Yellow-crowned Night-Heron is Ohio's rarest regularly breeding heron, with just a few colonies totaling from 12 to 15 nests. Listed as threatened in Ohio, this bird is not one that you would normally stumble upon. • Ohio lies at the extreme northern limit of this bird's range. Very small colonies have been present in Columbus and Dayton for years, and there are probably a few other nesting pairs here and there. In Ohio, the Yellow-crowned Night-Heron tends to be associated with large rivers, primarily the Great Miami and the Scioto.
• Despite its name, the Yellow-crowned Night-Heron commonly feeds by day as well as by night, though poor lighting is a standard criterion for suitable breeding and hunting habitat.
• A noninvasive way to observe Yellow-crowned Night-Herons at night is to shine a light on the floating yellow barrels strung across Columbus's Griggs Reservoir, just above the dam.

breeding

immature

ID: gray overall; black head with buffy white crown and white "cheek." *Breeding:* long, white head plumes. *In flight:* feet extend well beyond tail.
Size: *L* 24 in; *W* 3½ ft.
Habitat: in wetlands and along lowland rivers.
Nesting: singly or in colonies; in a tree or shrub near water; pair builds nest of heavy twigs and lines it with finer plant materials; pair incubates 2–4 pale greenish blue eggs for 21–25 days.
Feeding: stands and waits or wades slowly in shallow water to catch crabs, crayfish, other freshwater invertebrates, fish, frogs and insects; forages alone or in small groups.
Voice: a loud *quok!* that is less harsh and slightly higher in pitch than the Black-crowned Night-Heron's call.
Similar Species: *Black-crowned Night-Heron* (p. 90): black crown and back; shorter legs; thinner bill; immature has yellowish bill, shorter neck and legs and broader, smudgy breast streaks.
Best Sites: Griggs Reservoir (Columbus); generally expected only in Columbus and Dayton.

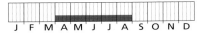

J F M A M J J A S O N D

BLACK VULTURE

Coragyps atratus

This avian scavenger is a southerner, but unlike Black Vultures in the south, which often frequent cities and dumps and generally associate with people, our birds are found largely in scenic and undeveloped areas of southern Ohio. • Although its few Ohio outposts represent the extreme northern limit of its range, this bird's population appears to be expanding, as evidenced by a relatively new colony established in the Holmes County region. • Black Vultures often associate with Turkey Vultures, but lack that species' soaring ability and must frequently flap their wings to stay aloft. They also lack the Turkey Vultures' keen olfactory sense, which comes in handy to locate the rotting carcasses that are the staple of both species' diets. Consequently, Black Vultures will often lurk around patrolling Turkey Vultures, exploiting their more advanced abilities to locate food, then following them to the often-hideous banquet. There, the Black Vultures' more aggressive nature enables them to shoulder the Turkey Vultures aside and take first dibs.

ID: all-black plumage with grayish head, legs and feet; whitish base to primaries.
Size: *L* 25 in; *W* 5 ft.
Habitat: forages over open country throughout its range, but tends to roost and nest in forested areas.
Nesting: in a cave, thicket, abandoned building or other sheltered site; no nest is built; female lays 2 creamy white to buff eggs in a large tree cavity or on the ground; both adults incubate the eggs and raise the young.
Feeding: carrion forms bulk of diet; also eats eggs, small reptiles and amphibians, small mammals and food waste from garbage dumps; occasionally some plant material.
Voice: generally silent; may hiss or grunt at nest sites or around communal food sources.
Similar Species: *Turkey Vulture* (p. 93): pink head; 2-tone wings (dark wing linings and light flight feathers); longer, narrower wings and tail; wings usually held in a shallow V-shape.
Best Sites: Clear Creek Valley (Hocking Co. and Fairfield Co.); Sugar Grove (roosts on large powerline towers on US Rte. 33); Ohio Brush Creek Valley (Adams Co.); Granville area (Licking Co.).

J F M A M J J A S O N D

92

TURKEY VULTURE

Cathartes aura

Turkey Vultures are unparalleled flyers and can exploit the slightest thermal, allowing them to drift effortlessly for hours. While airborne, these birds use their remarkable sense of smell to locate dead animals, which constitute nearly their entire diet. So keen is this ability that they can even find carcasses out of sight under the forest canopy. • Turkey Vultures specialize in eating small items that can be consumed rapidly. This helps to combat pirating by the more aggressive Black Vultures, which sometimes follow Turkey Vultures to food. • Turkey Vultures hide their nests in dark recesses such as hollow logs, cliffs and old buildings. Caution is advised upon discovery of a nest: the young will projectile-regurgitate their food at invaders as a defense mechanism. • Recent studies have shown that American vultures are more closely related to storks rather than to hawks and falcons as was previously thought. Molecular similarities with storks, plus the shared tendency to defecate on their own legs to cool down, strongly support this taxonomic reclassification.

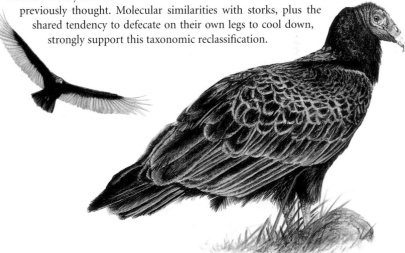

ID: all black; bare, red head. *Immature:* gray head. *In flight:* head appears small; silver gray flight feathers; black wing linings; wings are held in a shallow "V"; rocks from side to side while soaring.
Size: L 26–32 in; W 5½–6 ft.
Habitat: usually seen flying over open country, shorelines or roads.
Nesting: in a cave crevice or among boulders; rarely a hollow stump or log; no nest material is used; female lays 2 dull white eggs on bare ground; pair incubates the eggs for up to 41 days; young are fed by regurgitation.
Feeding: entirely on carrion (mostly mammalian); sometimes seen at road kills.

Voice: generally silent; occasionally produces a hiss or grunt if threatened.
Similar Species: *Golden Eagle* (p. 104) and *Bald Eagle* (p. 95): lack silvery gray wing linings; wings are held flat in flight; do not rock when soaring; heads are more visible in flight. *Rough-legged Hawk* (p. 103): dark morph has multi-banded tail with black subterminal band and whitish wing linings with dark bars. *Black Vulture* (p. 92): gray head; silvery tips on otherwise black wings; wings are held flat in flight.
Best Sites: statewide, particularly in rural areas; inexplicably returns to Hinckley "precisely" on March 15 every year.

J F M A M J J A S O N D

OSPREY

Pandion haliaetus

The Osprey is almost exclusively a fish-eater; in fact, many people know it as "Fish Hawk." It catches its underwater prey after a spectacular dive, sometimes from heights of up to 100 feet. After emerging from the water with the catch, the Osprey will always arrange the fish so that it points head forward, for optimal aerodynamics. • Ospreys nearly disappeared from North America when populations were severely reduced by unregulated use of agricultural pesticides in the 1950s and 1960s. These chemicals contaminated fish and negatively affected reproduction of some avian predators at the top of the food chain. By the early 1970s, only a handful of Ospreys were seen in Ohio each year. After the offending pesticides were banned, Osprey populations began to recover. In Ohio today, they are common migrants and the number of breeding pairs has steadily increased, making the Osprey a true environmental success story.

ID: dark brown upperparts; white underparts; dark eye line; light crown; yellow eyes. *Male:* all-white throat. *Female:* fine, dark "necklace." *In flight:* long wings are held in shallow "M"; dark "wrist" patches; brown and white tail bands.
Size: L 22–25 in; W 4½–6 ft.
Habitat: lakes and slowly flowing rivers and streams.
Nesting: on a treetop, usually near water; also on a specially made platform, utility pole or tower up to 100 ft high; massive stick nest is reused over many years; pair incubates 2–4 brown-blotched, yellowish eggs for about 38 days; both adults feed the young, but the male hunts more.

Feeding: dramatic, feet-first dives into water; small fish make up almost all of the diet.
Voice: series of melodious ascending whistles: *chewk-chewk-chewk;* also an often-heard *kip-kip-kip.*
Similar Species: unmistakable with careful observation. *Bald Eagle* (p. 95): larger; holds its wings straighter while soaring; larger bill with yellow base; yellow legs; clean white head and tail on otherwise dark body; lacks white underparts and dark "wrist" patches.
Best Sites: Deer Creek, Alum Creek and Mogadore reservoirs. *In migration:* large water bodies statewide.

| J | F | M | A | M | J | J | A | S | O | N | D |

BALD EAGLE

Haliaeetus leucocephalus

The Bald Eagle, our national symbol, is the largest breeding Ohio raptor and, like the Osprey, feeds primarily on fish. Sometimes an eagle will steal food from an Osprey, resulting in a spectacular aerial chase. • The Bald Eagle was severely affected by DDT and other pesticides, and numbers dropped precipitously in the 1950s and 1960s. By 1979, Ohio Bald Eagles had declined to just four breeding pairs. With the elimination of harmful pesticides, the Bald Eagle has made a dramatic comeback—88 Ohio nests produced 105 young in 2003; Ottawa County alone had 14 nests! A statewide survey in January 2004 found 352 wintering birds, an all-time high. • Bald Eagles do not mature until their fourth or fifth year, only then developing their characteristic white head and tail plumage. • A Bald Eagle nest in Vermilion, Ohio, which toppled in a storm in 1925 after continuous use for 35 years, was 12 feet high by 8 feet wide and weighed almost 2 tons!

immature

ID: white head and tail; dark brown body; yellow bill and feet; broad wings are held flat in flight. *1st-year:* dark overall; dark bill; some white in underwings. *2nd-year:* dark "bib"; white in underwings. *3rd-year:* mostly white plumage; yellow at base of bill; yellow eyes. *4th-year:* light head with dark facial streak; variable pale and dark plumage; yellow bill; paler eyes. *In flight:* broad wings are held flat.
Size: *L* 30–43 in; *W* 5½–8 ft.
Habitat: large lakes and rivers.
Nesting: usually in a tree bordering a lake or large river, but may be far from water; huge stick nest, up to 15 ft across, is reused for many years; pair incubates 1–3 white eggs for 34–36 days; pair feeds the young; young remain in nest until they can fly.

Feeding: captures waterbirds, small mammals and fish at the water's surface; frequently feeds on carrion.
Voice: thin, weak squeal or gull-like cackle: *kleek-kik-kik-kik* or *kah-kah-kah.*
Similar Species: adult is distinctive. *Golden Eagle* (p. 104): dark overall; golden nape; tail faintly banded with white; smaller head and bill; immature has white patch on wings and base of tail. *Osprey* (p. 94): similar to 4th-year Bald Eagle, but holds wings in an M-shape in flight, has dark "wrist" patches and dark bill, and is much smaller.
Best Sites: Killdeer Plains WA; Ottawa NWR; Mosquito Creek Reservoir. *In migration:* statewide.

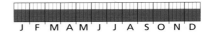

J F M A M J J A S O N D

NORTHERN HARRIER

Circus cyaneus

L ong known as "Marsh Hawk," the Northern Harrier is an extraordinary aerial acrobat. Britain's Royal Air Force was so impressed by this bird's maneuverability that it named its Harrier aircraft after this hawk. • Although identifying raptors at a distance tends to befuddle the casual birder, Harriers present no real problem. The slender body with its long tail, wings held in a slight "V" and a conspicuous white rump add up to a unique set of characteristics. • Northern Harriers generally hunt low over grasslands, continually tilting from side to side, as they seek mice and voles. Harriers are also unusual in that they use their keen hearing to help locate prey—a trait not known among other North American hawks. Harriers have an owl-like facial disc that increases sound perception and helps the birds triangulate on small animals. • The vast grasslands occupying reclaimed strip mines in southeastern Ohio have proven to be excellent habitat for Northern Harriers. In northern Ohio, this species is a rare grassland nester and is listed as endangered.

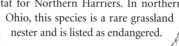

ID: long wings and tail; white rump; black wing tips. *Male:* blue gray to silver gray upperparts; white underparts; indistinct tail bands, except for 1 dark subterminal band. *Female:* dark brown upperparts; streaky, brown-and-buff underparts. *Immature:* rich reddish brown plumage; dark tail bands; streaked breast, sides and flanks.
Size: *L* 16–24 in; *W* 3½–4 ft.
Habitat: open country, including fields, wet meadows, cattail marshes and croplands.
Nesting: on the ground, often on a slightly raised mound in grass, cattails or tall vegetation; shallow depression or platform nest is lined with grass, sticks and cattails; female incubates 4–6 bluish white eggs for 30–32 days.
Feeding: hunts in low, rising and falling flights, often skimming the tops of vegetation; eats small mammals, birds, amphibians, reptiles and some invertebrates.
Voice: most vocal near the nest and during courtship, but generally quiet; high-pitched *ke-ke-ke-ke-ke-ke* near the nest.
Similar Species: *Rough-legged Hawk* (p. 103): broader wings; dark "wrist" patches; black tail with wide, white base; dark belly. *Red-tailed Hawk* (p. 102): lacks white rump and long, narrow tail.
Best Sites: The Wilds; Killdeer Plains WA; Deer Creek SP; large grasslands.

J F M A M J J A S O N D

SHARP-SHINNED HAWK

Accipiter striatus

Accipiters are "bird hawks"—designed for aerial maneuverability and for catching smaller birds on the wing. Short, rounded wings and a rudderlike tail enable quick directional changes, a decided advantage when pursuing songbirds through thickets and brush. • The Sharp-shinned Hawk is the smallest of the three North American accipiters. They are birds of large, unbroken forests, at least during breeding season. Sharpies will visit feeders—for the birds, not the seed—though they are not as frequent backyard raiders as are Cooper's Hawks. • Although most hawks exhibit sexual dimorphism, size difference is most pronounced in the Sharp-shinned Hawk, with males averaging only 57 percent of the size of females. • Sharpies are a common sight at prime Lake Erie viewing sites during March to April raptor migrations, and can be identified from afar by their habit of strafing and badgering larger, more sluggish fellow migrants such as Red-tailed Hawks.

immature

ID: short, rounded wings; long, straight, heavily barred, squared tail; dark barring on pale underwings; blue gray back; red horizontal bars on underparts; red eyes. *Immature:* brown overall; yellow eyes; vertical, brown streaking on breast and belly. *In flight:* flap-and-glide flier; very agile in wooded areas.
Size: *Male: L* 10–12 in; *W* 20–24 in. *Female: L* 12–14 in; *W* 24–28 in.
Habitat: dense to semi-open forests and large woodlots; wooded riparian corridors.
Nesting: in a conifer if available; builds a new stick nest each year or might remodel an abandoned crow nest; female incubates 4–5 brown-blotched, bluish white eggs for 34–35 days; male feeds the female during incubation.

Feeding: pursues small birds; rarely takes small mammals, amphibians and insects.
Voice: silent, except during the breeding season, when an intense and often repeated *kik-kik-kik-kik* can be heard.
Similar Species: *Cooper's Hawk* (p. 98): generally larger overall; rounder tail tip with broader terminal band; much heavier-headed in flight; more sluggish wingbeats; small male Cooper's can resemble female Sharp-shinned. *Merlin* (p. 106): pointed wings; 1 dark "tear streak"; brown streaking on buff underparts; dark eyes; usually darker, more obvious tail bands; rapid wingbeats.
Best Sites: *Breeding:* large forests of southeastern Ohio. *In migration* and *winter:* statewide.

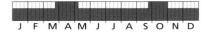

| J | F | M | A | M | J | J | A | S | O | N | D |

COOPER'S HAWK

Accipiter cooperii

The Cooper's Hawk is a common raider of backyard bird feeders, routinely barreling into yards and scattering the guests, often capturing a Northern Cardinal or other small bird. Even if not directly seen in action, this hawk's presence can often be detected by observing piles of its victim's feathers lying about. The Cooper's Hawk has adapted well to urbanization and now commonly nests within cities and towns. • The Cooper's Hawk is the classic "Chicken Hawk," formerly blamed by many farmers for pillaging their poultry. This persecution diminished the overall population significantly, and early ornithologists, who felt that accipiters were detrimental to songbird populations, abetted this attitude. Fortunately, modern ecology has taught us the importance of predators in the food chain. • In Ohio, the Cooper's Hawk is now common in all but the most intensively agricultural areas, and its numbers swell during migration.

immature

ID: short, rounded wings; long, straight, heavily barred, rounded tail; dark barring on pale undertail and underwings; squarish head; blue gray back; red, horizontal barring on underparts; red eyes; white terminal tail band. *Immature:* brown overall; dark eyes; vertical brown streaks on breast and belly. *In flight:* flap-and-glide flyer.
Size: *Male: L* 15–17 in; *W* 27–32 in. *Female: L* 17–19 in; *W* 32–37 in.
Habitat: mixed woodlands, riparian woodlands and woodlots; widespread and increasingly seen in urban yards with feeders.
Nesting: nest of sticks and twigs is built 20–65 ft up in a deciduous or coniferous tree; often near a stream or pond; might use an abandoned crow nest; female incubates 3–5 bluish white eggs for 34–36 days; male feeds the female during incubation.
Feeding: pursues prey in flight; eats mostly songbirds, squirrels and chipmunks; uses plucking post or nest for eating.
Voice: fast, woodpecker-like *cac-cac-cac-cac.*
Similar Species: *Sharp-shinned Hawk* (p. 97): smaller; squared tail; narrower terminal tail band. *Merlin* (p. 106): smaller; pointed wings; 1 dark "tear streak"; brown streaking on buff underparts; dark eyes; usually darker, more obvious tail bands; rapid wingbeats.
Best Sites: peak hawk movements along western L. Erie in March and April; large flights at Oak Openings MP and Maumee Bay SP.

J F M A M J J A S O N D

NORTHERN GOSHAWK

Accipiter gentilis

This agile and powerful predator will prey on any smaller animal it can over-take, dispatching its victim with powerful talons. The goshawk has even been known to chase quarry on foot should elusive prey disappear into dense thickets. Such is its reputation for ferocity that Attila the Hun's battle helmet was decorated with the figure of a goshawk. • The Northern Goshawk is cyclical—south-ward winter invasions occur every 10 or 11 years and correlate with population ebbs of snow-shoe hare and grouse, two principal prey species. For instance, over 1000 were recorded over Hawk Ridge, near Duluth, Minnesota on one day in October 1982. • In Ohio, Northern Goshawks are much scarcer, as we're on the southern edge of their win-tering range. There are only a few sight-ings annually, except in invasion years when as many as 15 to 20 have been reported. This species is also one of our most frequently misidentified raptors.

immature

ID: very large; rounded wings; long, banded tail with white terminal band; white "eye-brow"; dark crown; blue gray back; fine, gray, vertical streak-ing on pale breast and belly; gray barring on pale undertail and underwings; red eyes. *Immature:* brown overall; brown, vertical streaking on whitish breast and belly; brown barring on pale undertail and underwings; grayish yellow eyes.
Size: *Male: L* 21–23 in; *W* 3–3½ ft. *Female: L* 23–25 in; *W* 3½–4 ft.
Habitat: usually seen in open country with scattered woodlots.
Nesting: does not nest in Ohio.
Feeding: low foraging flights through forests; feeds primarily on large songbirds, grouse, rabbits and squirrels.

Voice: silent, except during the breeding season, when adults utter a loud, shrill, fast *kak-kak-kak-kak*.
Similar Species: *Cooper's Hawk* (p. 98) and *Sharp-shinned Hawk* (p. 97): much smaller; reddish breast bars; lack white "eyebrow" stripe; immatures lack speckled back and buffy underparts and have more even tail bands than immature goshawk. *Red-shouldered Hawk* (p. 100): juvenile is smaller, lacks prominent eye line and has narrower, more numerous tail bands.
Best Sites: most likely in northern third of state; Killdeer Plains WA; Maumee Bay SP; East Harbor SP.

J F M A M J J A S O N D

RED-SHOULDERED HAWK

Buteo lineatus

There are five subspecies of Red-shouldered Hawks, but the birds found in Ohio are the nominate group, *B. l. lineatus*, which is one of the showiest subspecies. No other buteo possesses the striking beauty of the adult, with its black-and-white-checked upperwings, rich rufous barring underneath and boldly banded tail. • Unlike its relative the Red-tailed Hawk, the Red-shouldered Hawk is largely a denizen of mature bottomland forests and not nearly so common. At one time, Red-shouldered Hawks were the most frequently occurring buteos in Ohio, but as forests were cleared, changing conditions favored the Red-tailed and caused dramatic reductions in Red-shouldered populations. Fortunately, forests are recovering and numbers may be on the rise again. Large numbers of Red-shouldered Hawks move through Ohio in migration, and a good March day along western Lake Erie may produce 150 birds passing overhead.

female incubates 2–4 darkly blotched, bluish white eggs for about 33 days; both adults raise the young.

Feeding: small mammals, birds, reptiles and amphibians are usually detected from a fence post, tree or telephone pole and caught in a swooping attack; may catch prey flushed by low flight.

Voice: repeated series of high *key-ah* notes; beware of Blue Jay imitations.

Similar Species: *Broad-winged Hawk* (p. 101): lacks reddish shoulders; wide, white tail bands; wings are broader, more whitish and dark-edged underneath. *Red-tailed Hawk* (p. 102): lacks barring on tail and light "windows" at base of primaries. *Northern Goshawk* (p. 99): juvenile is larger, has prominent eye line and broader, less numerous tail bands.

Best Sites: Cuyahoga Valley National Recreation Area; Conkle's Hollow SNP; Lake Hope SP.

ID: chestnut red shoulders on otherwise dark brown upperparts; reddish underwing linings; narrow, white bars on dark tail; barred, reddish breast and belly; reddish undertail coverts. *Immature:* large, brown streaks on white underparts; whitish undertail coverts. *In flight:* light and dark barring on underside of flight feathers and tail; white crescents or "windows" at base of primaries.

Size: *L* 19 in; *W* 3½ ft.

Habitat: mature deciduous and mixed forests, wooded riparian areas, swampy woodlands and large, mature woodlots.

Nesting: pair assembles a bulky nest of sticks and twigs, usually 15–80 ft up in a deciduous tree; nest is often reused;

J F M A M J J A S O N D

BROAD-WINGED HAWK

Buteo platypterus

The highly migratory Broad-winged Hawk winters in Central and South America and tends to migrate in groups called "kettles," with the greatest Ohio numbers occurring in the western Lake Erie vicinity. Spring kettles are usually small, with about 10 to 100 birds, though flocks numbering over 1200 have been seen on occasion. Fall produces the most spectacular flights—the biggest group recorded in Ohio was on September 18, 2002, when an estimated 20,000 birds were seen at Perrysburg. • This species nests in forests, where it spends much of its time below the canopy. In this habitat, its short, broad wings and highly flexible tail help it to maneuver through the dense growth. It is a common nester in the large forests of unglaciated Ohio. • The Broad-winged Hawk is often first detected by its plaintive, unbirdlike whistle. • Broad-winged Hawks are not hardy. Occasional winter reports are invariably misidentifications, and there are no indisputable winter records in Ohio.

ID: small size; broad, black and white tail bands; broad wings with pointed tips; heavily barred, rufous brown breast; dark brown upperparts. *Immature:* dark brown streaks on white breast, belly and sides; buff and dark brown tail bands. *In flight:* pale underwings are outlined with dark brown.

Size: *L* 14–19 in; *W* 32–39 in.

Habitat: *Breeding:* dense, mixed and deciduous forests and woodlots. *In migration:* escarpments and shorelines; also uses riparian and deciduous forests and woodland edges.

Nesting: usually in a deciduous tree, often near water; bulky stick nest is built in a crotch 20–40 ft up in a coniferous or deciduous tree; usually builds a new nest each year; mostly the female incubates

2–4 brown-spotted, whitish eggs for 28–31 days; both adults raise the young.

Feeding: swoops from a perch for small mammals, amphibians, insects and young birds; often has favored perches to which the bird repeatedly returns.

Voice: high-pitched, whistled *peeeo-wee-ee;* generally silent during migration.

Similar Species: *Other buteos* (pp. 100–03): lack broad banding on tail and broad, dark-edged wings with pointed tips. *Accipiters* (pp. 97–99): long, narrow tails with less distinct banding.

Best Sites: *Breeding:* Lake Hope SP; Shawnee SF. *In migration:* Oak Openings MP; Maumee Bay SP.

J F M A M J J A S O N D

RED-TAILED HAWK

Buteo jamaicensis

Our most common and ubiquitous hawks, Red-tailed Hawks are often conspicuous along highways, where they perch on posts, fences, trees and other obvious spots. Their extensive white underparts make them stand out at a distance. • Hunting Red-tailed Hawks soar high in the air in languid, circular patterns. Birders should familiarize themselves with the flight of this species, as it is the best method by which to distinguish this hawk from other raptors. • Migrants augment our resident populations, and good flight days in March along the western Lake Erie shore have produced over 400 birds. • Red-taileds range throughout North and Central America, and some authorities recognize up to 16 subspecies. Almost all Ohio birds are the typical eastern race (*B. j. borealis*), and the much paler, western "Krider's" subspecies is very rarely seen.

stick nest is usually added to each year; pair incubates 2–4 brown-blotched, whitish eggs for 28–35 days; male brings food to the female and the young.

Feeding: scans for food while perched or soaring; drops to capture prey; rarely hunts on foot; eats voles, mice, rabbits, chipmunks, birds, amphibians and reptiles; rarely takes large insects.

Voice: powerful, descending scream: *keeearrrr.*

Similar Species: *Rough-legged Hawk* (p. 103): white tail base; dark "wrist" patches on underwings; broad, dark terminal tail band. *Broad-winged Hawk* (p. 101): broadly banded tail; broader wings with pointed tips; lacks dark "belt." *Red-shouldered Hawk* (p. 100): pale wing "windows"; reddish wing linings and underparts; reddish shoulders.

Best Sites: almost anywhere, even in urban areas. *In migration:* large numbers in March at the "hawk tower" at Magee Marsh WA. *Winter:* Killdeer Plains; The Wilds.

ID: large, chunky, red tail; dark upperparts with some white highlights; dark brown band of streaks across belly. *Immature:* variable; lacks red tail; generally darker; band of streaks on belly. *In flight:* fan-shaped tail; white or occasionally tawny brown underside and underwing linings; dark edge on underside of wing; light underwing flight feathers with faint barring.

Size: *Male: L* 18–23 in; *W* 4–5 ft. *Female: L* 20–25 in; *W* 4–5 ft.

Habitat: open country with some trees; roadsides, fields, woodlots, mixed forests.

Nesting: in woodlands adjacent to open habitat; usually in a deciduous tree; bulky

J F M A M J J A S O N D

ROUGH-LEGGED HAWK

Buteo lagopus

When hunting, the Rough-legged Hawk often "wind-hovers" to scan the ground below, flapping to maintain a stationary position while facing upwind—a good clue to its identity. Another useful long-range identification tip is this hawk's habit of perching on the extreme top branch of a tree, kind of like an avian Christmas tree ornament.

• Rough-legged Hawks are best known in Ohio as winter visitors that prefer harsh, wind-swept grasslands and open country. There are two distinct color phases, the "light morph" and the "dark morph." In Ohio, light morphs predominate, outnumbering the dark birds by about ten to one. • As with other northern, tundra-breeding birds, wintering numbers in the south fluctuate from year to year, and correlate with cyclical fluctuations of prey.

• The name *lagopus*, meaning "hare's foot," refers to this bird's distinctive feathered legs, which are an adaptation for survival in cold climates.

dark morph ♀

♀

ID: white tail base with 1 wide, dark subterminal band; dark brown upperparts; pale flight feathers; legs are feathered to toes. *Light morph:* wide, dark abdominal "belt"; dark streaks on breast and head; dark "wrist" patches; light underwing linings. *Dark morph:* dark wing linings, head and underparts. *Immature:* lighter streaking on breast; bold belly band; buff leg feathers. *In flight:* frequently hovers; most birds show dark "wrist" patches.
Size: *L* 19–24 in; *W* 4–4½ ft.
Habitat: fields, meadows, reclaimed strip mines and agricultural croplands.

Nesting: does not nest in Ohio.
Feeding: soars and hovers while searching for prey; primarily eats small rodents; occasionally eats birds, amphibians and large insects.
Voice: alarm call is a catlike *kee-eer,* usually dropping at the end.
Similar Species: *Other buteos* (pp. 100–102): rarely hover; lack dark "wrist" patches and white tail base. *Northern Harrier* (p. 96): facial disc; obvious white rump; lacks dark "wrist" patches and dark belly band; longer, thinner tail lacks broad, dark subterminal band.
Best Sites: The Wilds; Killdeer Plains WA; Funk Bottoms WA.

J F M A M J J A S O N D

GOLDEN EAGLE

Aquila chrysaetos

The Golden Eagle is a species of big landscapes and wide-open spaces—mountain ranges, high plains and tundra. The biggest, fiercest raptor in North America, it is even able to bring down weakened, young deer. And the sight of one is unforgettable! • Golden Eagles are rare in Ohio, with about half a dozen migrants recorded annually. Because they are usually seen as "flybys" in migration and don't linger, they are one of the most difficult regularly occurring species for birders to get on their Ohio list. • With the passage of reclamation laws in 1974, thousands of acres of southeastern Ohio strip mines were converted to grasslands, creating vast, windswept, tundralike terrain unlike any other Ohio habitat. At least two Golden Eagles have been known to overwinter at The Wilds. This eagle has often been easy to find, and has provided many birders with their "life" Golden Eagle. It will be interesting to see if more eagles begin wintering in this habitat.

immature

ID: very large; brown overall with golden tint to neck and head; brown eyes; dark bill; brown tail has grayish white bands; yellow feet; fully feathered legs. *Immature:* takes five years to develop adult plumage; white tail base; white patch at base of underwing primary feathers. *In flight:* relatively short neck; long tail; long, large, rectangular wings; holds wings in a slight "V."
Size: L 30–40 in; W 6½–7½ ft.
Habitat: *In migration:* as fly-overs along western L. Erie shoreline and in Oak Openings region; occasionally along large rivers and grasslands.
Nesting: does not nest in Ohio.

Feeding: swoops down on prey from soaring flight; eats rabbits, rodents, foxes and occasionally young ungulates; often eats carrion.
Voice: generally quiet; rarely a short bark.
Similar Species: *Bald Eagle* (p. 95): longer neck; larger head and bill; shorter tail; soars with wings flat; immature lacks distinct, white underwing patches and tail base. *Turkey Vulture* (p. 93): naked, pink head; pale flight feathers; dark wing linings. *Dark-morph Rough-legged Hawk* (p. 103): much smaller; pale flight feathers; white tail base.
Best Sites: *In migration:* western L. Erie shoreline; Girdham Rd. sand dunes in Oak Openings MP. *Winter:* The Wilds.

J F M A M J J A S O N D

AMERICAN KESTREL

Falco sparverius

Long known as "Sparrow Hawk," the American Kestrel is the smallest and most common of our falcons, and the only one that regularly hovers. It hunts in open areas, and a kestrel perched on a telephone wire or fence post alongside an open field is a familiar sight throughout the region year-round. • Unlike many birds of prey, kestrels are largely insectivorous in warmer seasons and feed heavily on grasshoppers. They also adapt well to urban environments, and as our only non-owl cavity-nesting raptor, can often be enticed to use appropriate nest boxes. • A little-known aspect of the American Kestrel is the important role it has played in science. The kestrel was the principal species used to document accumulation of pesticides in birds of prey, and was the first falcon to be successfully reproduced via artificial insemination.

ID: 2 distinctive facial stripes. *Male:* rusty back; blue gray wings; blue gray crown with rusty "cap"; lightly spotted underparts. *Female:* rusty back, wings and breast streaking. *In flight:* frequently hovers; long, rusty tail; buoyant, indirect flight style.

Size: *L* 7½–8 in; *W* 20–24 in.

Habitat: open fields, riparian woodlands, woodlots, forest edges, grasslands and croplands.

Nesting: in a tree cavity (usually an abandoned woodpecker cavity) or nest box, in freeway signs or building eaves; mostly the female incubates 4–6 finely speckled, white to buff eggs for 29–30 days; both adults raise the young.

Feeding: swoops from a perch or from hovering flight; eats mostly insects and some small rodents, birds, reptiles and amphibians.

Voice: loud, shrill, often repeated *killy-killy-killy* when excited; female's voice is lower pitched.

Similar Species: *Merlin* (p. 106): only 1 facial stripe; less colorful; does not hover; flight is more powerful and direct. *Sharp-shinned Hawk* (p. 97): short, rounded wings; reddish barring on underparts; lacks facial stripes; flap-and-glide flight. *Mourning Dove* (p. 167): perched dove may look similar from a distance, but has thinner body and does not constantly bob its tail.

Best Sites: ubiquitous resident in open country statewide.

J F M A M J J A S O N D

MERLIN

Falco columbarius

The main weapons of this small falcon, like all its falcon relatives, are speed, surprise and sharp, daggerlike talons. The Merlin's sleek body, long, narrow tail and pointed wings provide the speed and agility required to snatch small birds out of the air. • Merlins were formerly known as "Pigeon Hawks"—not because they eat birds, which are their main prey, but because they somewhat resemble pigeons in flight. • Although only slightly larger than American Kestrels, Merlins generally appear significantly larger and heavier, especially in flight. They are very aggressive, and often badger larger raptors. • Merlins have been known to closely follow hunting Northern Harriers as they course low over grasslands. The harriers generally seek out small rodents, while the Merlins pick off the small birds inadvertently flushed by the harriers. • The Merlin is one of Ohio's few extirpated breeding birds. A small nesting population was probably present in the wilderness of pre-1900 northeastern Ohio—primarily Ashtabula County—but had disappeared by the 1930s.

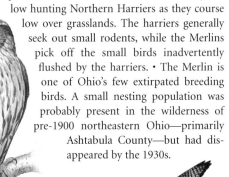

taiga form

ID: banded tail; heavily streaked underparts; 1 indistinct facial stripe; long, narrow wings and tail. *Male:* blue gray back and crown; rufous leg feathers. *Female:* brown back and crown. *In flight:* very rapid, shallow wingbeats.
Size: *L* 10–12 in; *W* 23–26 in.
Habitat: open fields; reclaimed strip mine grasslands; lakeshores; sometimes winters in large cemeteries.
Nesting: no longer known to nest in Ohio.
Feeding: overtakes smaller birds in flight; also eats rodents and large insects, such as grasshoppers and dragonflies; may also take bats.

Voice: loud, noisy, cackling cry: *kek-kek-kek-kek-kek* or *ki-ki-ki-ki;* calls in flight or while perched, often around the nest.
Similar Species: *American Kestrel* (p. 105): more colorful; 2 facial stripes; more buoyant, less direct flight style; often hovers. *Peregrine Falcon* (p. 107): much larger; well-marked, dark "helmet"; pale, unmarked upper breast; black flecking on light underparts. *Sharp-shinned Hawk* (p. 97) and *Cooper's Hawk* (p. 98): short, rounded wings; horizontal, reddish barring on breasts and bellies.
Best Sites: occasional at The Wilds and other large, reclaimed strip mine grasslands. *In migration:* western L. Erie shoreline. *Winter:* Spring Grove Cemetery (Cincinnati); Calvary Cemetery (Cleveland).

J F M A M J J A S O N D

PEREGRINE FALCON

Falco peregrinus

Peregrine Falcons are among the world's fastest predatory birds, reported to reach speeds in excess of 200 miles per hour in a stoop, or dive. The sight of a Peregrine rocketing into a flock of shorebirds and literally blowing one apart, so great is the force of the strike, is an unforgettable spectacle. • Peregrine means "wanderer"; these primarily arctic-nesting birds are highly migratory, with some birds traveling over 11,000 miles on their annual peregrinations. • Historically, Peregrine Falcons were limited in distribution by the availability of suitable nest sites—ledges on cliff faces. Ohio has no such habitat, and Peregrines are not native here as nesters. To aid in the recovery of Peregrine Falcons after they nearly became extirpated in North America in the 1960s, pairs were introduced to city skyscrapers and enticed to nest on high ledges. Peregrines readily adopted this new and artificial habitat, and can now be found in many major eastern U.S. cities. Pairs now reside in most big Ohio cities including Columbus, Cleveland, Toledo, Cincinnati and Dayton. These urban nesters can be either migratory or nonmigratory, depending on the pair.

ID: blue gray back; prominent, dark "helmet"; light underparts with fine, dark spotting and flecking. *Immature:* brown back; heavier breast streaks; gray feet and cere. *In flight:* pointed wings; long, narrow, dark-banded tail.

Size: *Male: L* 15–17 in; *W* 3–3½ ft. *Female: L* 17–19 in; *W* 3½–4 ft.

Habitat: lakeshores, river valleys, river mouths, urban areas and open fields.

Nesting: usually on a rocky cliff; may use a skyscraper ledge; no material is added, but the nest is littered with prey remains, leaves and grass; nest sites are often reused; mostly the female incubates 3–4 heavily speckled, creamy white to buff eggs for 32–34 days.

Feeding: high-speed, diving stoops; strikes birds with clenched feet in midair; prey is consumed on a nearby perch; primarily eats pigeons, waterfowl, shorebirds, flickers and larger songbirds; rarely takes small mammals or carrion.

Voice: loud, harsh, continuous *cack-cack-cack-cack-cack* near the nest site.

Similar Species: *Gyrfalcon* (p. 332): larger; longer tail; lacks dark "helmet." *Merlin* (p. 106): smaller; heavily streaked breast and belly; lacks prominent dark "helmet."

Best Sites: established pairs in cities are easily observed (some nests can be viewed via Internet "web cams"). *In migration:* "wild" migrants along L. Erie shoreline.

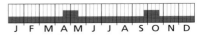

J	F	M	A	M	J	J	A	S	O	N	D

KING RAIL

Rallus elegans

John James Audubon described the King Rail, North America's largest rail, for the first time in 1835, but not much more than a century later, Richard Pough stated "the King Rail is becoming scarcer as more and more of its habitat is destroyed..." Once the most numerous breeding rail in Ohio, the King is now the rarest and is listed as endangered in the state, owing to our estimated 90 percent loss of wetlands. Probably less than 25 King Rail pairs breed in Ohio in most years. • King Rails can be surprisingly bold at times, and have been known to walk right up to birders attempting to lure them from the marshes with tape recordings. Perhaps the best way to detect these birds is to venture into the large western Lake Erie marshes at night, and listen for their calls. Listening to recordings to learn their calls is recommended, as rails are far more often heard than seen.

ID: long, slightly downcurved bill; black back feathers have buffy or tawny edges; cinnamon shoulders and underparts; strongly barred, black-and-white flanks; grayish brown "cheeks."
Immature: similar plumage pattern with lighter, washed-out colors.
Size: *L* 15 in; *W* 20 in.
Habitat: freshwater marshes, shrubby swamps, marshy riparian shorelines and flooded fields with shrubby margins.
Nesting: among clumps of grass or sedge just above the water or ground; male builds most of the platform nest, canopy and entrance ramp using marsh vegetation;

pair incubates 10–12 brown-spotted, pale buff eggs for 21–24 days.
Feeding: aquatic insects, crustaceans and occasionally seeds; forages in shallow water for small fish and amphibians, often in or near dense plant cover.
Voice: chattering call is 10 or fewer, evenly spaced *kek* notes.
Similar Species: *Virginia Rail* (p. 109): much smaller; brown back feathers; gray face; reddish bill.
Best Sites: large western L. Erie marshes such as Mallard Marsh WA; Big Island WA. *In migration:* may appear in any good-sized marsh.

J F M A M J J A S O N D

VIRGINIA RAIL
Rallus limicola

Rails may have inspired the expression "thin as a rail"—the narrow body of the Virginia Rail allows it to slip through the densest of wetland vegetation, particularly the thick cattail stands that it favors. • Virginia Rails often build extra "dummy" nests in their own territory, for unknown reasons. However, they have been observed destroying the nearby nests of other Virginia Rails, and these dummy nests may serve to throw raiders off the track. • The Virginia Rail often prefers to run from danger and only takes flight under serious harassment. When it does become airborne, it doesn't set any speed records as it flutters weakly along with its legs dangling limply below. It's hard to believe that the Virginia Rail migrates several hundred miles—and this always at night. • In Ohio, the Virginia Rail is still fairly common in the largest wetland complexes, and migrants may appear in almost any wetland. Surprisingly hardy, small numbers overwinter annually, except perhaps in the coldest winters.

ID: long, down-curved, reddish bill; gray face; rusty breast; barred flanks; chestnut brown wing patch; very short tail. *Immature:* much darker overall; pale bill.
Size: *L* 9–11 in; *W* 13 in.
Habitat: freshwater wetlands, especially cattail and bulrush marshes.
Nesting: concealed in emergent vegetation, usually suspended just over the water; loose basket nest is made of coarse grass, cattail stems or sedges; pair incubates 5–13 speckled, pale buff eggs for up to 20 days.
Feeding: probes into soft substrates and exposed mud near cover; gleans vegetation

for invertebrates, including beetles, snails, small snakes, spiders, earthworms, insect larvae and nymphs; also eats some pondweeds and seeds.
Voice: call is an often-repeated, telegraph-like *kidick, kidick*; also "oinks" and croaks.
Similar Species: *King Rail* (p. 108): much larger; dark legs; lacks reddish bill and gray face; juvenile is mostly pale gray. *Sora* (p 110): short, yellow bill; black face and throat. *Black Rail* (p. 334): very rare; chestnut nape; whitish back speckling; adult is similar in size and black color to Virginia Rail young.
Best Sites: good-sized wetlands with a lot of cattails; Springville Marsh SNP; Pipe Creek WA; Spring Valley WA.

J F M A M J J A S O N D

SORA

Porzana carolina

The Sora has the widest distribution and abundance of any North American rail and it can be quite common in Ohio. • Rails are very secretive as a rule, and the best way to detect Soras is to listen for their routinely delivered, ascending *kur-ree* call. A well-tossed stone splashed into a good wetland at the peak of migration might induce 20 or 30 Soras to cry their *keek* alarm note. • Soras are hunted in Ohio and in 30 other states. They weigh but 2½ ounces, though, so it is hardly worth the effort for a hunter to fill the daily limit of 25 birds. Hunting probably has little impact on overall numbers, as there aren't many rail hunters and the Sora lays an enormous clutch—up to 18 eggs that must be arranged in layers for the incubating bird to be able to cover them. • An interesting aspect of rails is their penchant for brood parasitism. One rail nest was found that contained a good variety of eggs: 9 King Rail, 7 Virginia Rail and 1 Sora.

breeding

ID: short, yellow bill; black face, throat and foreneck; gray neck and breast; long, greenish legs. *Immature:* no black on face; buffier with paler underparts; more greenish bill.
Size: *L* 8–10 in; *W* 14 in.
Habitat: wetlands with abundant emergent cattails, bulrushes, sedges and grasses.
Nesting: usually over water, but occasionally in a wet meadow under concealing vegetation; well-built basket nest is made of grass and aquatic vegetation; pair incubates 10–12 darkly speckled, buff or olive buff eggs for 18–20 days.

Feeding: gleans and probes for seeds, plants, aquatic insects and mollusks.
Voice: usual call is a clear, 2-note *kur-ree;* alarm call is a sharp *keek;* courtship song begins *or-Ah or-Ah,* descending quickly in a series of maniacal *weee-weee-weee* notes.
Similar Species: *Virginia Rail* (p. 109) and *King Rail* (p. 108): larger; long, downcurved bills; chestnut brown wing patches; rufous breasts. *Yellow Rail* (p. 334): rare; streaked back; tawny upperparts; white throat; white trailing edges of wings show in flight.
Best Sites: Magee Marsh WA; Big Island WA; Calamus Swamp. *In migration:* any marshy area.

J F M A M J J A S O N D

COMMON MOORHEN

Gallinula chloropus

Formerly known as the "Common Gallinule," this odd bird, with its variety of clucks, whines, cackles and grunts, accounts for many of the strange sounds heard in our marshes. Calling birds often inspire all of the resident Common Moorhens to erupt in a cacophony. William Brewster, writing in 1891, noted "The calls…were so varied and complex that it seems hopeless to attempt a full description…" • The Common Moorhen is still a fairly common breeder in the big western Lake Erie marshes, with possibly over 1000 pairs in total, but is rare or uncommon elsewhere. The record one-day count of 100 birds near Toledo on September 7, 1947, by the naturalist Louis W. Campbell won't be matched now—it's hard to find more than 10 on a good day. • As is true for some rails, Common Moorhens are hunted; fortunately, not many hunters pursue them.

breeding

ID: reddish forehead shield; yellow-tipped bill; gray black body; white streak on sides and flanks; long, greenish yellow legs. *Breeding:* brighter bill and forehead shield. *Juvenile:* paler plumage; duller legs and bill; white throat.
Size: *L* 12–15 in; *W* 21 in.
Habitat: primarily marshes, sometimes other wetland types such as buttonbush swamps.
Nesting: pair builds a platform nest or a wide, shallow cup of bulrushes, cattails and reeds in shallow water or along a shoreline; often built with a ramp leading to the

water; pair incubates 6–17 darkly spotted, buff-colored eggs for 19–22 days.
Feeding: eats mostly aquatic vegetation, berries, fruits, tadpoles, insects, snails, worms and spiders; may take carrion and eggs.
Voice: noisy in summer; various sounds include chickenlike clucks, screams, squeaks and a loud *cup;* courting males give a harsh *ticket-ticket-ticket.*
Similar Species: *American Coot* (p. 112): white bill and forehead shield; lacks white streak on flanks.
Best Sites: Spring Valley WA; western L. Erie marshes. *Breeding:* any large marsh. *In migration:* large and small wetlands.

J F M A M J J A S O N D

111

AMERICAN COOT

Fulica americana

Even though they are in the rail family, coots are very ducklike and, unlike most of their brethren, are not at all shy and retiring. Their lobed toes make these birds efficient swimmers, and they can even dive well, which enables them to reach the succulent submergent plants that are a dietary favorite. However, they are not above robbing Canvasbacks of coveted water celery when these more skilled divers pop to the surface. • In Ohio, American Coots are common to abundant migrants, and while the days of 10,000-bird flocks—like the one at Buckeye Lake in 1936—are over, we still get concentrations of up to several thousand at favored locales. These birds are also local nesters in the larger marshes, and overwinter regularly along Lake Erie and elsewhere if open water is available.

ID: gray black, duck-like bird; white, chickenlike bill with dark ring around tip; reddish spot on white forehead shield; long, green yellow legs; lobed toes; red eyes. *Immature:* lighter body color; darker bill and legs; lacks prominent forehead shield.

Size: *L* 13–16 in; *W* 24 in.

Habitat: shallow marshes, ponds, lakes and wetlands with open water and emergent vegetation.

Nesting: in emergent vegetation; pair builds a floating nest from cattails and grass; pair incubates 6–11 brown-spotted, buffy white eggs for 21–25 days; regularly produces 2 broods in a season.

Feeding: gleans the water's surface; sometimes dives, tips up or even grazes on land; eats aquatic vegetation, insects, snails, crayfish, worms, tadpoles and fish; may steal food from ducks.

Voice: calls frequently in summer, day and night: *kuk-kuk-kuk-kuk-kuk;* also grunts.

Similar Species: *Ducks* (pp. 44–69): all lack white, chickenlike bill and uniformly black body. *Common Moorhen* (p. 111): reddish forehead shield; yellow-tipped bill; white streak on flanks.

Best Sites: *In migration:* nearly any water body; large concentrations at Findlay Reservoir, western L. Erie marshes and many large reservoirs.

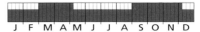

J F M A M J J A S O N D

SANDHILL CRANE

Grus canadensis

Deep, resonant, rattling calls announce the approach of a flock of migrating Sandhill Cranes long before they pass overhead. At first glance, the large, V-shaped flocks look similar to flocks of Canada Geese, but cranes often soar and circle, and their calls are unlike the honking of geese. • Sandhill Cranes were scattered nesters in northern Ohio until about 1900, then, as numbers declined, they became accidental visitors. Therefore, the discovery of a nesting pair in Wayne County in 1987 was a positive symbol of their comeback. Nesters have slowly increased, with three or four breeding locales now known. Migrant flocks have also increased, as evidenced by flocks totaling over 800 birds, seen throughout western Ohio in December 2003. Still, breeding Sandhill Cranes are listed as endangered in Ohio. • Occasionally, the Sandhill Crane's fascinating, elaborate courtship "dancing" display is observed among spring migrants.

nonbreeding

ID: very large; gray overall; long neck and legs; naked, red crown; long, straight bill; plumage is often stained rusty color from iron oxides in water. *Immature:* lacks red crown; reddish brown plumage may appear patchy. *In flight:* extends neck and legs; often glides, soars and circles.
Size: *L* 3½–4 ft; *W* 6–7 ft.
Habitat: *In migration:* agricultural fields and shorelines. *Breeding:* large wetland complexes.
Nesting: on a large mound of aquatic vegetation in the water or along a shoreline; pair incubates 2 brown-blotched, buff eggs

for 29–32 days; egg hatching is staggered; young fly at about 50 days.
Feeding: probes and gleans the ground for insects, soft-bodied invertebrates, waste grain, shoots and tubers; frequently eats small vertebrates.
Voice: loud, resonant, rattling *gu-rrroo gu-rrroo gurrroo.*
Similar Species: *Great Blue Heron* (p. 84): lacks red forehead patch; neck is folded back over shoulders in flight.
Best Sites: Deer Creek Reservoir; Killdeer Plains WA; Funk Bottoms WA; migrant flocks can occur throughout western and northern Ohio.

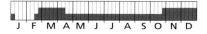

J F M A M J J A S O N D

BLACK-BELLIED PLOVER

Pluvialis squatarola

K nown as "Grey Plover" in the Old World, the Black-bellied Plover is cosmo-politan in distribution. • In breeding colors, this is one of our most striking shorebirds—"an aristocrat among shorebirds…" as put by A.C. Bent in 1929. Breeding-plumaged birds are unmistakable, but fall migrants closely resemble American Golden-Plovers, so care must be taken to separate the two species. • Like many plovers, Black-bellied Plovers are wary, and are often the first to sound an alarm when a threat appears, thus serving as sentries in mixed shorebird flocks. In migration, they can sometimes be heard delivering their plaintive, three-note, whistled call while passing overhead. • Black-bellied Plovers often feed at night—their large eyes are an adaptation for improving nocturnal vision. • This species normally occurs as single birds or in small flocks, but occasional groups of over 150 birds have been seen.

breeding

nonbreeding

ID: short, black bill; long, black legs. *Breeding:* black face, breast, belly and flanks; white under-tail coverts; white stripe leading from crown to "collar," neck and sides of breast; mottled, black-and-white back. *Nonbreeding:* mottled, gray brown upperparts; lightly streaked, pale underparts. *In flight:* black "wing pits"; whitish rump; white wing linings.
Size: *L* 11–13 in; *W* 29 in.
Habitat: primarily mudflats, also edges of reservoirs, marshes and sewage lagoons.
Nesting: does not nest in Ohio.

Feeding: run-and-stop foraging technique; eats insects, mollusks and crustaceans.
Voice: rich, plaintive, 3-syllable whistle: *pee-oo-ee.*
Similar Species: *American Golden-Plover* (p. 115): upperparts are mottled with gold; black undertail coverts in breeding plumage; lacks black "wing pits." *Red Knot* (p. 129): nonbreeding bird has longer bill, shorter legs and lacks black "wing pits."
Best Sites: large mudflats, but good locales vary from year to year; consistent areas include Big Island WA, Killdeer Plains WA and western L. Erie marshes.

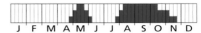

J F M A M J J A S O N D

AMERICAN GOLDEN-PLOVER

Pluvialis dominica

We are fortunate to still be able to observe the visually striking, breeding-plumaged American Golden-Plover. Historically, this species was tremendously abundant, with perhaps one of the largest populations of any bird worldwide. But this huge population was decimated by unregulated market hunting in the 1800s—on one spring day in 1821, for example, Louisiana hunters brought down over 48,000 birds at one location. Fortunately, after protective laws put an end to the slaughter, American Golden-Plovers eventually recovered something of their former abundance. • One of the greatest long-distance migrants, American Golden-Plovers travel up to 7000 miles round trip each year between their summer and winter ranges. • One of Ohio's greatest avian spectacles is a northbound flock of hundreds, or even thousands, of Golden-Plovers feeding in freshly plowed fields in April.

breeding

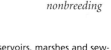

nonbreeding

ID: straight, black bill; long, black legs. *Breeding:* black face and underparts; S-shaped, white stripe from forehead to shoulders; dark upperparts speckled with gold and white. *Nonbreeding:* broad, pale "eyebrow"; dark streaking on pale neck and underparts; much less gold on upperparts. *Juvenile:* somewhat brighter than adult; white forehead. *In flight:* gray "wing pits"; holds wings briefly erect after alighting.
Size: *L* 10–11 in; *W* 26 in.
Habitat: favors freshly plowed fields; also sod farms, lakeshores and mudflats along

the edges of reservoirs, marshes and sewage lagoons.
Nesting: does not nest in Ohio.
Feeding: run-and-stop foraging technique; snatches insects, mollusks and crustaceans; also takes seeds and berries.
Voice: soft, melodious whistle: *quee, quee-dle.*
Similar Species: *Black-bellied Plover* (p. 114): white undertail coverts; whitish crown; conspicuous black "wing pits"; lacks gold speckling on upperparts.
Best Sites: large flocks occur in freshly plowed fields along western Ohio roads in April and May; Grand Lake St. Marys; Killdeer Plains WA; Big Island WA; Charlie's Pond.

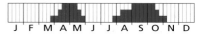

J F M A M J J A S O N D

115

SEMIPALMATED PLOVER

Charadrius semipalmatus

Like many other shorebirds, Semipalmated Plovers were decimated by unregulated market hunting, and by 1900 had become rare in many areas. The first
laws aimed at protecting migratory birds were promulgated in 1913, and the
Semipalmated Plover has once again become common. • This species is also thought
to be increasing in numbers because of the nesting habits of other birds. Arctic-
nesting geese, such as Snow, Ross's and Greater White-fronted geese, have recently
experienced population booms—the localized disturbance they create on their tundra nesting grounds creates suitable microhabitat that favors nesting Semipalmated
Plovers. • In Ohio, this is primarily a mudflat species, though Semipalmated Plovers
are versatile and can be found along lakeshores, in plowed fields and in other habitats. The largest concentrations occur along western
Lake Erie, with daily counts in excess of 900 birds.

nonbreeding

breeding

ID: *Breeding:* dark brown back; white breast with 1 black, horizontal band; long, orange legs; stubby, orange, black-tipped bill; white patch above bill; white throat and "collar"; black band across fore-head; small, white "eyebrow." *Nonbreeding:* duller; mostly dark bill. *Immature:* dark legs and bill; brown banding.
Size: *L* 7 in; *W* 19 in.
Habitat: mudflats, beaches, lakeshores and occasionally river edges.
Nesting: does not nest in Ohio.

Feeding: run-and-stop feeding, usually on shorelines and beaches; eats crustaceans, worms and insects.
Voice: crisp, high-pitched, 2-part, rising whistle: *tu-wee.*
Similar Species: *Killdeer* (p. 117): larger; 2 black bands across breast. *Piping Plover* (p. 335): much lighter upperparts; narrower breast band, incomplete in females and most males; lacks dark forehead band.
Best Sites: large mudflats, such as those at Hoover Reservoir, Sheldon Marsh SNP and Pipe Creek WA.

J F M A M J J A S O N D

KILLDEER

Charadrius vociferus

Because of its ability to adapt to human-modified landscapes, the Killdeer is probably more common now than ever. Its frequent presence around people and their haunts makes this species the best known of North American shorebirds. • Killdeer are not easily missed; in fact, they can be downright irritating to a birder attempting to stealthily move in on the mudflats for a closer look at the shorebirds. Killdeer make good sentinels; at the first sign of a threat, they alert everything within earshot with their loud, ringing cries. • The Killdeer frequently employs the "broken wing" act. An intruder venturing too near a nest—little more than a scrape in the gravel or dirt—is greeted by one of the adults, who cries piteously while dragging a wing and stumbling about as if injured. If the intruder is gullible enough to follow, the Killdeer will lead it on a circuitous route away from the nest until the feathered actor is satisfied all is well, at which point the bird is miraculously cured and flies away.

ID: long, dark yellow legs; white breast with 2 black bands; brown back; white underparts; brown head; white "eyebrow"; white patch above bill; black forehead band; rufous rump. *Immature:* downy; only 1 breast band.
Size: *L* 9–11 in; *W* 24 in.
Habitat: open ground, fields, lakeshores, sandy beaches, mudflats, gravel streambeds, wet meadows and grasslands; also urban areas and parks, often far from water.
Nesting: on open ground; in a shallow, usually unlined depression; pair incubates 4 heavily marked, creamy buff eggs for 24–28 days; occasionally raises 2 broods.

Feeding: run-and-stop feeder; eats mostly insects; also takes spiders, snails, earthworms and marine invertebrates.
Voice: loud and distinctive *kill-dee kill-dee kill-deer* and variations, including *deer-deer*.
Similar Species: *Semipalmated Plover* (p. 116): smaller; only 1 breast band. *Piping Plover* (p. 335): smaller; lighter upperparts; 1 breast band.
Best Sites: widespread in a variety of habitats including large lawns, mudflats, rivers, marshes and baseball fields. *In migration:* largest flocks form on big mudflats in fall.

J F M A M J J A S O N D

AMERICAN AVOCET

Recurvirostra americana

Any discovery of American Avocets in our state is noteworthy for two reasons: these birds are never common in Ohio, and they are one of the most spectacular of North American shorebirds, especially in their strikingly elegant breeding plumage. The American Avocet is the only avocet in the world that undergoes a yearly color change. • The long, upturned bill is not only a great field mark, but it allows avocets to exploit a peculiar feeding niche. These shorebirds sweep their bills through the muddy substrates of ponds and lakes like a scythe, extracting small animal life from the muck. As visibility is often nonexistent in the sediment-clouded water, extremely sensitive nerve endings at the bill's tip allow avocets to sense their prey by touch. Sometimes feeding groups will form lines, advancing through the water like a marching military regiment.

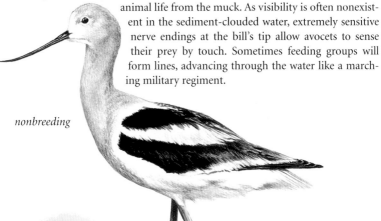

nonbreeding

breeding

ID: long, upturned, black bill; long, pale blue legs; black wings with wide, white patches; white underparts; female's bill is more upturned and shorter than male's. *Breeding:* peachy red head, neck and breast. *Nonbreeding:* gray head, neck and breast. *In-flight:* a "stick with wings"; long, skinny legs and neck; black-and-white wings.
Size: *L* 17–18 in; *W* 31 in.
Habitat: lakeshores, wetlands and exposed mudflats.
Nesting: does not nest in Ohio.

Feeding: sweeps its bill from side to side along the water's surface, picking up aquatic invertebrates and insects and occasionally seeds; male sweeps lower in the water than female; occasionally swims and tips up like a duck.
Voice: harsh, shrill *plee-eek plee-eek.*
Similar Species: generally unmistakable. *Black-necked Stilt* (p. 335): very rare; straight bill; mostly black head; pink legs.
Best Sites: larger wetland complexes throughout glaciated Ohio; Killdeer Plains WA; western L. Erie marshes; L. Erie dredge-spoil impoundments and harbors.

J F M A M J J A S O N D

GREATER YELLOWLEGS

Tringa melanoleuca

Greater Yellowlegs are one of the first shorebirds to arrive in spring, sometimes as early as late February or early March. Their loud, whistled calls warn other birds of danger while alerting observers to the Yellowlegs' presence before the birds are seen. • One of the classic identification problems that novice birders face is separating this species from its close relative, the Lesser Yellowlegs. The most important differences are structural—Greaters have almost twice the overall mass of Lessers, along with proportionally longer and thicker, slightly upturned bills. Some birders feel that the larger, knobby "knees" of Greater Yellowlegs are also a useful field mark. • A behavioral characteristic of Greater Yellowlegs that can aid in identification at a distance is their habit of foraging in deep water, sometimes up to their bellies and generally beyond the depth in which most other shorebirds can operate.

nonbreeding

breeding

ID: long, bright yellow legs; dark, 2-tone bill is slightly upturned and noticeably longer than head width. *Breeding:* brownish black back and upperwing; fine, dense, dark streaking on head and neck; dark barring on breast often extends onto belly; subtle, dark eye line; light lores. *Nonbreeding:* gray overall; fine streaks on breast; clear belly. *Juvenile:* warmer brown upperparts, marked with pale buff notches; clear, brown streaks on breast, flanks and undertail.
Size: *L* 13–15 in; *W* 28 in.
Habitat: almost all wetlands, including lakeshores, marshes, flooded fields and river shorelines.
Nesting: does not nest in Ohio.

Feeding: usually wades in water over its "knees"; sometimes sweeps its bill from side to side; primarily eats aquatic invertebrates, but will also eat small fish; occasionally snatches prey from the water's surface.
Voice: quick, whistled *tew-tew-tew*, usually 3 notes.
Similar Species: *Lesser Yellowlegs* (p. 120): smaller; straight bill is not noticeably longer than width of head; call is generally a pair of higher notes: *tew-tew*. *Willet* (p. 122): white wing bars; heavier, straighter bill; dark, greenish legs; clear, distinctive *will-will-willet* call.
Best Sites: large marshes, mudflats and flooded fields; suitable habitat can vary from year to year; Sheldon Marsh SNP; Gilmore Ponds Preserve; western L. Erie marshes.

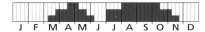

J F M A M J J A S O N D

LESSER YELLOWLEGS

Tringa flavipes

Yellowlegs—both species—were often referred to as "telltales" in the old days, because their loud cries of alarm, sounded at the first sign of a threat, alerted their less-observant shorebird companions to danger. • The Lesser Yellowlegs is noticeably more common in Ohio than the Greater, and can congregate in large numbers. As with most shorebirds, though, numbers vary from year to year depending upon the availability of habitat. Only recently have biologists begun to understand the importance of managing wetland habitats to accommodate peak movements of these long-distance migrants. Unfortunately, in many marshes, duck management strategies take precedence and shorebird habitat is given short shrift. Lately, though, some of the more progressive land management agencies such as Franklin County Metroparks have actively begun to manage wetlands specifically for migrating shorebirds—an example that will hopefully be followed by other agencies.

breeding

nonbreeding

ID: bright yellow legs; all-dark bill is not noticeably longer than width of head; brownish black back and upperwing; fine, dense, dark streaking on head, neck and breast; lacks barring on belly; subtle, dark eye line; light lores. *Nonbreeding:* similar, but gray overall.
Size: *L* 10–11 in; *W* 24 in.
Habitat: shorelines of lakes, rivers, marshes and ponds.
Nesting: does not nest in Ohio.
Feeding: snatches prey from the water's surface; frequently wades in shallow water; primarily eats aquatic invertebrates, but will also take small fish and tadpoles.
Voice: typically a high-pitched pair of *tew* notes.
Similar Species: *Greater Yellowlegs* (p. 119): larger; bill is slightly upturned and noticeably longer than width of head; *tew* call is usually given in a series of 3 notes. *Solitary Sandpiper* (p. 121): white eye ring; darker upperparts; paler bill; greenish legs.
Best Sites: Slate Run Metropark; Funk Bottoms WA; Hoover Reservoir; wetlands and flooded fields statewide.

J F M A M J J A S O N D

SOLITARY SANDPIPER

Tringa solitaria

The Solitary Sandpiper is well named, because it is usually encountered as a lone bird and seldom gathers in any numbers. • Solitary Sandpipers are very diverse in their choice of habitats, commonly living along streams, lakeshores, bogs, mudflats and in all manner of wetlands. They are quite confiding and normally can be closely approached, but when spooked, they flush with a whistled *peet weet* call that's quite similar to the Spotted Sandpiper's call. • The Solitary Sandpiper was first described in 1813 by ornithologist extraordinaire Alexander Wilson, but it wasn't until 1903 that the first nest of a Solitary was discovered. This was, in part, because the Solitary Sandpiper nests in remote northern muskeg bogs, but also because this bird picks very unusual nest locations—other birds' nests in trees! The mostly commonly used nests are those of the American Robin and the Rusty Blackbird, but it also uses the nests of many other species.

breeding

ID: white eye ring; short, green legs; dark, yellowish bill with black tip; spotted, brownish gray back; white lores; brownish gray head, neck and breast have fine white streaks; dark uppertail feathers with black-and-white barring on sides. *In flight:* dark underwings.
Size: *L* 7–9 in; *W* 22 in.
Habitat: wet meadows, sewage lagoons, muddy ponds, sedge wetlands, beaver ponds and wooded streams.
Nesting: does not nest in Ohio.
Feeding: stalks shorelines and wetlands, picking up aquatic invertebrates such as water boatmen and damselfly nymphs;

also gleans for terrestrial invertebrates; occasionally stirs the water with its foot to spook out prey.
Voice: high, thin *peet-wheet* or *wheat wheat wheat*.
Similar Species: *Lesser Yellowlegs* (p. 120): no eye ring; longer, bright yellow legs. *Spotted Sandpiper* (p. 123): smaller; shorter legs; incomplete eye ring; black-tipped, orange bill; spotted breast in breeding plumage. *Other sandpipers* (pp. 119–47): most have black bills and legs; lack white eye ring.
Best Sites: almost any stream or wetland; Big Island WA; Calamus Swamp; Mogadore Reservoir.

J F M A M J J A S O N D

121

WILLET

Catoptrophorus semipalmatus

Willets are not flashy birds while at rest, somewhat resembling squat, dumpy Greater Yellowlegs, but when they take flight, the conspicuous white wing flashes coupled with their loud, ringing cries make these birds stand out and leave no doubt as to their identity. • Virtually all Ohio Willets are of the western race, but there are at least two records that appear to be the eastern with more diminutive bills and legs, and darker, browner plumage. Breeding-plumaged easterns are heavily marked with bars and spots. • The scientific name of this species is cumbersome, but instructive. *Catoptrophorus* means "bearing mirrors," which refers to the bold, white wing stripes, and *semipalmatus* means "half-webbed," for the partially lobed toes. The Willet's toes allow it to better negotiate the mucky ooze of the mudflats that are its preferred habitat.

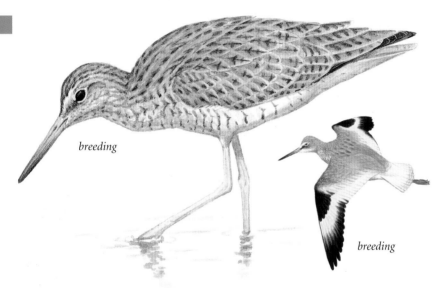

breeding

breeding

ID: plump; heavy, straight, black bill; light throat and belly. *Breeding:* dark streaking and barring overall. *In flight:* black-and-white wing pattern.
Size: *L* 14–16 in; *W* 26 in.
Habitat: wet fields and shorelines of marshes, lakes and ponds.
Nesting: does not nest in Ohio.
Feeding: feeds by probing muddy areas; also gleans the ground for insects; occasionally eats shoots and seeds.
Voice: loud, rolling *will-will-willet, will-will-willet;* also a repeated *wik wik* and *kreeliilii.*

Similar Species: *Marbled Godwit* (p. 127) and *Hudsonian Godwit* (p. 126): much longer, pinkish yellow bills with dark, slightly upturned tips; larger bodies; lack black-and-white wing pattern. *Greater Yellowlegs* (p. 119): long, yellow legs; slightly upturned bill; lacks black-and-white wing pattern.
Best Sites: marshes, mudflats and dredge-spoil impoundments along L. Erie; occasionally turns up in favorable inland habitats such as Hoover Reservoir and Killdeer Plains WA.

J F M A M J J A S O N D

SPOTTED SANDPIPER

Actitis macularia

Spotted Sandpipers are not only fairly common nesters throughout Ohio, but they also have one of the broadest breeding distributions of any North American sandpiper. • Spotted Sandpipers are easily recognized by their constant "teeter-totter" tail bobbing, and in flight, by their very shallow wingbeats that almost appear to quiver like a wire under tension. • A little known facet of the Spotted Sandpiper's ecology, and something that is quite rare in the bird world, is the phenomenon of role reversal. Female Spotteds return to breeding grounds first, aggressively court the males, mate, and then leave all the parental duties to the males. • A fascinating case of territorial aggression was documented by an entomologist who observed the largest North American dragonfly, the dragonhunter, relentlessly chase a Spotted Sandpiper until the bird finally dove under the water to escape.

breeding

nonbreeding

ID: teeters continuously. *Breeding:* white underparts heavily spotted with black; yellow orange legs; black-tipped, yellow orange bill; white "eyebrow." *Nonbreeding* and *juvenile:* pure white breast, foreneck and throat; brown bill; dull yellow legs. *In flight:* flies close to the water's surface with very rapid, shallow wingbeats; white upperwing stripe.
Size: *L* 7–8 in; *W* 15 in.
Habitat: shorelines, gravel beaches, ponds, marshes, alluvial wetlands, rivers, streams, swamps and sewage lagoons; occasionally in cultivated fields.
Nesting: usually near water; often under overhanging vegetation among logs or

under bushes; in a shallow depression lined with grass; almost exclusively the male incubates 4 heavily blotched, creamy buff eggs and raises the young.
Feeding: picks and gleans along shorelines for terrestrial and aquatic invertebrates; also snatches flying insects from the air.
Voice: sharp, crisp *eat-wheat, eat-wheat, wheat-wheat-wheat.*
Similar Species: *Solitary Sandpiper* (p. 121): complete eye ring; yellowish bill with dark tip; lacks breast spotting. *Other sandpipers* (pp. 119–47): most have black bills and legs; lack breast spotting.
Best Sites: gravel bars on streams and rivers; also rocky riprap along shores of lakes and reservoirs.

J F M A M J J A S O N D

123

UPLAND SANDPIPER
Bartramia longicauda

Upland Sandpipers had peaked in abundance in Ohio by the 1930s, when they were recorded as breeders in 76 counties. But as farming practices became increasingly inhospitable for birds, these sandpipers declined to the point where they are currently listed as threatened in Ohio. Now, they are generally recorded in only eight to ten counties each year. • Long known as "Upland Plover," the name was changed in 1973 as this bird is not a true plover. • Upland Sandpipers are obligate grassland birds that require extensive short-grass areas or a mosaic of open, fallow fields and hay meadows. Excellent areas in which to look for them are the large, grassy fields buffering airport runways, especially in the glaciated areas of Ohio. • The song of the Upland Sandpiper is an ethereal, mellow whistling—one of nature's most haunting sounds. • When alighting on a post or wire, Upland Sandpipers hold their wings extended upward momentarily, a very distinctive habit.

ID: small head; long neck; large, dark eyes; yellow legs; mottled brownish upperparts; lightly streaked breast, sides and flanks; white belly and undertail coverts; bill is about same length as head.
Size: *L* 11–12½ in; *W* 26 in.
Habitat: hayfields, ungrazed pastures, grassy meadows, abandoned fields, natural grasslands and airports.
Nesting: in dense grass or along a wetland; in a depression, usually with grass arching over the top; pair incubates 4 lightly spotted, pinkish to pale buff eggs for 22–27 days; both adults tend the young.
Feeding: gleans the ground for insects, especially grasshoppers and beetles.
Voice: courtship song is an airy, whistled *whip-whee-ee you;* alarm call is *quip-ip-ip.*
Similar Species: *Buff-breasted Sandpiper* (p. 140): shorter neck; larger head; daintier bill; lacks streaking on "cheek" and foreneck. *Pectoral Sandpiper* (p. 136): streaking on breast ends abruptly; smaller eyes; shorter neck; usually seen in larger numbers.
Best Sites: the entrance road to Bolton Field airport in Franklin Co. in April; fields in the western portion of Ottawa NWR.

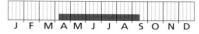

J F M A M J J A S O N D

WHIMBREL

Numenius phaeopus

Whimbrels, though not common, are the only regularly occurring curlew in Ohio. They are generally seen as scattered individuals or in small flocks, most often along the Lake Erie shore. They are often difficult to find because they rarely linger for more than a day and they tend to conceal themselves in vegetation. Thus, the flock of 233 that was recorded near Toledo on May 20, 1934, must have been an incredible sight. • Equally amazing was the individual—also seen near Toledo—that was most likely of the Siberian race *variegatus*. This July 1988 sighting is the only eastern North American record of this subspecies. • Whimbrels are powerful flyers, and some individuals fly nonstop from southern Canada to South America—a distance of over 1800 miles! • The scientific genus name *Numenius*, which means "new moon," refers to curlews' crescent-shaped bills.

ID: long, down-curved bill; striped crown; dark eye line; mottled brown body; long legs. *In flight:* dark underwings.
Size: *L* 17½ in; *W* 32 in.
Habitat: mudflats, sandy beaches, lakeshores, airports and flooded agricultural fields.
Nesting: does not nest in Ohio.
Feeding: probes and pecks for invertebrates in mud or vegetation; eats enormous amounts of berries in fall.
Voice: incoming flocks can be heard uttering a distinctive, rippling *bibibibibibi* long before they come into sight.

Similar Species: *Upland Sandpiper* (p. 124): smaller; straight bill; yellowish legs; lacks head markings. *Long-billed Curlew:* very rare; much larger overall; much longer bill; lacks bold striping on head and through eye.
Best Sites: flooded fields and wetlands in the vicinity of Ottawa NWR; also L. Erie shoreline dredge-spoil impoundments such as those at Lorain and Huron; Maumee Rapids.

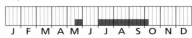

J F M A M J J A S O N D

HUDSONIAN GODWIT

Limosa haemastica

At the time of the earliest surveyors of North American natural history, Hudsonian Godwits were considered quite rare. Since that time, overall numbers have increased markedly. • Severe overhunting nearly eliminated the Hudsonian Godwit in the late 1800s. Its elliptical migration route—west of the Mississippi in northbound migration, east of that river while southbound—is virtually identical to the route used by the ill-fated Eskimo Curlew *(Numenius borealis)*. Although the Eskimo Curlew population was decimated to a point where the species could not recover and is now considered extinct, the Hudsonian Godwit did eventually rebound—the total population now stands at about 50,000 birds. • Most of our birds are seen as singles or in small groups of southbound juveniles in late September and October.

breeding

nonbreeding

ID: long, yellow orange bill with dark, slightly upturned tip; black tail; long, blue black legs. *Breeding:* heavily barred, chestnut red underparts; dark, grayish upperparts; male is more brightly colored. *Nonbreeding:* grayish upperparts; whitish underparts may show a few short, black bars. *Juvenile:* dark gray brown upperparts fringed with pale buff; well-marked scapulars and tertials. *In flight:* white rump; black "wing pits" and wing linings.
Size: *L* 14–15½ in; *W* 29 in.
Habitat: flooded fields, marshes, mudflats and lakeshores.
Nesting: does not nest in Ohio.
Feeding: walks into deeper water than most shorebirds but rarely swims; probes deeply into water or mud; eats mollusks, crustaceans, insects and other invertebrates; also picks earthworms from plowed fields.
Voice: usually quiet in migration; sometimes a sharp, rising *god-WIT!*
Similar Species: *Marbled Godwit* (p. 127): larger; mottled brown overall; lacks white rump. *Greater Yellowlegs* (p. 119): shorter, all-dark bill; bright yellow legs; lacks white rump. *Long-billed Dowitcher* (p. 142) and *Short-billed Dowitcher* (p. 141): much smaller; shorter legs; straight, all-dark bills; yellow green legs; mottled, rust brown upperparts in breeding plumage.
Best Sites: western L. Erie, particularly in the vicinity of Ottawa NWR; occasional on large mudflat complexes inland, such as Hoover Reservoir and Funk Bottoms WA.

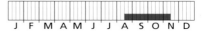

J F M A M J J A S O N D

MARBLED GODWIT

Limosa fedoa

The Marbled Godwit's genus name *Limosa*, meaning "mud," is very descriptive, since this bird is usually seen in muddy haunts in Ohio—often with its long bill stuck face-deep into the mud, probing for food. • Marbled Godwits are rare here, and are normally seen as single birds, sometimes in spring but mostly in fall. The largest recorded flock had 16 birds. • This species is one of our largest regularly occurring shorebirds, and this gigantic sandpiper is unlikely to be missed on the mudflats. • Marbled Godwits have very interesting population dynamics. The breeding range is divided into three very distinct and well-separated populations—Alaska, the Great Plains and along James Bay in Canada. Furthermore, they have three distinct wintering grounds. With this kind of geographic isolation, perhaps in the future there will be three species of Marbled Godwits.

nonbreeding

breeding

ID: long, yellow orange bill with dark, slightly upturned tip; long neck and legs; mottled buff brown plumage is darkest on upperparts; long, blue black legs. *In flight:* cinnamon wing linings.
Size: *L* 16–20 in; *W* 30 in.
Habitat: flooded fields, wet meadows, marshes, mudflats and lakeshores.
Nesting: does not nest in Ohio.
Feeding: probes deeply in soft substrates for worms, insect larvae, crustaceans and mollusks; picks insects from grass; may also eat the tubers and seeds of aquatic vegetation.
Voice: loud, ducklike, 2-syllable squawks: *co-rect co-rect* or *god-wit god-wit.*

Similar Species: *Hudsonian Godwit* (p. 126): smaller; white rump; nonbreeding is grayer. *Greater Yellowlegs* (p. 119): shorter, all-dark bill; bright yellow legs. *Long-billed Dowitcher* (p. 142) and *Short-billed Dowitcher* (p. 141): much smaller; straight, all-dark bills; white rump wedges; yellow green legs.
Best Sites: best sought from July to September along western L. Erie in the vicinity of Ottawa NWR; occasionally at large mudflats in the interior such as Hoover Reservoir, Funk Bottoms WA and Killdeer Plains WA.

J F M A M J J A S O N D

RUDDY TURNSTONE

Arenaria interpres

Boldly patterned, breeding-plumaged Ruddy Turnstones are unmistakable and are one of our showiest shorebirds. Even in comparatively drab, nonbreeding plumage, they are distinctive and should not befuddle even the casual birder. • Ruddy Turnstones are very much birds of the shoreline; the best places to seek them are Lake Erie beaches and shoreline riprap. They are hardy birds, with at least three overwintering records, all from the Cleveland area. • The Ruddy Turnstone is well named, as it uses its stout, spadelike bill to flip pebbles, shells and mud clods in its search for food. On occasion, an ambitious turnstone will dig a substantial excavation, rooting about like a feathered piglet as it slings sand in the pursuit of subterranean invertebrates. • These powerful flyers are virtually global in distribution and nest at some of the highest latitudes of any shorebird.

♂ *breeding*

nonbreeding

ID: white belly; black "bib" curves up to shoulder; stout, black, slightly upturned bill; orange red legs. *Breeding:* ruddy upperparts (female is slightly paler); white face; black "collar"; dark, streaky crown. *Nonbreeding:* brownish upperparts and face.

Size: *L* 9½ in; *W* 21 in.

Habitat: beaches and rocky riprap, lakeshores, sometimes marshes and mudflats.

Nesting: does not nest in Ohio.

Feeding: probes under and flips rocks, weeds and shells for food items; picks, digs and probes for invertebrates from the soil or mud; also eats crabs, berries, seeds, spiders and carrion.

Voice: clear, staccato, rattling *cut-a-cut* alarm call and lower, repeated contact notes.

Similar Species: *Other sandpipers* (pp. 119–47): all lack the Ruddy Turnstone's bold patterning and flashy wing markings in flight. *Plovers* (pp. 114–17): equally bold plumage but in significantly different patterns; more inconspicuous wing bars.

Best Sites: L. Erie beaches at Crane Creek SP and Maumee Bay SP, also flooded fields in that region; occasionally at large inland mudflats.

J F M A M J J A S O N D

RED KNOT
Calidris canutus

R ed Knots are one of our most beautiful spring sandpipers when they are in breeding plumage, yet they are drab and inconspicuous in nonbreeding plumage. • Red Knots are one of the world's greatest long-distance migrants. Nesting at the most northerly reaches of the globe, they travel to southern South America to winter—a round trip passage of up to 19,000 miles! To fuel themselves for this arduous journey, tremendous numbers of Red Knots stage at strategic Atlantic coast mudflats, where they gorge themselves primarily on horseshoe crab eggs. Densities of roosting birds there can exceed 20 per square meter. • In Ohio, Red Knots are much scarcer, and scattered individuals are seen mostly in fall along Lake Erie. Occasional small groups are observed; the largest was a group of 150 birds in Ottawa County on May 26, 1956.

breeding

nonbreeding

ID: chunky, round body; greenish legs. *Breeding:* rusty face, breast and underparts; brown, black and buff upperparts. *Nonbreeding:* pale gray upperparts; white underparts with some faint streaking on upper breast; faint barring on rump. *Immature:* buffy wash on breast; scaly-looking back. *In flight:* white wing stripe.
Size: *L* 10½ in; *W* 23 in.
Habitat: lakeshores, marshes, mudflats and plowed fields.
Nesting: does not nest in Ohio.
Feeding: gleans shorelines for insects, crustaceans and mollusks; probes soft substrates, creating lines of small holes.
Voice: usually silent; low, monosyllabic *knut* reminiscent of its name.

Similar Species: *Long-billed Dowitcher* (p. 142) and *Short-billed Dowitcher* (p. 141): much longer bills, at least 1½ times longer than width of head; barring under tail and on flanks; white "V" on rump and tail and on trailing edge of wings. *Buff-breasted Sandpiper* (p. 140): light buff; finer, shorter bill; dark flecking on sides. *Peeps* (pp. 131–35): much smaller; most have black legs; only the *Sanderling* (p. 130) and the very rare *Curlew Sandpiper* (p. 335) show reddish coloration on undersides in breeding plumage.
Best Sites: dredge-spoil impoundments and large mudflats bordering L. Erie such as Sheldon Marsh SNP and in the vicinity of Ottawa NWR; occasionally at large inland mudflats such as Killdeer Plains WA and Funk Bottoms WA.

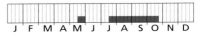

J F M A M J J A S O N D

129

SANDERLING

Calidris alba

The Sanderling is the classic wave-chasing beach sandpiper, darting into the damp sand on the heels of a receding wave to snatch small invertebrates before the next wave crashes onto shore. • The scientific name *alba* means "white," and refers to these birds in nonbreeding plumage, when they are very pale overall. Breeding-plumaged Sanderlings are a rich, rufous brown. • Sanderlings are most reliably found on beaches and should be sought along Lake Erie. Normally, they are seen in small flocks, sometimes with Ruddy Turnstones, but occasionally larger groups are encountered. • Increased recreational use of beaches and the associated disturbances may be detrimental to Sanderlings, but the North American population still numbers about 300,000 birds. • The arctic-nesting Sanderling is a true globetrotter, appearing on every continent in migration.

breeding

nonbreeding

ID: straight, black bill; black legs; white underparts; white wing bar; pale rump. *Breeding:* dark spotting or mottling on rusty head and breast. *Nonbreeding:* pale gray upperparts; black shoulder patch (often concealed).
Size: *L* 7–8½ in; *W* 17 in.
Habitat: sandy shores and large mudflats.
Nesting: does not nest in Ohio.
Feeding: gleans shorelines for insects, crustaceans and mollusks; probes repeatedly, creating a line of small holes in the sand or mud; often seen picking through large windrows of zebra mussel shells that litter L. Erie beaches.

Voice: flight call is a sharp *kip*.
Similar Species: *Dunlin* (p.-138): larger and taller; darker; slightly downcurved bill. *Red Knot* (p. 129): much larger; gray-barred, whitish rump; breeding adult has unstreaked, reddish belly. *Least Sandpiper* (p. 133): smaller and darker; yellowish legs; lacks rufous breast in breeding plumage. *Western Sandpiper* (p. 132) and *Semipalmated Sandpiper* (p. 131): sandy upperparts in nonbreeding plumage; lack rufous breast in breeding plumage.
Best Sites: beaches at Crane Creek SP, Maumee Bay SP and East Harbor SP; small numbers at large, inland mudflats such as those at Hoover Reservoir, Killdeer Plains WA and Funk Bottoms WA.

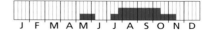

J F M A M J J A S O N D

SEMIPALMATED SANDPIPER

Calidris pusilla

Peeps—so dubbed for their similar call notes—are small sandpipers in the genus *Calidris*, and in Ohio include the Semipalmated, Least, Western, Baird's and White-rumped sandpipers. Identifying these small shorebirds is particularly vexing to novice birders, but with practice and experience, separating the peeps usually isn't too difficult. One advantage is that they are often easy to study out on the open mudflats, as opposed to treetop-seeking, foliage-obscured warblers. • In terms of sheer numbers, the Semipalmated Sandpiper is our most abundant peep, and is generally common anywhere suitable mudflats form. • The Semipalmated Sandpiper is one species that birders should strive to become intimately familiar with, as it is an excellent benchmark bird by which to judge and compare other small shorebirds.

breeding

nonbreeding

ID: short, straight, black bill; black legs. *Breeding:* mottled upperparts; slight rufous tinge on ear patch, crown and scapulars; faint streaks on upper breast and flanks. *Nonbreeding:* white "eyebrow"; gray brown upperparts; white underparts with light brown wash on sides of upper breast. *Juvenile:* similar to breeding adult, but with smudgier markings on upper breast; often washed warm buff; white line above eye; dark brown lores and ear coverts. *In flight:* narrow, white wing stripe; white rump is split by black line.
Size: *L* 5½–7 in; *W* 14 in.
Habitat: mudflats and the shores of ponds and lakes.
Nesting: does not nest in Ohio.
Feeding: probes soft substrates and gleans for aquatic insects and crustaceans.

Voice: flight call is a harsh *cherk;* sometimes a longer *chirrup* or chittering alarm call.
Similar Species: *Least Sandpiper* (p. 133): yellowish legs; darker upperparts. *Western Sandpiper* (p. 132): longer, slightly down-curved bill; bright rufous wash on crown and ear patch, sometimes on back (breeding). *Sanderling* (p. 130): pale gray upperparts; blackish trailing edge on flight feathers in nonbreeding plumage. *White-rumped Sandpiper* (p. 134): larger; white rump; folded wings extend beyond tail. *Baird's Sandpiper* (p. 135): larger; longer bill; folded wings extend beyond tail.
Best Sites: large mudflat complexes in western L. Erie marshes and inland; Killdeer Plains WA; Big Island WA; Funk Bottoms WA; C.J. Brown and Hoover reservoirs; any good-sized mudflat or flooded field is likely to host migrants.

J	F	M	A	M	J	J	A	S	O	N	D

131

WESTERN SANDPIPER

Calidris mauri

Distinguishing Western Sandpipers from Semipalmated Sandpipers is one of the few consistently difficult identification problems that routinely bedevil Ohio birders. • Western Sandpipers have a very limited breeding range that encompasses a small part of western Alaska and eastern Siberia, and the primary migratory corridor is along the Pacific coast. They are virtually unknown in Ohio in the spring, and in fall they migrate later and in much smaller numbers than the Semipalmated Sandpiper. Westerns typically occur in small flocks—the largest recorded group is 75 birds. After September, any sandpiper that resembles either the Semipalmated or Western is likely to be a Western. • Most Ohio Westerns are juveniles, which in general have longer, more drooped bills, brighter rusty scapulars and paler faces and heads than Semipalmateds.

nonbreeding

nonbreeding

ID: slightly down-curved, black bill; black legs. *Breeding:* rufous patches on crown, ear and scapulars; V-shaped streaking on upper breast and flanks; pale underparts. *Nonbreeding:* white "eyebrow"; gray brown upperparts; white underparts; streaky, light brown wash on upper breast. *In flight:* narrow, white wing stripe; white rump is split by black line.
Size: *L* 6–7 in; *W* 14 in.
Habitat: flooded fields, lakeshores and mudflats.
Nesting: does not nest in Ohio.
Feeding: gleans and probes mud; often wades in deeper water than Semipalmated Sandpiper; occasionally submerges its head; primarily eats aquatic insects, worms and crustaceans.
Voice: flight call is a high-pitched *cheep.*

Similar Species: *Semipalmated Sandpiper* (p. 131): see above. *Least Sandpiper* (p. 133): smaller; yellowish legs; darker breast wash; lacks rufous patches. *White-rumped Sandpiper* (p. 134): larger; white rump; folded wings extend beyond tail; lacks rufous wing patches. *Baird's Sandpiper* (p. 135): larger; folded wings extend beyond tail; lacks rufous patches. *Dunlin* (p. 138): larger; longer bill is thicker at base and droops at tip; black belly in breeding plumage; grayer, unstreaked back in nonbreeding plumage. *Sanderling* (p. 130): nonbreeding plumage shows pale gray upperparts, blackish trailing edge on flight feathers and bold, white upperwing stripe in flight.
Best Sites: mudflats along the western basin of L. Erie; large inland mudflats at C.J. Brown and Hoover reservoirs, Funk Bottoms WA, Big Island WA and Killdeer Plains WA.

J F M A M J J A S O N D

LEAST SANDPIPER

Calidris minutilla

The Least Sandpiper is the world's smallest shorebird, tipping the scales at an average weight of less than 1 ounce. This species is a common and ubiquitous part of Ohio's mudflat avifauna, and is second only to the Semipalmated Sandpiper in terms of abundance among the peeps. • Least Sandpipers are fairly easily recognized, as they are the only peeps with yellow legs, but beware of exceptionally muddy individuals. • Although the various species of peeps often occur together on the same mudflat, they generally segregate into distinct habitat niches, which can offer identification clues. Least Sandpipers and Semipalmated Sandpipers—which shouldn't be difficult to separate—tend to forage on open, saturated mud, often near open water, with the Leasts ranging into the drier locales. Western Sandpipers and White-rumped Sandpipers can be found wading and feeding in shallow water, and Baird's Sandpipers usually don't associate directly with other peeps, but feed among dry vegetation above the open, muddy zones.

breeding

ID: *Breeding:* black bill; yellowish legs; dark, mottled back; buff brown breast, head and nape; light breast streaking; prominent, white "V" on back. *Nonbreeding:* much duller; prominent, streaked breast band; often lacks back stripes. *Immature:* similar to breeding adult, but with faintly streaked breast.
Size: *L* 5–6½ in; *W* 13 in.
Habitat: sandy beaches, lakeshores, mudflats and wetland edges.
Nesting: does not nest in Ohio.
Feeding: probes or pecks for insects, crustaceans, small mollusks and occasionally seeds.

Voice: high-pitched *kreee.*
Similar Species: *Semipalmated Sandpiper* (p. 131): black legs; lighter upperparts; rufous tinge on crown, ear patch and scapulars. *Western Sandpiper* (p. 132): slightly larger; black legs; lighter breast wash in all plumages; rufous patches on crown, ear and scapulars in breeding plumage. *Other peeps* (pp. 131–35): all are larger; dark legs.
Best Sites: any mudflat or flooded field is likely to host migrants.

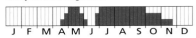

J F M A M J J A S O N D

WHITE-RUMPED SANDPIPER

Calidris fuscicollis

The white rump of this species is diagnostic, but can be surprisingly difficult to see. However, White-rumped Sandpipers have a distinctive, attenuated profile, with the wings projecting noticeably beyond the tail. Their overall body length is at least an inch longer than that of our other peeps, except for the Baird's Sandpiper. • White-rumps are uncommon at best in Ohio, and are generally seen as individuals or very small flocks. The largest concentration was 200 birds on May 23, 1971, in Ottawa County—an exceptional number. • Ornithologist Richard Pough, writing in 1959, noted that White-rumped Sandpipers were the most abundant shorebirds wintering in Argentina, and were also common in certain parts of the Arctic, but were inexplicably rare in the interior U.S. We now know the reason: most White-rumps fly over Ohio and much of the rest of the country, making their long journey in a few tremendous bursts. Some of these nonstop flights might last 60 hours and cover 1800 miles.

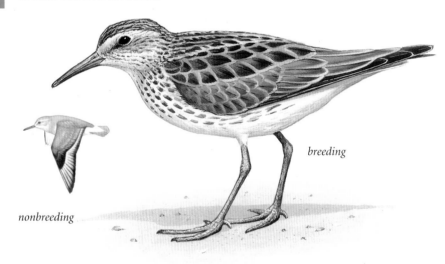

breeding

nonbreeding

ID: black legs and bill; wings extend well beyond tail. *Breeding:* mottled brown-and-rufous upperparts; streaked breast, sides and flanks. *Nonbreeding:* mottled gray upperparts; white "eyebrow." *Immature:* black upperparts edged with white, chestnut and buff. *In flight:* white rump; dark tail; indistinct wing bar.
Size: *L* 7–8 in; *W* 17 in.
Habitat: lakeshores, marshes, sewage lagoons, reservoirs and flooded fields.
Nesting: does not nest in Ohio.

Feeding: often feeds in standing water; gleans mud for insects, crustaceans and mollusks.
Voice: flight call is a characteristic, squealing *tzeet*, higher than any other peep.
Similar Species: *Other peeps* (pp. 131–35): all have dark line through rump. *Baird's Sandpiper* (p. 135): lacks clean white rump; breast streaking does not extend onto flanks.
Best Sites: large wetland-mudflat ecosystems, particularly those along western L. Erie; also any mudflat in western Ohio.

J F M A M J J A S O N D

BAIRD'S SANDPIPER

Calidris bairdii

Baird's Sandpipers travel from one end of the globe to the other between their winter and summer ranges. These sandpipers follow an elliptical migration route—northbound birds pass west of the Mississippi River, while many southbound migrants filter through the eastern U.S. Consequently, Baird's Sandpipers are accidental in Ohio in spring, but small numbers are regularly encountered in fall. • Baird's Sandpipers typically shun the company of other "peeps," preferring to forage in much drier, sparsely vegetated habitats, as are found along the upper reaches of the mudflat community. Sometimes, Baird's even appear on sod farms or on the grassy expanses that buffer airports. • Virtually all Baird's seen in Ohio are juveniles, which are readily distinguished by the buffy breast, the buff-colored, "scaly" edgings on the back feathers, and long wings that extend well beyond the tail.

nonbreeding

nonbreeding

ID: black legs and bill; faint, buff brown breast speckling; folded wings extend beyond tail. *Breeding:* black, diamondlike pattern on back and wing coverts. *Juvenile:* brighter and browner; back appears "scaly."
Size: *L* 7–7½ in; *W* 17 in.
Habitat: sandy beaches, mudflats, shortgrass meadows and wetland edges.
Nesting: does not nest in Ohio.
Feeding: gleans aquatic invertebrates, especially larval flies; also eats beetles and grasshoppers; rarely probes.
Voice: soft, rolling *kriit kriit.*
Similar Species: "scaly" back and long wings are fairly distinctive. *White-rumped Sandpiper* (p. 134): clean white rump; breast streaking extends onto flanks; more streaked head and back. *Pectoral Sandpiper* (p. 136): much more robust; dark breast streaks end abruptly at edge of white belly. *Least Sandpiper* (p. 133): smaller; yellowish legs. *Western Sandpiper* (p. 132) and *Sanderling* (p. 130): lack buffy breast speckling. *Semipalmated Sandpiper* (p. 131): smaller; shorter bill; lacks streaked breast in nonbreeding plumage.
Best Sites: suitable habitat locations vary yearly; consistent locales include C.J. Brown and Hoover reservoirs, western L. Erie marshes and wetlands, and sometimes dredge-spoil impoundments bordering L. Erie.

| J | F | M | A | M | J | J | A | S | O | N | D |

PECTORAL SANDPIPER

Calidris melanotos

Pectoral Sandpipers are harbingers of spring in Ohio, returning in numbers by early March, with some enthusiastic individuals occasionally arriving in late February. They are also one of our most common shorebirds, and can be expected statewide. When abundant, appropriate habitat is available, large numbers can occur. • Pectoral Sandpipers are unusual among *Calidris* shorebirds in that they exhibit extreme sexual dimorphism—males may be 25 to 30 percent larger than females. • In an interesting case of hybridism, Pectorals have crossbred with the Curlew Sandpiper, producing the "Cox's Sandpiper," which is known from three Australian specimens. Shorebird enthusiasts occasionally report Cox's Sandpipers in other locales, such as Massachusetts, but definitive identification of these other records is dubious.

breeding

ID: brown breast streaks end abruptly at edge of white belly; white under-tail coverts; black bill with slightly downcurved tip; long, yellow legs; mottled upperparts; may have faintly rusty, dark crown and back; folded wings extend beyond tail. *Juvenile:* less spotting on breast; broader, white feather edges on back form 2 white "V"s.
Size: *L* 8½ in; *W* 18 in (female is noticeably smaller).
Habitat: lakeshores, marshes, mudflats and flooded fields or pastures; short-grass airport buffers.
Nesting: does not nest in Ohio.

Feeding: probes and pecks for small insects, mainly flies, but also beetles and some grasshoppers; may also take small mollusks, crustaceans, berries, seeds, moss, algae and some plant material.
Voice: sharp, short, low *krrick krrick*.
Similar Species: *Sharp-tailed Sandpiper:* most similar, but extremely rare in Ohio; juveniles are brighter buff with prominent eye line and rusty "cap," but lack densely streaked breast. *Peeps* (pp. 131–35): much smaller; lack well-defined, dark "bib" and yellow legs. *Ruff* (p. 335): rare; larger; longer legs; lacks cleanly demarcated "bib."
Best Sites: Funk Bottoms WA; Big Island WA; western L. Erie marshes; grassy airport buffers.

J F M A M J J A S O N D

PURPLE SANDPIPER

Calidris maritima

This rare bird is one of the most sought-after shorebirds to regularly appear in Ohio. Purple Sandpipers can only be expected along the shoreline of Lake Erie, where they visit rocky breakwaters, riprap and occasionally beaches. In spite of this well-defined habitat, they can be surprisingly hard to spot, as their cryptic plumage blends well with the mossy, wave-washed rocks that they frequent. • Purple Sandpipers are extremely hardy, wintering in the most brutal conditions of any North American shorebird. They generally don't appear in Ohio until November, and then stay until freeze-up, sometimes overwintering in milder years. • For birders, Purple Sandpiper pursuit often takes place in harsh, very cold weather, and involves peering at ice-encrusted rocks while Common Goldeneyes and Red-breasted Mergansers swim nearby—a stark contrast to most shorebirding, which is done on sun-warmed mudflats. Normally only one or two Purple Sandpipers are encountered, but occasionally, little groups of three to six are seen.

nonbreeding

ID: long, slightly drooping, black-tipped bill with orange yellow base; orange yellow legs; dull streaking on breast and flanks. *Nonbreeding:* unstreaked, gray head, neck and upper breast form "hood"; gray spots on white belly. *Juvenile:* streaking on head; chestnut brown, white and buff feather edgings on upperparts.

Size: *L* 9 in; *W* 17 in.

Habitat: sand and gravel beaches, rocky shorelines, piers and breakwaters.

Nesting: does not nest in Ohio.

Feeding: food is found visually and is snatched while moving over rocks and sand; eats mostly mollusks, insects, crustaceans and other invertebrates; also eats a variety of plant material.

Voice: call is a soft *prrt-prrt*.

Similar Species: *Peeps* (pp. 131–35): all lack bicolored bill, yellow orange legs and unstreaked, gray "hood" in nonbreeding plumage. *Ruddy Turnstone* (p. 128): distinctive, dark brown patches on chest; less "scaly" back; lacks breast and flank streaks.

Best Sites: breakwaters at Mentor Headlands, Lorain Harbor, Avon L. and most of the protected harbors and dredge-spoil impoundments along the central basin of L. Erie.

J F M A M J J A S O N D

DUNLIN

Calidris alpina

Spring Dunlins return in early April, resplendent in their breeding plumage, sporting a prominent black belly and a bright rusty back, the latter earning them the former name "Red-backed Sandpiper." • This is one of the most abundant and widespread shorebirds in Ohio and in much of North America. The hardy Dunlin is also the only commonly occurring sandpiper regularly seen into December. • This broadly ranging shorebird has been split into nine subspecies based on regional variation, three of which comprise North American populations. Ohio birds are the subspecies *hudsonia,* which breeds in the Arctic along western Hudson Bay. • This abundant shorebird was once rare owing to rampant overhunting; its unwary nature, dense flocks and propensity to return to downed comrades made Dunlins easy pickings. Now, two primary predators are the Merlin and the Peregrine Falcon.

nonbreeding   *breeding*

ID: slightly downcurved, black bill; black legs. *Breeding:* black belly; streaked, white neck and underparts; rufous wings, back and crown. *Nonbreeding:* pale gray underparts; brownish gray upperparts; light brown streaking on breast and nape. *Juvenile:* buffy head and breast; mantle and scapulars fringed chestnut brown and white; whitish "V" on back. *In flight:* white wing stripe.
Size: *L* 7½–9 in; *W* 17 in.
Habitat: mudflats and the shores of ponds, marshes and lakes; occasionally seen in pastures or sewage lagoons.
Nesting: does not nest in Ohio.
Feeding: gleans and probes for aquatic crustaceans, worms, mollusks and insects, often wading in water up to belly.

Voice: flight call is a grating *cheezp* or *treezp.*
Similar Species: black belly in breeding plumage is distinctive. *Western Sandpiper* (p. 132) and *Semipalmated Sandpiper* (p. 131): smaller; nonbreeding plumage is browner overall; bill tip of female is less downcurved. *Least Sandpiper* (p. 133): smaller; darker upperparts; yellowish legs. *Stilt Sandpiper* (p. 139): larger; longer bill; yellowish green legs. *Curlew Sandpiper* (p. 335): very rare; very similar in nonbreeding plumage, but has longer, more downcurved bill, grayer back and white rump.
Best Sites: western L. Erie marshes and vicinity; Killdeer Plains, Big Island and Funk Bottoms wildlife areas inland; any flooded field or mudflat.

J F M A M J J A S O N D

STILT SANDPIPER

Calidris himantopus

By becoming familiar with the feeding niches and habits of shorebirds, birders can often gain useful clues to species identities. The Stilt Sandpiper normally forages belly-deep in water—often in association with yellowlegs and dowitchers—submerging its head to probe the bottom. • Stilts are very rare spring visitors, but become more frequent in fall migration, and are normally seen in small groups of 20 or fewer. • Novice birders are sometimes surprised to learn that southbound shorebirds appear very early in Ohio. Stilts provide a typical chronology. The earliest birds, which return in early July, are failed breeders that lack the time to attempt renesting in the short arctic summer. Successful postbreeding females follow them a few weeks later, and males arrive a week after that. Finally come the juveniles, which comprise virtually all of our late fall shorebird migrants.

breeding

nonbreeding

ID: long, greenish legs; long bill droops slightly at end. *Breeding:* chestnut red "ear" patch; white "eyebrow"; striped crown; streaked neck; barred underparts. *Nonbreeding:* less conspicuous, white "eyebrow"; dirty white neck and breast; white belly; dark brownish gray upperparts. *Juvenile:* dark brown upperparts, fringed with rufous or light buff; buff wash on throat and breast; faintly streaked, white belly. *In flight:* white rump; legs trail behind tail; no wing stripe.
Size: *L* 8–9 in; *W* 18 in.
Habitat: lakeshores, reservoirs, marshes and flooded fields.
Nesting: does not nest in Ohio.
Feeding: probes deeply in shallow water; eats mostly invertebrates; occasionally picks insects from the water's surface or the ground; also eats seeds, roots and leaves.
Voice: soft, rattling *querp* or *kirr* in flight; clearer *whu.*
Similar Species: *Greater Yellowlegs* (p. 119) and *Lesser Yellowlegs* (p. 120): straight bills; yellow legs; lack red "ear" patch of breeding adult, blotchy back feathers of nonbreeding adult or chestnut mantle of juvenile. *Dunlin* (p. 138): shorter, black legs; dark rump; whitish wing bar. *Long-billed Dowitcher* (p. 142) and *Short-billed Dowitcher* (p. 141): shorter legs; less "scaly" backs; longer bills; white upper rumps.
Best Sites: Ottawa NWR, Magee Marsh WA and other western L. Erie marshes; Hoover and C.J. Brown reservoirs inland.

J F M A M J J A S O N D

BUFF-BREASTED SANDPIPER

Tryngites subruficollis

irders need sharp eyes to spot Buff-breasted Sandpipers lurking in the drier upper reaches of mudflats where their mousy coloration blends with dried vegetation. These are very much sandpipers of short vegetation, from their arctic breeding grounds to their wintering grounds on the pampas of Argentina. • In Ohio, the Buff-breast is strictly a fall bird, and normally occurs as singles or a few scattered birds. • Buff-breasted Sandpipers are the only North American shorebirds that use a "lek" mating system, similar to prairie-chickens. Males gather communally and engage in elaborate courtship displays, using a characteristic "raised wing" display posture. The male will hold one wing perfectly vertical, flashing his prominent, white underwing like a flag. The females gather around the posturing males and select the most impressive specimen with which to mate. These studs may win the affection of multiple females, as they are unabashed polygamists.

breeding

ID: unpatterned, buffy face and foreneck; large, dark eyes; very thin, straight, black bill; buff underparts; small spots on crown, nape, breast, sides and flanks; "scaly" look to back and upperwings; yellow legs. *In flight:* pure white underwings; no wing stripe.
Size: *L* 7½–8 in; *W* 18 in.
Habitat: drier, sparsely vegetated areas of mudflats; also sod farms, mowed fields, airports and sometimes golf courses.
Nesting: does not nest in Ohio.
Feeding: gleans the ground and shorelines for insects, spiders and small crustaceans; may eat seeds.

Similar Species: *Upland Sandpiper* (p. 124): more boldly streaked breast; longer neck; smaller head; larger bill; streaking on "cheek" and foreneck. *Pectoral Sandpiper* (p. 136): grayer brown back and breast; white on belly; white ovals at sides of tail. *Ruff* (p. 335): very rare; larger and heavier; sloping forehead; less tidy markings; greener legs; white ovals at sides of tail. *Baird's Sandpiper* (p. 135): much smaller; shorter dark legs; shorter neck.
Best Sites: largest numbers in vicinity of Ottawa NWR; also inland at Hoover Reservoir, Killdeer Plains WA and Big Island WA.

J F M A M J J A S O N D

SHORT-BILLED DOWITCHER

Limnodromus griseus

Distinguishing the two dowitcher species has confounded ornithologists since these birds were discovered and is still a problem for many. It wasn't until 1950 that the two dowitchers were awarded full species status. • The Short-billed Dowitcher is by far the more common of the two in Ohio, and is a regular component of our spring and fall mudflat ecosystems. Fall numbers can be in the thousands at favorable locales. • Both dowitcher species typically forage in deeper water and rapidly probe the substrate with a motion like a sewing machine, habits that render them very distinctive. • Compounding the Short-billed Dowitcher identification problem is the division of the species into three subspecies. Fortunately our birds are all of the race *hendersoni*, so we aren't plagued with the entire suite of variations.

breeding

nonbreeding

ID: straight, long, dark bill; white "eyebrow"; chunky body; yellow green legs. *Breeding:* white belly; dark spotting on reddish buff neck and upper breast; prominent dark barring on white sides and flanks. *Nonbreeding:* dirty gray upperparts; dirty white underparts. *Juvenile:* chestnut-edged crown and back; barred or striped tertials; no streaking or spotting on underparts. *In flight:* white wedge on rump and lower back.

Size: *L* 11–12 in; *W* 19 in.

Habitat: lakeshores, mudflats, reservoirs, marshes and flooded fields.

Nesting: does not nest in Ohio.

Feeding: wades in shallow water or on mud, probing deeply into the substrate with a repeated up-down bill motion; eats aquatic invertebrates, including insects, mollusks, crustaceans and worms; may feed on seeds, aquatic plants and grasses.

Voice: generally silent; flight call is a mellow, repeated *tututu, toodulu* or *toodu.*

Similar Species: *Long-billed Dowitcher* (p. 142): very little white on belly; dark spotting on neck and upper breast; black-and-white barring on red flanks in breeding plumage; alarm call is a high-pitched *keek. Red Knot* (p. 129): much shorter bill; unmarked, red breast in breeding plumage; nonbreeding birds lack barring on tail and white wedge on back in flight. *Wilson's Snipe* (p. 143): heavy streaking on neck and breast; bicolored bill; light median stripe on crown; shorter legs.

Best Sites: largest numbers on western L. Erie; also mudflats and flooded fields statewide.

J F M A M J J A S O N D

LONG-BILLED DOWITCHER
Limnodromus scolopaceus

Separating the two dowitcher species is a conundrum for many birders and identifications are often clouded with uncertainty. In Ohio, different migration timing may help. In spring, Long-billeds tend to be earlier and are rare, passing through by May, prior to the Short-billed Dowitcher invasion. In fall, Long-billeds migrate later—mostly juveniles are seen from late September through November. Any dowitcher sighted after October is much more likely to be a Long-billed. • Spring birds are easier to separate visually: Long-billeds are more heavily barred below and the center of the breast is never spotted. • The real identification problem lies with nonbreeding- and juvenile-plumaged birds in fall. A helpful structural characteristic involves wing length—Long-billed wings fall slightly short of the tail; Short-billed wings extend slightly beyond the tail. Long-billeds also show a sharp demarcation between the gray upper and white lower breast, and the tertial and scapular feathers have pale, narrow, buff edgings. • Calls are perhaps the most diagnostic way to separate these two species. Long-billeds, which call more frequently, give a single sharp *keek*, whereas Short-billeds emit a mellow, whistled *tu-tu-tu*.

breeding

nonbreeding

ID: very long, straight, dark bill; dark eye line; white "eyebrow"; chunky body; yellow green legs. *Breeding:* black-and-white barring on reddish underparts; some white on belly; dark, mottled upperparts. *Nonbreeding:* gray overall; dirty white underparts. *In flight:* white wedge on rump and lower back.
Size: *L* 11–12½ in; *W* 19 in.
Habitat: lakeshores, ponds, reservoirs, shallow marshes and mudflats.
Nesting: does not nest in Ohio.
Feeding: probes in shallow water and mudflats with a repeated up-down bill motion; frequently plunges its head underwater;

eats shrimps, snails, worms, larval flies and other soft-bodied invertebrates.
Voice: alarm call is a loud, high-pitched *keek,* occasionally given in series.
Similar Species: *Short-billed Dowitcher* (p. 141): see above. *Red Knot* (p. 129): much shorter bill; unmarked, red breast in breeding plumage; nonbreeding birds lack barring on tail and white wedge on back in flight. *Wilson's Snipe* (p. 143): shorter legs; bicolored bill; heavy streaking on neck and breast; light median stripe on crown.
Best Sites: western L. Erie attracts the greatest numbers; Ottawa NWR; large mudflat complexes inland such as at Killdeer Plains WA, Big Island WA and Hoover Reservoir.

| J | F | M | A | M | J | J | A | S | O | N | D |

WILSON'S SNIPE

Gallinago delicata

The Wilson's Snipe is one of North America's most abundant sandpipers—overall population estimates are as high as 5 million—but most people would never know. This bird is normally very secretive, spending its time foraging in low, wet areas among vegetation that blends with its cryptic coloration. The 500 birds seen on September 29, 1987, at Ottawa National Wildlife Refuge is evidence of how common the snipe can be in migration. • Wilson's Snipe is a very rare breeder in Ohio, with perhaps only half a dozen pairs most years. The most consistent breeding site is Irwin Prairie State Nature Preserve, where visitors can observe the spectacular courtship flights of displaying males. They become aerial acrobats, coursing high aloft and then dropping to the ground in irregular fits and spurts, producing a hollow, tremulous booming from air rushing through their tail feathers. • Long known as "Common Snipe," this superspecies was split in 2002 into the North American "Wilson's Snipe" and the Eurasian "Common Snipe."

ID: long, sturdy, bicolored bill; relatively short legs; heavily striped head, back, neck and breast; dark eye stripe; dark barring on sides and flanks; unmarked, white belly. *In flight:* quick zigzags on takeoff.
Size: *L* 10½–11½ in; *W* 18 in.
Habitat: damp to wet cattail and bulrush marshes, sedge meadows, bogs and fens, and even soggy fields and cattle pastures; rare in winter, especially southward.
Nesting: usually in dry grass, often under vegetation; nest is made of grass, moss and leaves; female incubates 4 darkly marked, olive buff to brown eggs for 18–20 days; both parents raise the young, often splitting the brood.
Feeding: probes soft substrates for larvae, earthworms and other soft-bodied invertebrates; also eats mollusks, crustaceans, spiders, small amphibians and some seeds.
Voice: eerie, accelerating courtship song is produced in flight: *woo-woo-woo-woo-woo-woo;* often sings *wheat wheat wheat* from an elevated perch; alarm call is a nasal *scaip.*
Similar Species: *Long-billed Dowitcher* (p. 142) and *Short-billed Dowitcher* (p. 141): longer legs; all-dark bills; lack heavy striping on head, back, neck and breast; usually seen in flocks. *American Woodcock* (p. 144): unmarked, buff underparts; yellowish bill; light bars on black crown and nape.
Best Sites: Killdeer Plains WA; Big Island WA; western L. Erie marshes.

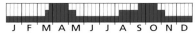

J F M A M J J A S O N D

AMERICAN WOODCOCK

Scolopax minor

Ecologically distinct from all other species, "Timberdoodles" are perhaps our most bizarre shorebirds. American Woodcocks are inhabitants of scruffy, successional woodlands where consistently damp soil is present. They can be hardy and have overwintered where seepages provide constantly moist soil. • Normally, woodcocks are quite secretive, but in spring, courting males come out of their shells and provide one of our most interesting avian spectacles. Returning in late February to early March, males promptly begin to woo females with an ornate ritual. Males launch themselves from a small dancing ground and spiral high into the air, producing distinctive twittering sounds with their wings. Then, as they quickly drop to earth, a characteristic chirping sound emanates from the primary feathers. Once back on the ground, they resume their nasal *peent* calls. Sometimes beginning birders confuse this latter call with that of the Common Nighthawk, a species that doesn't return to Ohio until early May.

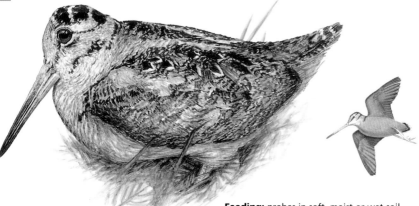

ID: very long, sturdy bill; very short legs; large head; short neck; chunky body; large, dark eyes; unmarked, buff underparts; pale bars on black crown and nape. *In flight:* rounded wings; makes a twittering sound when flushed from cover.
Size: *L* 11 in; *W* 18 in.
Habitat: moist woodlands and brushy thickets adjacent to grassy clearings or abandoned fields.
Nesting: on the ground in woods or overgrown fields; female digs a scrape and lines it with dead leaves and other debris; female incubates 4 pinkish buff eggs, blotched with brown and gray, for 20–22 days; female tends the young.

Feeding: probes in soft, moist or wet soil for earthworms and insect larvae; also takes spiders, snails, millipedes and some plant material, including seeds, sedges and grasses.
Voice: nasal *peent;* during courtship dance male produces high-pitched, twittering, whistling sounds.
Similar Species: *Wilson's Snipe* (p. 143): heavily striped head, back, neck and breast; dark barring on sides and flanks. *Long-billed Dowitcher* (p. 142) and *Short-billed Dowitcher* (p. 141): all-dark bills; longer legs; lack pale barring on dark crown and hindneck; usually seen in flocks in open areas.
Best Sites: any area with a combination of wet fields and thickets; great places to observe displays are Irwin Prairie SNP, Resthaven WA and Spring Valley WA.

J F M A M J J A S O N D

WILSON'S PHALAROPE

Phalaropus tricolor

Wilson's Phalarope is the only nonmarine species of the three phalaropes, and the most common in Ohio. • Several unusual habits distinguish phalaropes, including sexual role-reversal, which is very rare in the bird world. The female is larger, more colorful and more aggressive than the male. Once the female lays the eggs, all parental duties fall to the hapless male, who must incubate the eggs, guard the nest and watch over the young after they hatch. • This species probably historically bred in Ohio's wet prairies, virtually all of which have been destroyed. Therefore, it was a real success story when the Ohio Division of Wildlife's massive wetland restoration project at Big Island Wildlife Area attracted a nesting pair in 2000 and again in 2002.

nonbreeding ♂ *breeding* ♀

ID: dark, needlelike bill; white "eyebrow," throat and nape; light underparts; black legs. *Breeding female:* gray "cap"; chestnut brown on sides of neck; black eye line extends down side of neck and onto back. *Breeding male:* duller overall; dark "cap." *Nonbreeding:* all-gray upperparts; white "eyebrow"; gray eye line; white underparts; dark yellowish or greenish legs. *Juvenile:* dark brown upperparts appear "scaly"; buffy sides of breast; pinkish yellow legs.
Size: *L* 9–9½ in; *W* 17 in.
Habitat: marshes, lakeshores, ponds and mudflats.
Nesting: often near water; well concealed in a depression lined with grass and other vegetation; male incubates 4 brown-blotched, buff eggs for 18–27 days; male rears the young.
Feeding: swims in tight, spinning circles to stir up prey, then picks aquatic insects, worms and small crustaceans from the water's surface or just below it; on land makes short jabs to pick up invertebrates.
Voice: deep, grunting *work work* or *wu wu wu*, usually given on the breeding grounds.
Similar Species: *Red-necked Phalarope* (p. 146): rufous stripe down side of neck in breeding plumage; dark nape and line behind eye in nonbreeding plumage. *Red Phalarope* (p. 147): reddish neck, breast and underparts in breeding plumage; dark nape and broad, dark line behind eye in nonbreeding plumage; rarely seen away from L. Erie. *Stilt Sandpiper* (p. 139) and *Lesser Yellowlegs* (p. 120): much different foraging behavior.
Best Sites: western L. Erie marshes; also large interior mudflat complexes such as Funk Bottoms WA, Killdeer Plains WA and Hoover Reservoir.

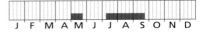

J F M A M J J A S O N D

RED-NECKED PHALAROPE

Phalaropus lobatus

R ed-necked Phalaropes are mostly pelagic, spending up to nine months of the year at sea. They swim like little ducks and often engage in a feeding behavior called "spinning." Whirling rapidly in circles, swimming phalaropes create an upwelling of water that pulls small animals within the bird's reach. • In Ohio, this is an uncommon species at best, and is far more likely to be seen in fall. Usual sightings are of singles or a few birds, and rarely flocks of 20 or more. Red-necked Phalaropes are not hardy and are rarely seen after September. • Although still an abundant bird overall, ornithologists are alarmed over the mysterious disappearance of the huge migratory flocks that would stop over in the Bay of Fundy on the coasts of Nova Scotia and New Brunswick, Canada. Until the early 1980s, up to 3 million Red-necked Phalaropes congregated there; by the early 1990s they had disappeared from this locale.

nonbreeding

ID: thin, black bill; long, gray legs; lobed toes. *Breeding female:* chestnut brown stripe on neck and throat; white "chin"; blue black head; incomplete, white eye ring; white belly; 2 rusty buff stripes on each upperwing. *Breeding male:* white "eyebrow"; less intense colors than female. *Nonbreeding:* white underparts; dark nape; black "cap"; broad, dark band from eye to ear; whitish stripes on blue gray upperparts. *Juvenile:* buff-and-black upperparts; white underparts.
Size: *L* 7 in; *W* 15 in.
Habitat: open water bodies including ponds, lakes, marshes and sewage lagoons; flooded fields.
Nesting: does not nest in Ohio.
Feeding: swims in tight, spinning circles to stir up prey, then picks aquatic insects, worms and small crustaceans from the

water's surface or just below it; on land makes short jabs to pick up invertebrates.
Voice: often noisy in migration; soft *krit krit krit.*
Similar Species: *Wilson's Phalarope* (p. 145): very long, thin bill; female has gray "cap" and black eye line extending down side of neck and onto back in breeding plumage; lacks prominent eye stripe in nonbreeding plumage. *Red Phalarope* (p. 147): noticeably thicker bill, normally pale at base; all-red neck, breast and underparts in breeding plumage; in nonbreeding plumage has uniformly pale gray back and unmarked, white underwings.
Best Sites: western L. Erie marshes; numerous records at Ottawa NWR, Cleveland lakefront, Mosquito Creek L. and Killdeer Plains WA; may appear at any large mudflat complex in the glaciated region.

J F M A M J J A S O N D

RED PHALAROPE

Phalaropus fulicarius

The most pelagic of the three phalarope species, Red Phalaropes spend most of their lives on open ocean waters beyond the sight of land. European whalers dubbed them "Bowhead Birds" for their habit of congregating in large numbers over feeding bowhead whales, which stir up small crustaceans upon which the phalaropes feed. In fact, whalers used this species as a guide for locating whales. • This is the rarest of the three phalarope species in Ohio, with four to six single-bird sightings in most years. Every five years or so, mini-invasions will result in a doubling of the normal number of sightings. • Red Phalaropes are very late migrants to Ohio, with most records from November. They sometimes linger into January, though they normally do not overwinter. The vast majority are seen in the eastern basin of Lake Erie, from Huron eastward, where the birds are found lurking in sheltered waters of the lake near stone breakwaters.

nonbreeding

ID: thicker bill than other phalaropes. *Breeding female:* chestnut red throat, neck and underparts; white face; black crown and forehead; black-tipped, yellow bill. *Breeding male:* mottled brown crown; duller face and underparts. *Nonbreeding:* white head, neck and underparts; blue gray upperparts; mostly dark bill; black nape; broad, dark patch extending from eye to ear. *Juvenile:* similar to nonbreeding adult, but buff-colored overall; dark streaking on upperparts.
Size: *L* 8½ in; *W* 17 in.
Habitat: open waters of L. Erie; rarely on large interior lakes.
Nesting: does not nest in Ohio.
Feeding: small crustaceans, mollusks, insects and other invertebrates; rarely takes

vegetation or small fish; gleans from the water's surface, usually while swimming in tight, spinning circles.
Voice: calls include a shrill, high-pitched *wit* or *creep* and a low *clink clink*.
Similar Species: *Red-necked Phalarope* (p. 146): smaller; thinner bill; breeding bird lacks all-red underparts; nonbreeding bird has pale white stripes on upperwing and dark stripes on leading edge of lower wing. *Wilson's Phalarope* (p. 145): much longer, needlelike bill; breeding bird lacks all-red underparts; nonbreeding bird lacks dark "mask."
Best Sites: breakwaters along L. Erie; Huron Municipal Pier; Avon Lake Power Plant; Cleveland Lakefront Park and Conneaut Harbor.

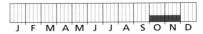

J F M A M J J A S O N D

POMARINE JAEGER

Stercorarius pomarinus

Pomarine Jaegers are pelagic for most of the year and are quite rare inland. Small numbers pass through the Great Lakes, and in a normal year three to five birds are seen along Lake Erie in Ohio waters. • Jaegers are easy to recognize as a group, but separating the three species is much trickier. Compounding identification difficulties is the fact that nearly all jaegers seen here are subadults that lack the diagnostic tail streamers and plumage characters of adult birds. Timing of migration may offer a clue: the Long-tailed Jaeger, by far our rarest species, passes through in late August to September, the Parasitic Jaeger from late September to October, and the Pomarine Jaeger from November to early January.

*breeding
light morph*

*juvenile
intermediate morph*

ID: *Adult* (almost never seen in Ohio): long, central, twisted tail feathers; black "cap." *Light morph:* dark, mottled breast band, sides and flanks; dark vent. *Dark morph:* dark body except for white on wing. *Juvenile:* central tail feathers extend just past tail; white at base of upperwing primaries; variable dark barring on underwings and underparts; lacks black "cap." *In flight:* wings are wide at base of body; powerful, steady wingbeats; white flash at base of underwing primaries.
Size: *L* 20–23 in; *W* 4 ft.
Habitat: L. Erie; very rare on inland reservoirs.
Nesting: does not nest in Ohio.
Feeding: snatches fish from the water's surface while in flight; chases down small birds; may also take small mammals and nestlings; pirates food from gulls.

Voice: generally silent; may give a sharp *which-yew,* a squealing *weak-weak* or a squeaky, whistled note in migration.
Similar Species: jaegers of all species in Ohio almost always juveniles. *Parasitic Jaeger* (p. 336): long, thin, pointed tail; very little white on upperwing primaries; short, sharpened tail streamers; lacks mottled sides and flanks; juvenile has barred underparts. *Long-tailed Jaeger* (p. 336): very long, thin, pointed tail; very little white on upperwing primaries; lacks very dark vent and dark, mottled breast band, sides and flanks; juvenile has stubby, spoon-shaped tail streamers and solid dark markings on throat and upper breast.
Best Sites: L. Erie shore from Huron eastward; Cleveland Lakefront Park; Avon Lake Power Plant; Conneaut Harbor; best on cold, windy days when a lot of gulls are moving.

J	F	M	A	M	J	J	A	S	O	N	D

LAUGHING GULL

Larus atricilla

This beautiful little gull's numbers were decimated back in the days of unregulated market hunting and the Laughing Gull was virtually eliminated from many of its traditional strongholds. The primary reason that this gull was hunted was to provide feathers, and in some cases entire birds, for decorating women's hats. Fortunately, Laughing Gulls have made a tremendous comeback in the last 100 years, and are once again common along the Atlantic and Gulf coasts. In Ohio, they are relatively rare but are increasing, with about 10 reports annually. • Most sightings are along Lake Erie, but Laughing Gulls sometimes appear on inland reservoirs and large rivers. Single birds are normally encountered, but small groups appear occasionally. • There are at least two Ohio records of Laughing Gull and Ring-billed Gull hybrids—origins unknown.

nonbreeding

breeding

ID: *Breeding:* black head; broken, white eye ring; red bill. *Nonbreeding:* white head with some pale gray bands; black bill. *Juvenile:* variable, brown to gray and white overall; broad, black, subterminal tail band. *2nd-year:* white neck and underparts; dark gray back; black-tipped wings; black legs.
Size: *L* 15–17 in; *W* 3 ft.
Habitat: shorelines and open water of lakes and rivers.
Nesting: does not nest in Ohio.
Feeding: omnivorous; gleans insects, small mollusks, crustaceans, spiders and small fish from the ground or water while flying, wading, walking or swimming; may steal food from other birds; may eat the eggs and nestlings of other birds.
Voice: loud, high-pitched, laughing call: *ha-ha-ha-ha-ha-ha.*

Similar Species: *Franklin's Gull* (p. 150): smaller and more diminutive overall; shorter, slimmer bill; white on outer primaries; nonbreeding normally has much more extensive "hood"; juvenile has unmarked white underwings. *Black-headed Gull* (p. 336) and *Bonaparte's Gull* (p. 152): Black-headed is very rare; orange or reddish legs; slimmer bill (Bonaparte's has black bill); paler mantle; white wedge on upper leading edge of wing; black "hood" on breeding adult does not extend over nape. *Little Gull* (p. 151): much smaller; paler mantle; reddish legs; dainty black bill; no eye ring; lacks black wing tips.
Best Sites: anywhere along L. Erie shore; occasionally large inland reservoirs such as C.J. Brown and East Fork Lake, or large rivers such as the Scioto.

J F M A M J J A S O N D

FRANKLIN'S GULL

Larus pipixcan

Sometimes referred to as "Prairie Pigeon," Franklin's Gull frequents prairie sloughs and marshes in the interior western U.S. and Canada, where it forms enormous nesting colonies. Franklin's Gulls are not nearly so numerous in Ohio, with only small numbers passing through, primarily in late fall. Highly migratory, they leave the U.S. entirely in winter, traveling to the Pacific coast of South America. • Inveterate plow-followers and frequent companions of many a western farmer, Franklin's Gulls glean freshly exposed insects from the wakes of tractors, a behavior that they occasionally engage in here. • The Franklin Gull's small size and delicate, buoyantly graceful flight is very different from the powerful, aggressive appearance of many of the larger gulls. • The species was discovered in 1823 on an expedition to northwestern Canada led by British explorer Sir John Franklin, and was named in his honor.

nonbreeding

breeding

ID: dark gray mantle; broken, white eye ring; white underparts. *Breeding:* black head; orange red bill and legs; breast may have pinkish tinge. *Nonbreeding:* white head; dark patch on side of head. *In flight:* black crescent on white wing tips; pure white underwings.

Size: *L* 13–15 in; *W* 3 ft.

Habitat: agricultural fields, marshlands, river and lake shorelines.

Nesting: does not nest in Ohio.

Feeding: very opportunistic; gleans agricultural fields and meadows for grasshoppers and insects; often catches dragonflies, mayflies and other flying invertebrates in midair; also eats small fish and some crustaceans.

Voice: mewing, shrill *weeeh-ah weeeh-ah* while feeding and in migration.

Similar Species: *Laughing Gull* (p. 149): larger; longer, heavier bill; less extensive "hood" in breeding plumage; lacks pure white underwings. *Little* (p. 151), *Bonaparte's* (p. 152) and *Black-headed* (p. 336) *gulls:* lack prominent, white eye crescents and bold, white primary tips.

Best Sites: western L. Erie, particularly vicinity of Ottawa NWR; irregular along L. Erie and in the glaciated western half of state, particularly around reservoirs.

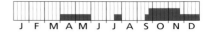

J F M A M J J A S O N D

LITTLE GULL

Larus minutus

Intrepid adventurer John Franklin put his stamp on the gull world, not only with the discovery of his namesake gull, but also with the Little Gull, which was first found in North America on his 1819 to 1820 expedition. • This European species—the world's smallest gull—is something of a mystery in that it seems to be a relatively recent colonizer on this continent. It probably arrived as a transatlantic wanderer from western Europe, and has slowly built up its numbers in North America, a pattern seemingly being followed by the Black-headed Gull. • The first discovery of nesting Little Gulls in the New World was in 1962 on Lake Ontario. There have now been nearly 1000 documented nestings, but none in Ohio as of yet. • Ohio Little Gulls are almost exclusively seen on Lake Erie, and are virtually always associated with large flocks of migrant Bonaparte's Gulls. The best way to find a Little Gull is to visit harbors and power plant outlets hosting large concentrations of Bonaparte's, and start sorting through the birds. Perseverance often leads to a Little Gull reward.

nonbreeding

nonbreeding

ID: white neck, rump, tail and underparts; gray back and wings; orange red feet and legs. *Breeding:* black head; dark red bill. *Nonbreeding:* black bill; dark "ear" spot and "cap." *Immature:* pinkish legs; brown and black in wings and tail. *In flight:* white wing tips and trailing edge of wing; dark underwings.

Size: *L* 10–11 in; *W* 24 in.

Habitat: almost always open waters of L. Erie; the few inland records have been from large reservoirs.

Nesting: does not yet nest in Ohio.

Feeding: gleans insects from the ground or from the water's surface while flying, wading, walking or floating; may also take small mollusks and fish, crustaceans, marine worms and spiders.

Voice: repeated *kay-ee* and a low *kek-kek-kek*.

Similar Species: 2nd-year and adult birds are unlikely to be mistaken, as the sooty black underwings and upperwings that lack dark pigment (slightly darkened in 2nd-year birds) are shared only by the very rare *Ross's Gull* (2 records).

Best Sites: Huron Municipal Pier; Avon Lake Power Plant; Cleveland Lakefront Park; Eastlake Power Plant; Conneaut Harbor.

J F M A M J J A S O N D

BONAPARTE'S GULL

Larus philadelphia

Few Ohio gulls are as showy as the breeding-plumaged Bonaparte's Gull, which passes through in small flocks in spring, beginning in early to mid-March. It begins acquiring its beautiful dark gray "hood" in late March, so spring flocks are often composed of both white-headed and hooded birds. • Unlike many of their larger, cruder brethren, Bonaparte's Gulls do not scavenge at dumps or in fast-food restaurant parking lots; rather, they are accomplished fishers, dipping agilely to the water's surface to snap up small piscine prey. • Enormous hordes of Bonaparte's Gulls congregate along eastern Lake Erie from November to early January. Some birds may overwinter on Lake Erie, their numbers depending on the severity of the winter. Scattered birds may be seen in any month of the year. • Bonaparte's Gulls are unique in the gull world for being arboreal nesters. Breeding in the open boreal forests of Canada and Alaska, Bonaparte's construct nests in the boughs of conifers, sometimes 20 feet above the ground.

nonbreeding

breeding

ID: black bill; gray mantle; white underparts. *Breeding:* gray black head; white eye ring; orange legs. *Nonbreeding:* white head; dark "ear" patch. *In flight:* white forewing wedge; black wing tips.
Size: *L* 11½–14 in; *W* 33 in.
Habitat: large lakes, rivers and marshes; flooded fields in spring.
Nesting: does not nest in Ohio.
Feeding: dabbles and tips up for aquatic invertebrates, small fish and tadpoles; gleans the ground for terrestrial invertebrates; also captures insects in the air.

Voice: scratchy, soft *ear ear* while feeding.
Similar Species: *Black-headed Gull* (p. 336): rare; larger; sooty black underwing primaries. *Little Gull* (p. 151): similar to 1st-winter Bonaparte's, but has much less prominent, black, M-shaped bars on upperwing and lacks dusky "cap."
Best Sites: spectacular numbers in late fall and early winter at Huron Municipal Pier, Lorain Harbor, Avon Lake Power Plant, Cleveland Lakefront SP, Eastlake Power Plant and Conneaut Harbor; migratory birds also at any good-sized water body.

J F M A M J J A S O N D

RING-BILLED GULL

Larus delawarensis

The Ring-billed Gull is our most common and widespread gull, and is easily the most common gull away from Lake Erie. It is the gull of mall parking lots, fast-food joints and other urban scavenging habitats. • John James Audubon called the Ring-billed Gull the "Common American Gull," which it undoubtedly was 200 years ago. By the early 20th century, however, Ring-billed Gull populations had been decimated by unregulated market hunting, primarily to provide feathers for hats, and this bird had become quite rare in Ohio. Fortunately, it is an adaptable species and the population has rebounded fantastically—fall concentrations on Lake Erie can reach 100,000 birds at favored locales! • Ring-billed Gulls are omnivores, meaning that they will eat anything. This trait, coupled with their extreme mobility, has allowed them to quickly exploit new sources of food. • This species, along with the Herring Gull, nests in Ohio in a few scattered colonies near Lake Erie.

nonbreeding

breeding

ID: white head; yellow bill and legs; black ring around bill tip; pale gray mantle; yellow eyes; white underparts. *Immature:* gray back; brown wings and breast. *In flight:* black wing tips with a few white spots.
Size: *L* 18–20 in; *W* 4 ft.
Habitat: *Breeding:* sparsely vegetated natural and artificial islands, open beaches and breakwaters. *In migration* and *winter:* lakes, rivers, landfills, fields, sometimes urban areas.
Nesting: colonial; in a shallow scrape on the ground lined with plants, debris, grass

and sticks; pair incubates 2–4 brown-blotched, gray to olive eggs for 23–28 days.
Feeding: gleans the ground for garbage, spiders, insects, rodents, earthworms, grubs and some waste grain; scavenges for carrion; surface tips for aquatic invertebrates and fish.
Voice: high-pitched *kakakaka-akakaka;* also a low, laughing *yook-yook-yook*.
Similar Species: black bill ring is diagnostic for adults. *Herring Gull* (p. 154): larger; pinkish legs; red spot near tip of lower mandible; lacks bill ring. *Mew Gull* (p. 336): accidental; different structure; more delicate, less fierce appearance; dark eyes; darker mantle; lacks bill ring.
Best Sites: L. Erie; any sizable lake or river.

J F M A M J J A S O N D

153

HERRING GULL

Larus argentatus

The Herring Gull is a complex species with nine subspecies, and it also freely hybridizes with several other species of gulls. However, only one subspecies, *smithsonianus*, breeds in North America, making matters a bit simpler. • Herring Gulls take four years to attain the classic, clean, gray-and-white adult colors, whereas many other species take two or three years. With Herring Gulls, generally speaking, first-year birds look mostly dark brown, second-years are dirty brown with gray in the mantle and a bicolored bill, third-years are similar to adults, but have a black-tipped bill and retain brownish pigment in some of the feathers. • Herring Gulls are most numerous along Lake Erie where there are scattered nesting colonies and become increasingly scarce inland the farther one gets from the lake.

breeding

nonbreeding

ID: large gull; yellow bill; red spot on lower mandible; pale eyes; light gray mantle; pink legs. *Breeding:* white head; white underparts. *Nonbreeding:* white head and nape washed with brown. *Immature:* mottled brown overall. *In flight:* white-spotted, black wing tips.

Size: *L* 23–26 in; *W* 4 ft.

Habitat: large lakes, wetlands, rivers, landfills and urban areas.

Nesting: singly or colonially, often with other gulls; on the ground or sometimes on a gravel rooftop; in a shallow scrape lined with plant material and sticks; pair incubates 3 darkly blotched, olive to buff eggs for 31–32 days.

Feeding: surface-tips for aquatic invertebrates and fish; gleans the ground for insects and worms; scavenges human food

waste at landfills; eats other birds' eggs and young.

Voice: loud, bugling *kleew-kleew;* also an alarmed *kak-kak-kak.*

Similar Species: *Ring-billed Gull* (p. 153): smaller; black bill ring; yellow legs; immature has much grayer back and pinkish bill. *Glaucous Gull* (p. 158) and *Iceland Gull* (p. 156): paler mantles; lack black in wings; immatures generally whiter or paler buff. *Lesser Black-backed Gull* (p. 157) and *Great Black-backed Gull* (p. 159): much darker mantles; immatures more discretely marked with brown.

Best Sites: anywhere along L. Erie at any season; spectacular flocks in winter at power plants such as Avon Lake or Eastlake; inland, small numbers wherever large flocks of Ring-billed Gulls occur, primarily in late fall, winter and early spring.

J F M A M J J A S O N D

154

THAYER'S GULL

Larus thayeri

The concept of a species is somewhat arbitrary—a human-devised method of classifying organisms so that we can better catalog them. Sometimes, as with Thayer's Gull, very similar "species" within a genus may not warrant full species status. Recent work suggests that Thayer's Gull is really just a subspecies of the Iceland Gull, and differences are merely the result of subtle changes in appearance over a broad geographic area. • Thayer's Gulls are strictly Lake Erie winter birds in Ohio—rare but regular visitors, usually seen as single birds among large gatherings of Herring Gulls, but occasionally as six or more together under favorable circumstances. • Most sightings are of first-year birds, which are actually the easiest to recognize in the field—they have frosty brown plumage with a neat checkerboard pattern on the back, and very pale undersides on the primaries.

nonbreeding

nonbreeding

ID: white-spotted, dark gray wing tips; dark eyes; yellow bill with red spot at tip of lower mandible; dark pink legs. *Breeding:* clean white head, neck and upper breast. *Nonbreeding:* brown-flecked head, neck and upper breast. *Immature:* variable, mottled white-and-brown plumage; 1st-year has black bill; 2nd-year has black ring around dusky bill.
Size: *L* 22–25 in; *W* 4½ ft.
Habitat: open water of L. Erie, normally where gizzard shad concentrations occur.
Nesting: does not nest in Ohio.
Feeding: gleans from the water's surface while in flight; eats small fish, crustaceans, mollusks, carrion and human food waste.

Voice: various raucous and laughing calls are given, much like the Herring Gull's *kak-kak-kak*.
Similar Species: *Herring Gull* (p. 154): black wing tips; light eyes; darker mantle. *Iceland Gull* (p. 156): light eyes; more white than dark gray on wing tips. *Glaucous Gull* (p. 158): larger; longer, heavier bill; light eyes; pure white wing tips; lighter pink legs. *Ring-billed Gull* (p. 153): smaller; dark ring on yellow bill; yellow legs. *Lesser Black-backed Gull* (p. 157): darker mantle; black wing tips; yellow feet.
Best Sites: winter concentrations at warm water outlets of L. Erie power plants; Avon Lake Power Plant; Cleveland Lakefront Park; Eastlake Power Plant.

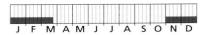

J F M A M J J A S O N D

ICELAND GULL

Larus glaucoides

The hardy arctic Iceland Gull is generally a rare winter visitor along Lake Erie. In mild winters, there are few reports, but in brutal winters, when the lake freezes, these gulls are much easier to find. • Birders often refer to the Iceland Gull and the Glaucous Gull as "white-winged gulls," referring to their very pale plumage and normally completely white wings in all plumages. This lack of dark pigmentation makes Icelands easy to spot among the massive packs of Herring Gulls and Ring-billed Gulls. • Most Iceland Gulls winter well to the north of Ohio; in fact, it is suspected that many overwinter on arctic polynas—large open leads among the sea pack ice, possibly created by warmer current upwellings. • The "Kumlien's" subspecies of Iceland Gull is occasionally reported—look for small, dark gray spots on the outer primaries to identify this bird.

nonbreeding

nonbreeding

ID: white overall; pale gray mantle; white wings; pink legs; relatively small bill with red spot. *1st-year:* pale chocolate brown; completely pale wings. *2nd-year:* ivory-colored overall with pale wings. *3rd-year:* similar to adult, but with brownish streaking on head and sometimes on wings and mantle; black mark on bill.
Size: *L* 22 in; *W* 4½ ft.
Habitat: landfills, harbors and open water on large lakes and rivers.
Nesting: does not nest in Ohio.
Feeding: scavenges at landfills and in harbors; eats mostly fish; may also take crustaceans, mollusks, carrion, seeds and human food waste.
Voice: high, screechy calls; much less bugling than other large gulls.
Similar Species: *Glaucous Gull* (p. 158): larger; appears fiercer and more powerful; significantly bigger bill; prominent orbital ridge over eye; flatter head. *Thayer's Gull* (p. 155): 1st-year similar to 1st-year Iceland, but tends to be slightly darker with neat checkerboard pattern on back.
Best Sites: among large L. Erie winter gull flocks; Avon Lake and Eastlake power plants; Cleveland Lakefront Park; Conneaut Harbor; Maumee River Rapids.

J F M A M J J A S O N D

LESSER BLACK-BACKED GULL

Larus fuscus

This European species is a recent addition to the North American avifauna, and is rapidly increasing. Reports slowly increased following the first continental U.S. record on September 9, 1934, in New Jersey. After the mid-1970s, Lesser Black-backed Gull populations began to increase dramatically—birds had been seen in 31 eastern states by 1994. The first Ohio report was in the winter of 1977, when at least one bird was seen near Cleveland. Lesser Black-backed Gulls have been reported annually since then, and are now easily found along Lake Erie in winter. Inland reports, as well as nonwinter records, are increasing. • All of our records are of the subspecies *graellsii*, but birders should also watch for the much darker-mantled subspecies *intermedius*, which has been reported on occasion in the U.S. The *intermedius* subspecies approaches the Great Black-backed Gull in dark coloration, with little contrast between the dark back and dark primary feathers.

nonbreeding

nonbreeding

ID: *Breeding:* dark gray or black mantle; mostly black wing tips; yellow bill with red spot on lower mandible; yellow eyes; yellow legs; white head and underparts. *Nonbreeding:* brown-streaked head and neck. *Immature:* dark or light eyes; black or pale bill with black tip; various plumages with varying amounts of gray on upperparts and brown flecking over entire body.

Size: *L* 20½ in; *W* 4½ ft.

Habitat: landfills, harbors and open water on large lakes and rivers.

Nesting: does not nest in Ohio.

Feeding: eats fish, crustaceans, mollusks, insects, small rodents, carrion, seeds and human food waste; scavenges at landfills and harbors.

Voice: screechy call is like a lower-pitched version of the Herring Gull's.

Similar Species: conspicuously darker mantle than any other regular Ohio gull except Great Black-backed. *Great Black-backed Gull* (p. 159): much larger; pink legs; darker gray mantle; lacks strongly brown-streaked head and neck in non-breeding plumage.

Best Sites: among large L. Erie winter gull flocks, such as form at Cleveland Lakefront SP, Avon Lake and Eastlake power plants; anywhere along L. Erie; occasional at large inland reservoirs.

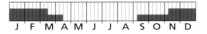

J F M A M J J A S O N D

GLAUCOUS GULL

Larus hyperboreus

This gull is our most common white-winged gull and usually outnumbers the Iceland Gull by as much as ten to one. The Glaucous Gull superficially resembles the Iceland, but is normally much larger, with a bigger bill and a pronounced orbital ridge over the eye that lends a fierce appearance. Other than the Great Black-backed Gull, the Glaucous Gull is the largest gull to visit Ohio. • The Glaucous Gull is highly predatory and reigns supreme on the avian food chain on its arctic breeding grounds. It is a prolific pilferer of other birds' eggs and young, and even sometimes adult birds—a birder watching gulls at Lorain once observed a careless European Starling snatched and swallowed whole by a Glaucous Gull. • The Glaucous Gull occasionally hybridizes with the Herring Gull, producing "Nelson's Gull," which is sometimes reported in Ohio.

nonbreeding

nonbreeding

ID: *Breeding:* relatively long, heavy, yellow bill with red spot on lower mandible; pure white wing tips; flattened crown profile; yellow eyes; pink legs; white underparts; very pale gray mantle. *Nonbreeding:* brown-streaked head, neck and breast. *Immature:* dark eyes; pale, black-tipped bill; various plumages have varying amounts of brown flecking on body.

Size: *L* 27 in; *W* 5 ft.

Habitat: landfills, harbors and open water on large lakes and rivers.

Nesting: does not nest in Ohio.

Feeding: predator, pirate and scavenger; eats mostly fish, crustaceans, mollusks and some seeds; feeds on carrion and at landfills.

Voice: high, screechy calls similar to Herring Gull's *kak-kak-kak.*

Similar Species: *Iceland Gull* (p. 156): smaller; smaller bill; rounder head; slightly darker mantle; primaries project farther beyond tail on sitting birds. *Herring Gull* (p. 154): smaller; black wing tips; darker mantle.

Best Sites: typical L. Erie winter gull spots, such as Avon Lake Power Plant, Lorain Harbor, Cleveland Lakefront SP and Eastlake Power Plant; harder winters bring more birds; only consistent inland locale is Maumee River Rapids.

J F M A M J J A S O N D

GREAT BLACK-BACKED GULL

Larus marinus

As with some other species of large gulls, Great Black-backed Gulls have increased dramatically during the past several decades. Before 1930, they were rare on Lake Erie; now groups numbering in the hundreds may be seen at times. • This is the largest gull in North America—individuals may weigh more than 3½ pounds (a Herring Gull weighs about 2½ pounds). The massive, black-mantled adults easily stand out among the smaller, paler Ring-billed Gulls and Herring Gulls. Even subadults aren't too hard to identify, as their large size and powerful proportions are unlike other species. • Great Black-backeds are most frequent in winter, particularly along eastern Lake Erie. Inland, they are accidental, but sightings are increasing. The number of summering birds along Lake Erie is increasing, too, but as of yet there are no Ohio breeding records.

nonbreeding

nonbreeding

ID: very large gull; all white except for gray underwings and black mantle; pale pinkish legs; light-colored eyes; large, yellow bill with red spot on lower mandible. *Nonbreeding:* may have faintly streaked nape. *Immature:* variable, mottled grayish brown, white and black; black bill or black-tipped, pale bill; blackish mantle with neat checkerboard pattern on wings.
Size: *L* 30 in; *W* 5½ ft.
Habitat: open water on large lakes and rivers.
Nesting: does not nest in Ohio.
Feeding: opportunistic feeder; finds food by flying, swimming or walking; eats fish,

eggs, birds, small mammals, berries, carrion, mollusks, crustaceans, insects and other invertebrates, as well as human food waste; scavenges at landfills.
Voice: a harsh *kyow*.
Similar Species: adult is distinctive. *Lesser Black-backed Gull* (p. 157): much smaller; paler mantle; yellow legs; non-breeding bird has heavy brown streaking on head.
Best Sites: along L. Erie, with numbers increasing farther eastward; large concentrations in winter at Cleveland, Lorain, Conneaut and Ashtabula; rocky islet on north side of State Route 2 bridge over Sandusky Bay.

J F M A M J J A S O N D

SABINE'S GULL

Xema sabini

Joseph Sabine named this beautiful gull in honor of his brother, Sir Edward Sabine, the English astronomer who discovered this species on an 1818 expedition to the Arctic. • Sabine's Gulls are highly migratory arctic breeders that winter in tropical seas off South America and Africa, and normally migrate at sea; only small numbers pass through the Great Lakes. They are strictly fall migrants in Ohio. • The best chance of seeing a Sabine's Gull is to carefully watch Lake Erie from a good vantage point on days of blustery north winds in late September and October. • Virtually all of our birds are juveniles, but are distinctive with their bold, tricolored wing pattern that is visible at great distances. A closer look will reveal the slightly forked tail. • Sabine's Gulls are not hardy, therefore a bird that overwintered in Cleveland in 1989 was amazing, and one of very few North American winter records.

nonbreeding

nonbreeding

ID: yellow-tipped, black bill; dark gray mantle; black feet. *Breeding:* dark, slate gray hood trimmed with black. *Nonbreeding:* white head; dark gray nape. *In flight:* tricolored wing: gray at base, then white, then black at tip; shallowly forked tail. *Juvenile:* more common than adult in Ohio; brown in areas where adult is gray; black-tipped tail.
Size: *L* 13–14 in; *W* 3 ft.
Habitat: L. Erie; accidental on lakes and large rivers.
Nesting: does not nest in Ohio.

Feeding: gleans the water's surface while swimming or flying; eats mostly insects, fish and crustaceans.
Voice: ternlike *kee-kee;* not frequently heard in migration.
Similar Species: unmistakable in any plumage by combination of striking M-shaped wing pattern in bold white, brown (on juveniles) and black, slightly forked tail, and small size.
Best Sites: wherever large numbers of Bonaparte's Gulls gather along L. Erie, particularly sites that offer a good vista of the lake; Huron Municipal Pier; Avon Lake Power Plant; Sherod Park (Vermilion); Eastlake Power Plant; Mentor Headlands.

J F M A M J J A S O N D

BLACK-LEGGED KITTIWAKE

Rissa tridactyla

The Black-legged Kittiwake was dubbed "Frost Gull" by New Englanders, because the arrival of migrant kittiwakes marked the onset of harsh winter weather. That name would even be appropriate here in Ohio; the normal window of passage, which is almost entirely along Lake Erie, is from late October through December. Even though they arrive late in the year, these birds rarely overwinter.
• Although the Black-legged Kittiwake is one of the world's most abundant gulls, with a population estimated at several million, it is a rare migrant through Ohio. These birds are among the most pelagic of gull species away from their breeding grounds, spending most of their time at sea well out of sight of land. • Virtually all of our birds are juveniles, which are probably even more distinctive than adults. Normally only single birds are observed. • Kittiwakes often capture prey by diving below the water's surface—the only gull species to regularly do so.

1st year

nonbreeding

ID: *Nonbreeding:* gray nape; dark gray smudge behind eye; black wing tips; small, entirely yellow bill; slightly indented tail. *Immature:* black bill; wide, black "half-collar"; dark ear patch. *In flight:* solid black triangles on wing tips; immature has black "M" on upper forewing from wing tip to wing tip and black terminal tail band.
Size: *L* 16–18 in; *W* 3 ft.
Habitat: open water on L. Erie; rarely large interior lakes and rivers.
Nesting: does not nest in Ohio.
Feeding: dips to the water's surface to snatch prey; may plunge under the water's surface or glean from the surface while swimming; prefers small fish; also takes crustaceans, insects and mollusks.
Voice: calls are *kittewake* and *kekekek*.
Similar Species: almost all Ohio birds are juveniles, which are easily told from other species by a combination of bold, black, M-pattern on upperwings, broad, black "collar" and truncated to slightly notched, black-tipped tail.
Best Sites: areas of large gull congregations along L. Erie, such as Huron Municipal Pier, Cleveland Lakefront SP, Avon Lake Power Plant, Eastlake Power Plant and Conneaut Harbor.

J F M A M J J A S O N D

CASPIAN TERN

Sterna caspia

The Caspian Tern is the world's largest tern and one of the most cosmopolitan, occurring on every continent but Antarctica. These terns—named for the Caspian Sea, where the first specimens were collected—resemble gulls, with huge bills, broad wings and a propensity for soaring on fixed wings, quite unlike their smaller, graceful tern brethren. • North America's Caspian Terns have an odd, disjunct breeding range, nesting in six well-separated regions. The closest colonies to Ohio are along Lakes Huron and Michigan, but these birds have never been found breeding along Lake Erie. Small groups regularly pass through Ohio's interior in both spring and fall, but the greatest numbers occur along Lake Erie where flocks numbering into the hundreds have been reported on occasion. • In fall passage, adults are often accompanied by their offspring, who continue to pester their parents for food. The long-suffering adults provide the longest period of parental care of any tern, enduring the freeloading young for several months.

breeding

ID: *Breeding:* black "cap"; heavy, red orange bill with faint black tip; light gray mantle; black legs; shallowly forked tail; white underparts; long, frosty, pointed wings; dark gray on underside of outer primaries. *Nonbreeding:* black "cap" streaked with white.

Size: *L* 19–23 in; *W* 4–4½ ft.

Habitat: wetlands and shorelines of large lakes and rivers.

Nesting: does not nest in Ohio.

Feeding: hovers over water and plunges headfirst after small fish, tadpoles and aquatic invertebrates; also feeds by swimming and gleaning at the water's surface.

Voice: low, harsh *ca-arr;* loud *kraa-uh;* juveniles answer with a high-pitched whistle.

Similar Species: large size and huge red bill make confusion with other terns unlikely.

Best Sites: large mudflats and wetlands buffering L. Erie, such as Sheldon Marsh SNP and Magee Marsh WA; coastal dredge-spoil impoundments, such as Huron Municipal Pier, Lorain Harbor and Conneaut Harbor; most larger inland reservoirs host small numbers in migration.

J F M A M J J A S O N D

COMMON TERN

Sterna hirundo

Both male and female Common Terns perform aerial courtship dances, and for most pairs, the nesting season commences when the female accepts her suitor's gracious fish offerings. • Common Terns were slaughtered in huge numbers in the late 1800s for the millinery trade; the snowy white feathers—and sometimes whole birds—were used to decorate hats. The enactment of the 1918 Migratory Bird Treaty Act put an end to the wanton killing of most birds, and by the 1930s, Common Tern populations had rebounded remarkably. • The dramatic expansion of nesting Herring Gulls has displaced most Common Tern breeding colonies in Ohio. Efforts by the Ohio Division of Wildlife to induce terns to use artificial nesting platforms on Lake Erie have had positive results and should be applauded as a wildlife-management success story.

breeding

nonbreeding

ID: *Breeding:* black "cap"; thin, red, black-tipped bill; red legs; white rump; white tail with gray outer edges; white underparts. *Nonbreeding:* black nape; lacks black "cap." *In flight:* shallowly forked tail; long, pointed wings; dark gray wedge near lighter gray upperwing tips.

Size: *L* 13–16 in; *W* 30 in.

Habitat: *Breeding:* natural and artificial islands, breakwaters and beaches. *In migration:* large lakes, wetlands and rivers.

Nesting: primarily colonial; usually on an island with nonvegetated, open areas, or on an artificial nesting platform; in a small scrape lined sparsely with pebbles, vegetation or shells; pair incubates 1–3 variably marked eggs for up to 27 days.

Feeding: hovers over the water and plunges headfirst after small fish and aquatic invertebrates, sometimes from heights of 30 ft or more.

Voice: high-pitched, drawn-out *keee-are* is most commonly heard at colonies, but also in foraging flights.

Similar Species: *Forster's Tern* (p. 164): gray tail with white outer edges; upper primaries have silvery look; broad, black eye band in nonbreeding plumage. *Arctic Tern:* very rare; all-red bill; deeply forked tail; upper primaries lack dark gray wedge; grayer underparts. *Caspian Tern* (p. 162): much larger overall; much heavier, red orange bill; very dark primary underwing patch.

Best Sites: especially in fall along L. Erie at Sheldon Marsh SNP, Bayshore Power Plant and Lorain Harbor; small numbers at larger inland reservoirs; nesting platforms at Pipe Creek WA.

J F M A M J J A S O N D

FORSTER'S TERN

Sterna forsteri

Forster's Tern doesn't breed in Ohio, but it nests as close as southeastern Michigan. Forster's is a marsh bird and selects floating mats of vegetation or a muskrat lodge as a nest site. Breeding areas are large, mixed-emergent marshes at least 50 acres in size; the Cedar Point National Wildlife Refuge near Toledo would be an excellent locale to discover our first breeding record of Forster's Tern. • Most terns are known for their ability to catch fish in dramatic headfirst dives, but the Forster's excels at snatching flying insects in midair. • Common Terns and Forster's Terns, which are both common migrants and often occur together, particularly in fall, are often confused. The disconnected, dark ear patches help to identify juvenile and nonbreeding Forster's. Breeding Forster's in spring are trickier to separate from Commons, but can be told by their glimmering white upper primaries—Commons have a dark wedge on the upperwing—and by their larger orange, rather than reddish, bill and whiter underparts. The Forster's tail projects beyond the wings when the bird is at rest.

nonbreeding

breeding

ID: *Breeding:* black "cap" and nape; thin, orange, black-tipped bill; orange legs; light gray mantle; pure white underparts; white rump. *Nonbreeding:* black band through eyes; lacks black "cap." *In flight:* forked, gray tail; long, pointed wings.
Size: *L* 14–16 in; *W* 31 in.
Habitat: mixed-emergent marshes, lakes and large rivers.
Nesting: does not nest in Ohio.
Feeding: hovers above the water and plunges headfirst after small fish and aquatic invertebrates; catches flying insects and snatches prey from the water's surface.
Voice: flight call is a short, nasal *keer keer;* also a grating *tzaap.*
Similar Species: *Common Tern* (p. 163): see text above. *Caspian Tern* (p. 162): much larger overall; much heavier, red orange bill. *Arctic Tern:* very rare; stubby red legs; gray underparts; white tail with gray outer edges.
Best Sites: L. Erie coastal mudflats, marshes, harbors and dredge-spoil impoundments, such as Maumee Bay SP, Ottawa NWR and Huron Municipal Pier; small numbers at large inland reservoirs.

J F M A M J J A S O N D

BLACK TERN

Chlidonias niger

Black Terns were once a common and fascinating part of western Lake Erie's enormous mixed-emergent marshes, breeding in numbers in every sizable wetland. A group of 1500 was seen in fall migration near Toledo on August 26, 1944; flocks of a few dozen would be newsworthy now. Today, Black Terns are rare migrants and even rarer nesters—listed as endangered in Ohio, a few pairs still nest in good years. • Although a variety of factors may have played a role in the Black Tern's demise, significant increases in aggressive, invasive plants such as purple loosestrife and *Phragmites* have reduced floristic diversity in wetlands, degrading Black Tern nesting habitat. • The Black Tern, with its dark coloration and diminutive size, is quite unlike other North American terns.

breeding

nonbreeding

ID: *Breeding:* black head and underparts; gray back, tail and wings; white undertail coverts; black bill; reddish black legs. *Nonbreeding:* white underparts and forehead; molting fall birds may be mottled with brown. *In flight:* long, pointed wings; shallowly forked tail.
Size: *L* 9–10 in; *W* 24 in.
Habitat: shallow, freshwater cattail marshes, wetlands and lake edges.
Nesting: loosely colonial; flimsy nest of dead plant material is built on floating vegetation, a muddy mound or a muskrat house; pair incubates 3 darkly blotched, olive to pale buff eggs for 21–22 days.
Feeding: snatches insects from the air, tall grass and the water's surface; also eats small fish.
Voice: greeting call is a shrill, metallic *kik-kik-kik-kik-kik;* typical alarm call is *kreea.*
Similar Species: *Other terns* (pp. 162–64): all are light in color, not dark.
Best Sites: large wetlands of L. Erie's western basin: Ottawa NWR, Magee Marsh WA, Sheldon Marsh SNP; may also appear on almost any interior wetland or lake.

J F M A M J J A S O N D

ROCK PIGEON

Columba livia

Originally native to North Africa, Europe and parts of Asia, the Rock Pigeon was brought to North America by European colonists by the early 1600s. The species is now established continent-wide. • As the name implies, wild Rock Pigeons occupied cliffs and rocky promontories in their native range. Skyscrapers, silos, barns and bridges provide a similar alternative, and no city, town or good-sized farm is without these birds. • Probably no other wild bird displays such a staggering range of color variation, although certain "types" predominate; the classic "blue-bar" (light gray body, two black bars) represents the "wild" phenotype. • Although much maligned by birders and the citizenry in general, Rock Pigeons have made tremendous contributions to science—the literature on them could fill a library. Much of what we know about avian orientation, color genetics and flight mechanics stems from pigeon research. In the course of endocrinology studies with Rock Pigeons, the pituitary hormone prolactin was discovered. • Formerly known as "Rock Dove," the commonly used "pigeon" name became official when this species' name was changed to Rock Pigeon in 2003.

ID: color is highly variable (iridescent blue gray, red, white or tan); usually has white rump and orange feet; dark-tipped tail. *In flight:* holds wings in deep "V" while gliding.
Size: *L* 12–13 in; *W* 28 in.
Habitat: urban areas, railroad yards, agricultural areas; occasional birds "revert" to wild habits and nest on large sandstone cliffs in southeastern Ohio.
Nesting: on ledges of barns, cliffs, bridges, buildings and towers; flimsy nest of sticks, grass and assorted vegetation; pair incubates 2 white eggs for 16–19 days; pair feeds the young "pigeon milk"; may raise broods year-round.
Feeding: gleans the ground for waste grain, seeds and fruits; occasionally eats insects.
Voice: soft, cooing *coorrr-coorrr-coorrr.*
Similar Species: unmistakable; even most nonbirders instantly recognize Rock Pigeons. If you struggle with this one, turn in your binoculars immediately.
Best Sites: common year round in cities and towns and near farms.

J F M A M J J A S O N D

MOURNING DOVE

Zenaida macroura

Mourning Dove populations have increased tremendously since European settlement, as conversion of the extensive and relatively unbroken eastern deciduous forest to farmland, pastures and towns created better habitats for this open-country bird. This is in sharp contrast to a related species, the Passenger Pigeon *(Ectopistes migratorius)*, which became extinct by 1900 after the great forests were cleared and uncontrolled hunting took its toll. • The Mourning Dove is one of the most abundant North American birds today; continent-wide breeding-bird surveys record it as the second most common species, after the Red-winged Blackbird. • The Mourning Dove is the leading game bird in North America, with up to 70 million harvested annually—more than the total harvest of all other migratory game birds combined. However, its prolific reproduction rate allows populations to remain high in spite of hunting and other mortality factors. • Two of the five subspecies of Mourning Dove occur in Ohio; the widespread *carolinensis*, and in far northwestern Ohio, the longer-billed, paler *marginella* of the western U.S.

ID: buffy, gray brown plumage; small head; long, white-trimmed, tapering tail; sleek body; dark, shiny patch below ear; dull red legs; dark bill; pale rosy underparts; black spots on upperwing.
Size: *L* 11–13 in; *W* 18 in.
Habitat: open or riparian woodlands, woodlots, forest edges, agricultural and suburban areas and open parks; has benefited from human-induced habitat change.
Nesting: occasionally on the ground or in the fork of a shrub or tree; female builds a fragile, shallow, platform nest from twigs supplied by male; pair incubates 2 white eggs for 14 days; young are fed "pigeon milk."
Feeding: gleans the ground and vegetation for seeds; visits feeders.
Voice: mournful, soft, slow *oh-woe-woe-woe*.
Similar Species: *Rock Pigeon* (p. 166): stockier; white rump; shorter tail; iridescent neck. Be mindful of the following rare doves: *White-winged Dove:* prominent, white wing patch. *Ringed Turtle-Dove* and *Eurasian Collared-Dove:* paler; sandy-colored; conspicuous, black neck "collar"; lack dark spots on wings. *Common Ground-Dove:* much smaller; rufous underwings.
Best Sites: ubiquitous and widespread year round; large concentrations can form in agricultural areas in fall.

J F M A M J J A S O N D

BLACK-BILLED CUCKOO

Coccyzus erythropthalmus

This is the less common of our two cuckoo species, although getting a good handle on populations is difficult, as Black-billed Cuckoos are shy, retiring and unwilling to show themselves. They might even be largely nocturnal in summer, as evidenced by their frequent nighttime singing. • Both Yellow-billed Cuckoos and Black-billed Cuckoos frequent scruffy, successional woodlands, though Black-billeds may prefer larger, more unbroken forests. • Many more cuckoos are heard than seen, and typical songs are easily distinguished. Beware of abbreviated calls and notes, which can be hard to discern from those of the Yellow-billed Cuckoo. • Both of our cuckoos occasionally engage in nest parasitism, like the Brown-headed Cowbird. This behavior is most likely to occur in years of peak food abundance when reproductive rates may be higher. Young cuckoos quickly outgrow the host bird's own young, and eventually push them from the nest.

ID: brown upperparts; white underparts; long, white-spotted undertail; downcurved, dark bill; reddish eye ring. *Juvenile:* buff eye ring; may have buff tinge on throat and undertail coverts.
Size: *L* 11–13 in; *W* 17½ in.
Habitat: dense second-growth woodlands, shrubby areas and thickets; often in tangled riparian areas and abandoned farmlands with low, deciduous vegetation and adjacent open areas.
Nesting: in a shrub or small deciduous tree; flimsy nest of twigs is lined with grass and other vegetation; occasionally lays eggs in other birds' nests; pair incubates

2–5 blue green, occasionally mottled eggs for 10–14 days.
Feeding: gleans hairy caterpillars from leaves, branches and trunks; also eats other insects and berries.
Voice: fast, repeated *cu-cu-cu* or *cu-cu-cu-cu-cu;* also a series of *ca, cow* and *coo* notes.
Similar Species: *Yellow-billed Cuckoo* (p. 169): stouter appearance; yellow bill; rufous tinge to primaries; larger, more prominent, white undertail spots; lacks red eye ring.
Best Sites: *Breeding:* northeastern Ohio, as at Cuyahoga Valley NRA and Pymatuning SP. *In migration:* Magee Marsh WA bird trail; Green Lawn Cemetery.

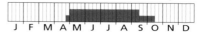

J F M A M J J A S O N D

YELLOW-BILLED CUCKOO

Coccyzus americanus

Yellow-billed cuckoos are fairly common, probably nesting in every county and greatly outnumbering Black-billed Cuckoos. Although not small—they are bigger than Blue Jays—they skulk furtively in dense growth and are hard to see. Their loud series of distinctive, hollow notes carries long distances, and is how these birds are usually detected. • Oldtimers referred to the Yellow-billed Cuckoo as "Rain Crow" for this species' alleged tendency to sing more before storms, though its skill as a weather forecaster is unproven. • The year-to-year fluctuation in Yellow-billed Cuckoo numbers is directly related to food abundance. This bird is heavily dependent on species like eastern tentworm caterpillars and cicadas, and in boom years the cuckoo quickly takes advantage. Because the Yellow-billed takes but 17 days from egg laying to fledging of young, multiple broods are possible.

ID: olive brown upperparts; white underparts; down-curved bill with black upper mandible and yellow lower mandible; yellow eye ring; long tail with large, white spots on underside; rufous tinge on primaries.

Size: *L* 11–13 in; *W* 18 in.

Habitat: semi-open deciduous habitats; dense tangles and thickets at the edges of orchards, urban parks and black locust groves in reclaimed strip mines; sometimes woodlots.

Nesting: on a horizontal branch in a deciduous shrub or small tree, within 7 ft of the ground; builds a flimsy platform of twigs lined with roots and grass; pair incubates 3–4 pale bluish green eggs for 9–11 days.

Feeding: gleans insect larvae, especially hairy caterpillars, from deciduous vegetation; also eats berries, small fruits, small amphibians and occasionally the eggs of small birds.

Voice: long series of deep, hollow *kuks,* slowing near the end: *kuk-kuk-kuk-kuk kuk kop kow kowlp kowlp.*

Similar Species: *Black-billed Cuckoo* (p. 168): all-black bill; lacks rufous tinge on primaries; less prominent, white undertail spots; red rather than yellow eye ring; juvenile has buff eye ring and may have buff wash on throat and undertail coverts.

Best Sites: black locust thickets in reclaimed strip mines such as at Crown City WA and Woodbury WA; thickly overgrown fields and woodland edges statewide.

J F M A M J J A S O N D

BARN OWL

Tyto alba

The Barn Owl has one of the broadest distributions of any terrestrial bird in the world, and is the widest ranging owl on the globe. But, like several other species of birds that colonized Ohio relatively recently, Barn Owls have undergone a "boom and bust" cycle. First appearing in the state in the mid-1800s after extensive forest-clearing created suitable habitat, populations peaked in the early 20th century, with sightings in 84 counties by 1935. However, by 1990, numbers had collapsed to one or two dozen pairs statewide. • Barn Owls are cavity nesters, and the availability of cavities is a limiting factor; recent efforts to erect nest boxes have increased numbers locally. Although still listed as threatened in Ohio, perhaps 50 pairs are now known to nest in the state. • Strictly nocturnal, Barn Owls have developed incredibly acute night vision and excellent hearing that allows them to triangulate on prey in total darkness.

ID: heart-shaped, white facial disc; dark eyes; pale bill; golden brown upperparts spotted with black and gray; creamy white, black-spotted underparts; long legs; white undertail and underwings.

Size: *L* 12½–18 in; *W* 3½ ft.

Habitat: roosts and nests in cliffs, hollow trees, barns and other unoccupied buildings; increasingly uses nest boxes; requires open areas with abundant vole populations for hunting, including agricultural fields, pastures and marshy meadows.

Nesting: in a natural or artificial cavity, often in a sheltered, secluded hollow in a building; may dig a hole in a dirt bank or use an artificial nest box; no nest is built; female incubates 3–8 whitish eggs for 29–34 days; male feeds incubating female.

Feeding: eats mostly small mammals, especially rodents; also takes snakes, lizards, birds and large insects; rarely takes frogs and fish.

Voice: calls include harsh, raspy screeches and hisses; also makes metallic, bill-snapping sounds.

Similar Species: no other Ohio owl is likely to be confused with the Barn Owl.

Best Sites: because of the Barn Owl's very rare status, and the fact that almost all known nests are on private lands, often in barns, directions to sites should not be publicized; these birds can sometimes be discovered by searching dense pine plantings located near suitable foraging habitat.

J F M A M J J A S O N D

EASTERN SCREECH-OWL

Megascops asio

These are easily our most numerous owls, although the uninitiated would never know it. Eastern Screech-Owls are strictly nocturnal, concealing themselves in cavities or dense tangles of vegetation during the day. They occur almost everywhere in Ohio—even very urban areas—and can be surprisingly common: a 1981 Toledo-area Christmas bird count recorded 112 birds in an area only 15 miles in diameter. • Eastern Screech-Owls respond quite readily to whistled imitations of their calls, and sometimes several owls will quickly appear to investigate the fraudulent perpetrator. • There are two distinct color morphs—red and gray. Both morphs are common in southern Ohio, with grays more frequent to the north. Very rarely, an intermediate brown morph occurs. • Erecting nest boxes in suitable habitat can often entice Eastern Screech-Owls to nest. They will even adopt Wood Duck nest boxes. • Screech-owls have one of the most varied diets of any owl, capturing a broad array of small animals, and sometimes even snagging small fish from creeks.

red morph

ID: short "ear" tufts; reddish or grayish overall; dark breast streaking; yellow eyes; pale grayish bill. **Size:** *L* 8–9 in; *W* 20–22 in. **Habitat:** a wide variety of woodlands, open parklands, cemeteries and neighborhoods with scattered trees, brushy pastures and thickly overgrown riparian areas.
Nesting: in a natural cavity or artificial nest box; no lining is added; female incubates 4–5 white eggs for about 26 days; male brings food to the female during incubation.
Feeding: small mammals, earthworms, fish, birds and insects, including moths in flight; feeds at dusk and at night.

Voice: horselike "whinny" whistle that rises and falls; rarely "screeches."
Similar Species: *Northern Saw-whet Owl* (p. 177): long, reddish streaks on white underparts; lacks "ear" tufts. *Long-eared Owl* (p. 175): much longer, slimmer body; longer, closer-set "ear" tufts; rusty facial disc; grayish, brown white body. *Great Horned Owl* (p. 172): much larger; lacks vertical breast streaks.
Best Sites: Wood Duck boxes and natural tree cavities along Magee Marsh WA bird trail; also locate young, scruffy woods with trees of varying ages, interspersed with fields and a stream nearby, and either whistle the call or play a tape.

J F M A M J J A S O N D

GREAT HORNED OWL

Bubo virginianus

The Great Horned Owl is our largest owl and a fierce hunter that sits at the top of the avian food chain. Ninety percent of its prey are mammals, up to the size of groundhogs and (better keep Fluffy inside) domestic cats. • These owls are common throughout the state in a wide range of habitats, even surprisingly in urban areas where they will catch Norway rats. • They begin nesting in January or February, and the female's superb, insulating down keeps eggs at about 98° F, even in subzero temperatures. As they often use Red-tailed Hawk nests, incubating Great Horned Owls can sometimes be spotted by their "ears" poking over the top of the nest. • Birders often locate these owls via enraged, mobbing packs of crows, which despise the owls. Olfactory senses can also help locate them— Great Horneds are one of few predators that capture skunks, and they are apparently not bothered by the perfuming they receive.

ID: yellow eyes; tall "ear" tufts set wide apart on head; fine, horizontal barring on breast; facial disc is outlined in black and is often rusty orange in color; white "chin"; heavily mottled gray, brown and black upperparts; overall plumage varies from light gray to dark brown.
Size: *L* 18–25 in; *W* 3–5 ft.
Habitat: fragmented forests, agricultural areas, woodlots, meadows, riparian woodlands, wooded suburban parks and the wooded edges of landfills and town dumps.
Nesting: usurps nests of larger hawks, American Crows and Great Blue Herons; about 50 percent of Ohio birds nest in broken-off tree snags or cavities; adds little or no material to the nest; mostly the female incubates 2–3 dull whitish eggs for 28–35 days.
Feeding: mostly nocturnal; also hunts at dusk or by day in winter; swoops from a perch; eats small mammals, birds, snakes, amphibians and even fish.
Voice: 4–6 deep hoots during breeding season: *hoo-hoo-hoooo hoo-hoo* or *eat-my-food, I'll-eat you;* male gives higher-pitched hoots; pairs often heard singing "duets" in fall and winter.
Similar Species: *Long-eared Owl* (p. 175): smaller; thinner; vertical breast streaks; "ear" tufts are close together. *Eastern Screech-Owl* (p. 171): much smaller; vertical breast streaks. *Short-eared Owl* (p. 176) and *Barred Owl* (p. 174): no "ear" tufts.
Best Sites: Killdeer Plains WA (several pairs); Green Lawn Cemetery; statewide in open country with scattered woodlots.

J F M A M J J A S O N D

SNOWY OWL
Bubo scandiacus

Most owls rely on cryptic plumage and nocturnal habits to avoid detection, not so the Snowy Owl. This diurnal hunter is bold, conspicuous and relatively fearless, perching in obvious spots and often permitting close approaches. Walking up to one on its wintering grounds isn't a problem, but they are highly aggressive on their arctic nesting territory. Snowies are known to attack people and even wolves that venture too near a nest. • Ohio gets a few Snowy Owls almost every winter along Lake Erie. Cyclical irruptions every four to six years, primarily owing to the periodic collapse of northern lemming populations, bring larger numbers. There is no doubt that this 2-ounce rodent is a major food source— a Snowy Owl consumes over 1500 annually. • Few birds generate as much excitement among birders, and even among nonbirders, as do Snowy Owls. A cooperative individual near Wilmington in Clinton County during the winter of 2000 was visited by hundreds of people and was the subject of numerous newspaper articles and TV spots, becoming one of the most celebrated Ohio birds ever.

ID: predominantly white; yellow eyes; black bill and talons; no "ear" tufts. *Male:* almost entirely white, with very little dark flecking. *Female:* prominent dark barring or flecking on breast and upperparts. *Immature:* heavier barring than adult female; almost all Ohio birds are immatures.
Size: *L* 20–27 in; *W* 4½–6 ft.
Habitat: open country, including croplands, meadows, airports and lakeshores; often perches on fence posts, breakwaters, buildings and utility poles.

Nesting: does not nest in Ohio.
Feeding: swoops from a perch, often punching through the snow, to take mice, voles, grouse, hares, weasels and rarely songbirds and waterbirds.
Voice: quiet in winter; breeding males produce an incredible booming hoot that carries for miles across the open tundra.
Similar Species: no other owl in the region is mostly white and lacks "ear" tufts.
Best Sites: mostly along L. Erie; Burke Lakefront Airport in Cleveland; Bayshore Power Plant near Toledo; Mentor Headlands.

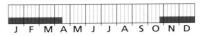

J F M A M J J A S O N D

BARRED OWL

Strix varia

Few nocturnal sounds are as spectacular—or as frightening to the uninitiated—as the hooting, screaming and caterwauling of courting Barred Owls. • Prior to European settlement, when Ohio was 95 percent forested, this was undoubtedly the most common owl. While still common in the unglaciated hill country and in much of southwestern Ohio, Barred Owls have been eliminated from large areas because of deforestation. • Barred Owls are strictly a woodland species, preferring older-growth stands with plenty of large dead and dying trees, which often have cavities suitable for nest sites. In Ohio, Barred Owls are found in large, bottomland forests along rivers and streams, and swamp woods. These areas have fast-growing trees such as sycamores, beech and maples, which quickly grow to a size where large cavities form. Many upland forests are overmanaged for silviculture purposes to eliminate cavity trees, which make for bad timber. • To find Barred Owls, locate good-looking habitat, then return at night to listen for them. Their calls are easily imitated; even a crude rendition may provoke an answer. If the mimic persists, one or two owls will sometimes fly in for a close look.

ID: dark eyes; horizontal barring around neck and upper breast; vertical streaking on belly; pale bill; no "ear" tufts; mottled, dark gray brown plumage.

Size: *L* 17–24 in; *W* 3½–4 ft.

Habitat: mature deciduous forests, especially in mature stands with plenty of cavity trees near swamps, streams and lakes.

Nesting: in a natural tree cavity, broken treetop or rarely an abandoned stick nest; adds little material to the nest; female incubates 2–3 white eggs for 28–33 days; male feeds the female during incubation.

Feeding: nocturnal; swoops from a perch to pounce on prey; eats mice, voles and squirrels; also eats amphibians and smaller birds.

Voice: most characteristic of all owls; loud, hooting, rhythmic, laughing call mostly in spring, also throughout the year: *Who cooks for you? Who cooks for you all?*

Similar Species: *Barn Owl* (p. 170): much paler; prominent, heart-shaped face.

Best Sites: ravines around the north end of Alum Creek Reservoir; Lake Hope SP; Shawnee SF; most large forests.

J F M A M J J A S O N D

LONG-EARED OWL

Asio otus

Because Long-eared Owls are a secretive species, it is probable that many more of these rare winter residents and migrants occur in Ohio than are ever detected. Completely nocturnal, Long-eareds hunt over open fields, but roost in dense conifers, grapevine tangles and pin oaks during the day. A key to locating them is to find suitable hunting meadows, then look for appropriate roosting cover. Carefully looking through the dense vegetation may reveal up to 20 owls, as Long-eareds often roost communally. They can be hard to spot; not only do they hide in thick vegetation, but they can also compress their feathers and appear very much like a broken-off stick. • Long-eared Owls are secretive with good reason—hawks and other owls frequently prey on them. • This species is a very rare and sporadic breeder across the northern half of Ohio.

ID: long, relatively close-set "ear" tufts; slim body; vertical belly markings; rusty brown facial disc; mottled brown plumage; yellow eyes; white around bill.

Size: L 13–16 in; W 3–4 ft.

Habitat: woodlots, particularly with dense pin oaks, overgrown riparian woodlands and hedgerows; also isolated conifer groves in meadows, fields, cemeteries, farmyards or parks.

Nesting: often in an abandoned hawk or crow nest; female incubates 2–6 white eggs for 26–28 days; male feeds female during incubation.

Feeding: nocturnal; flies low, pouncing on prey from the air; eats mostly voles and mice; occasionally takes shrews, moles, small rabbits, birds and amphibians.

Voice: breeding call is a low, soft, ghostly *quoo-quoo;* alarm call is *weck-weck-weck;* also various shrieks, hisses, whistles, barks, hoots and dovelike *coo;* seldom heard in Ohio.

Similar Species: *Great Horned Owl* (p. 172): much larger; "ear" tufts farther apart; stout body; rounder face. *Short-eared Owl* (p. 176): bolder, more complete black markings on upperwing; lacks long "ear" tufts, but very similar in flight. *Eastern Screech-Owl* (p. 171): much shorter, stout body; shorter, wider-set "ear" tufts.

Best Sites: broadcasting locations of Long-eared Owl roosts is not advisable; frequent human contact can cause birds to abandon favored sites; look for these owls in thickly overgrown woodlots near L. Erie during spring migration, and in fields overgrown with red cedar in southwestern Ohio, such as at Caesar Creek SP or Chaparral Prairie SNP; Cedar Bog State Memorial regularly hosts wintering birds.

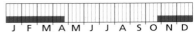

J F M A M J J A S O N D

175

SHORT-EARED OWL

Asio flammeus

The partly diurnal Short-eared Owl resembles a giant moth as it leisurely courses over the grasslands and meadows that are its favored hunting habitats. Its odd, distinctive, deep and slow wingbeats coupled with its erratic, low-flying hunting technique mean that the Short-eared Owl can be identified at great distances. • Short-eareds are one of the world's most broadly distributed owls, occurring on every continent but Australia. In Ohio, Short-eareds are primarily winter residents; their numbers are cyclical and they can occur in large concentrations when food—mainly the meadow vole—is plentiful. For instance, a huge, reclaimed strip mine grassland in Muskingum County hosted over 80 owls in the winter of 1999; the next year only a few were present. • Short-eared Owls have always been rare and erratic nesters in Ohio, but with the advent of enormous grasslands created in the course of strip mine reclamation, breeding bird numbers may increase. • Migrants may appear almost anywhere, even in very urban areas.

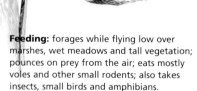

ID: yellow eyes set in black sockets; heavy, vertical streaking on buff belly; straw-colored upperparts; inconspicuous, short "ear" tufts. *In flight:* dark "wrist" crescents; deep wingbeats; long wings.

Size: *L* 13–17 in; *W* 3–4 ft.

Habitat: open areas, including grasslands, wet meadows, marshes, fields and airports.

Nesting: on the ground in an open area; a slight depression is sparsely lined with grass; female incubates 4–7 white eggs for 24–27 days; male feeds the female during incubation.

Feeding: forages while flying low over marshes, wet meadows and tall vegetation; pounces on prey from the air; eats mostly voles and other small rodents; also takes insects, small birds and amphibians.

Voice: generally quiet; produces a soft *toot-toot-toot* during breeding season; also squeals and barks like a small dog.

Similar Species: *Long-eared Owl* (p. 175) and *Great Horned Owl* (p. 172): long "ear" tufts; rarely hunt during the day.

Best Sites: Killdeer Plains, Magee Marsh and Big Island wildlife areas; large, reclaimed strip mine grasslands, such as The Wilds; also Crown City, Tri-Valley and Woodbury wildlife areas.

J F M A M J J A S O N D

NORTHERN SAW-WHET OWL

Aegolius acadicus

The tiny Northern Saw-whet Owl is an opportunistic hunter, taking whatever it can, whenever it can. If temperatures are below freezing and prey is abundant, this small owl may choose to catch more than it can eat in a single sitting. The extra food is usually stored in a tree, where it quickly freezes. When hunting efforts fail, a hungry saw-whet will return to thaw out its frozen cache by "incubating" the food as if it were a clutch of eggs! • Very secretive, saw-whets are most often discovered roosting in conifers or dense grapevine tangles. They are incredibly tame and can often be approached closely. A good way to detect them is by looking for "whitewash"—buildups of excrement—under possible roosting habitat. Like the Long-eared Owl, the vast majority of Northern Saw-whet Owls no doubt go undetected in Ohio. • This bird's name is derived from its peculiar whistled song, said to mimic the sound of a saw being whetted (sharpened). Unfortunately, this sound is unlikely to be heard here, as migrant and wintering birds are silent, and the Northern Saw-whet Owl is an extremely rare breeder.

ID: small body; large, rounded head; pale, unbordered facial disc; dark bill; vertical, rusty streaks on underparts; white-spotted, brown upperparts; white-streaked forehead; short tail. *Immature:* white patch between eyes; rich brown head and breast; buff brown belly.
Size: *L* 7–9 in; *W* 17–22 in.
Habitat: pure and mixed coniferous and deciduous forests.
Nesting: in an abandoned woodpecker cavity or natural hollow in a tree; female incubates 5–6 white eggs for 27–29 days; male feeds the female during incubation.
Feeding: swoops down from a perch; eats mostly mice and voles; also eats larger insects, songbirds, shrews, moles and occasionally amphibians; may cache food.
Voice: whistled, evenly spaced notes repeated about 100 times per minute: *whew-whew-whew-whew;* continuous and easily imitated.
Similar Species: smallest Ohio owl; when seen, bird is generally at close range and not likely to be confused with any other species.
Best Sites: Magee Marsh WA bird trail; pines near Gordon Park Impoundment; Green Lawn Cemetery. *Winter:* more common southward; successional fields overgrown with red cedars in SW Ohio, such as at Caesar Creek SP.

J F M A M J J A S O N D

COMMON NIGHTHAWK

Chordeiles minor

Sometimes called "Bullbats" for their batlike, erratic flight, Common Nighthawks are a ubiquitous part of our urban environment. Human-made nesting sites largely dictate this choice of habitat—nighthawks like to nest on flat, gravel-covered rooftops. These birds return to Ohio in the first week of May, and become quite conspicuous at dusk, hawking about in the sky to catch insects and frequently giving their loud *peent* call. • The male's spectacular courtship display involves a steep, rapid drop toward the ground from a great height, culminating in a loud booming sound as the bird pulls out of the dive and the air rushes through the primary feathers. • Tremendous flocks of nighthawks can pass over in late August and early September, with sometimes up to 5000 birds observed in a few hours. This is just the beginning of their journey, though, as Common Nighthawks winter in southern South America.

ID: cryptic, mottled plumage; barred underparts. *Male:* white throat. *Female:* buff throat. *In flight:* bold, white "wrist" patches on long, pointed wings; shallowly forked, barred tail; erratic flight.

Size: *L* 8½–10 in; *W* 24 in.

Habitat: *Breeding:* on flat rooftops; sometimes in recently logged woods, barren fields or large rocky areas. *In migration:* anywhere large numbers of flying insects can be found; usually roosts in trees, often near water.

Nesting: on bare ground; no nest is built; female incubates 2 well-camouflaged eggs for about 19 days; both adults feed the young.

Feeding: feeds primarily at dawn or dusk; catches insects in flight; eats mosquitoes, blackflies, midges, beetles, flying ants, moths and other flying insects; may fly around street lights to catch prey attracted to the light.

Voice: frequently repeated, nasal *peent;* also makes a deep, hollow *vroom* with its wings during courtship flight.

Similar Species: *Whip-poor-will* (p. 180) and *Chuck-will's-widow* (p. 179): less common; shorter, rounder wings; rounded tails; lack white "wrist" patches. *American Woodcock* (p. 144): nasal calls are often confused with those of Common Nighthawk in late February and March.

Best Sites: around bright lights at night on baseball fields and around tops of stadiums; skies at dusk in any urban area.

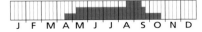

J F M A M J J A S O N D

CHUCK-WILL'S-WIDOW

Caprimulgus carolinensis

There should be no problem hearing a Chuck-will's-widow—it often sings its name incessantly—but seeing a "Chuck" is much more difficult, as this species is strictly nocturnal. • The most famous site, and probably the easiest place to locate Chuck-will's-widow, is along the southern reaches of Ohio Brush Creek in Adams County, which was also where the Chuck was first recorded in Ohio, back in 1932. There are occasional reports from elsewhere in southern Ohio, and this species is probably overlooked to some degree. • Migrants can rarely appear almost anywhere, particularly during May migration. These well-camouflaged birds are usually spotted in frequently birded areas, perched perpendicularly on a branch. • This species is the rarest of the Ohio *Caprimulgids*, which include the Common Nighthawk and Whip-poor-will. Perched birds can look similar, but Chuck-will's-widow is larger, with a warm rufous coloration, and lacks a white wing spot and dark crown stripe.

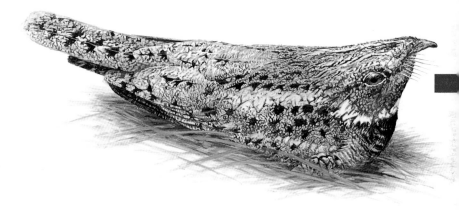

ID: mottled, brown-and-buff body with overall reddish tinge; pale brown to buff throat; whitish "necklace"; dark breast; long, rounded tail. *Male:* inner edges of outer tail feathers are white.

Size: *L* 11–13 in; *W* 24–26 in.
Habitat: deciduous riparian and upland woodlands interspersed with open fields.
Nesting: on bare ground; no nest is built; female incubates 2 heavily blotched, creamy white eggs for about 21 days and raises the young alone.
Feeding: catches insects on the wing; eats beetles, moths and other large flying insects.

Voice: 3 loud, whistled notes often paraphrased as *chuck-will's-widow.*
Similar Species: *Whip-poor-will* (p. 180): smaller; grayer overall; "necklace" contrasts with black throat; male shows much more white in tail feathers; female's dark tail feathers are bordered with buff on outer tips. *Common Nighthawk* (p. 178): smaller; forked tail; white patches on wings; male has white throat; female has buff throat.
Best Sites: along roads parallel to Ohio Brush Creek in Adams Co.; a few can be found in Hocking Co., near junction of Starner and Cantwell Cliffs roads, just south of Clear Creek Valley.

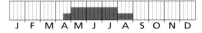

J F M A M J J A S O N D

WHIP-POOR-WILL

Caprimulgus vociferus

Woe to the sleepers who have one of these odd Goatsuckers set up territory outside their tent or window. "Whips" sing their loud, ringing song incessantly, up to 59 times per minute, and on occasion a bird can deliver a thousand renditions nonstop. • The colloquial name "Goatsucker" dates back to at least Aristotle, who wrote, "Flying to the udders of she-goats, it sucks them, and thus gets its name." He was probably referring to the European Nightjar *(C. europaeus)*, but the name has been applied to the entire family. Fortunately, better ornithologists than Aristotle disproved this myth long ago. • Whip-poor-wills and other members of the family are strictly insectivorous, catching insects on the wing at night. To aid in the capture of prey, they have long stiff hairs around the bill, known as rictal bristles, which help funnel prey into the mouth. • Whip-poor-wills have declined considerably in the last century, and are now most common in unglaciated hill country, with scattered populations elsewhere.

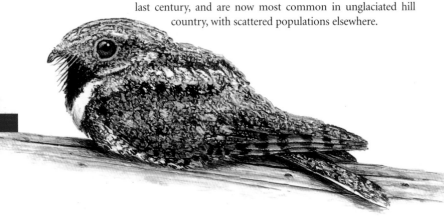

ID: mottled, gray brown overall with black flecking; reddish tinge on rounded wings; black throat; long, rounded tail. *Male:* white "necklace"; white outer tail feathers. *Female:* buff "necklace."

Size: *L* 9–10 in; *W* 16–20 in.

Habitat: open deciduous woodlands, sometimes in pine plantations; often along forest edges.

Nesting: on the ground, sometimes in leaf or pine needle litter; no nest is built; mostly the female incubates 2 whitish eggs, blotched with brown and gray, for 19–21 days; both adults raise the young.

Feeding: almost entirely nocturnal; catches insects in flight, often high in the air; eats mosquitoes, blackflies, midges, beetles and other flying insects; partial to moths; some grasshoppers are taken and swallowed whole.

Voice: far-carrying and frequently repeated *whip-poor-will,* with emphasis on the *will;* most often uttered at dusk and through the night.

Similar Species: *Common Nighthawk* (p. 178): less rufous in plumage; longer, pointed wings with white "wrist" patches; shallowly forked, barred tail. *Chuck-will's-widow* (p. 179): larger; pale brown to buff throat; whitish "necklace"; darker breast; more reddish overall; much less white on male's tail; different call.

Best Sites: in appropriate habitat in most of unglaciated Ohio; along roads in and around Clear Creek MP, Zaleski Forest, Crown City WA and Lake Katherine SNP; also Oak Openings MP region.

J F M A M J J A S O N D

CHIMNEY SWIFT

Chaetura pelagica

No aircraft can compare with the flying prowess of the Chimney Swift. Consider this: a Chimney Swift weighs 23 grams, yet has a wingspan of 14 inches, encompassing about 25 square inches. Thus, each square inch of wing surface supports less than 1 gram. In comparison, a Boeing 727 airliner would have to have a wingspan of about 42,000 feet, or nearly 8 miles, to match the wing-surface to weight ratio of a swift. • The Chimney Swift is much more adept aerially than terrestrially. It has very weak feet and can't even land or perch horizontally; rather, it must alight on a vertical surface and prop itself up with its stiff, bristle-tipped tail. Once airborne, though, the swift is the perfect flying machine. • Fall migrants sometimes form amazing roosts of hundreds or thousands of individuals; as dusk falls, these swifts resemble a tornado of birds as they form a vortex and begin funneling into chimneys.

ID: brown overall; slim body; long, thin, pointed wings; squared tail. *In flight:* rapid wingbeats; boomerang-shaped profile; erratic flight pattern.

Size: *L* 5–5½ in; *W* 12–13 in.

Habitat: forages over cities and towns; roosts and nests in chimneys; rarely nests in tree cavities in remote areas.

Nesting: often colonial; nests in the interior of a chimney, in a tree cavity or in the attic of an abandoned building; pair fixes a half-saucer nest of short, dead twigs to a vertical wall with saliva; pair incubates 4–5 white eggs for 19–21 days; both adults feed the young.

Feeding: flying insects are swallowed whole during continuous flight.

Voice: rapid, chattering call is given in flight: *chitter-chitter-chitter;* also gives a rapid staccato series of *chip* notes.

Similar Species: *Swallows* (pp. 213–18): broader, shorter wings; smoother flight pattern; most have forked or notched tails.

Best Sites: urban areas statewide; foraging birds wander widely into the countryside; large roosts form at favored, large chimneys in late summer and fall; locations vary from year to year.

J F M A M J J A S O N D

181

RUBY-THROATED HUMMINGBIRD

Archilochus colubris

Hummingbirds are the helicopters of the bird world, possessing extraordinary powers of flight that enable them to fly in any direction. One of these little birds weighs about the same as a nickel, can beat its wings up to 80 times a second, and in full locomotion its heart rate accelerates to 1200 beats per minute. • Ruby-throated Hummingbirds are the only breeding "hummers" in the eastern U.S. and are common throughout Ohio, though Rufous Hummingbird sightings are increasing here. • Filling a niche similar to bees, Ruby-throats feed primarily on nectar extracted from flowers via their specially adapted long bills, and pollinate the flowers in return. Ruby-throated Hummingbirds' time in Ohio coincides with the peak of our wildflower season. • These fascinating birds can easily be lured to yards and gardens with sugarwater feeders or by planting hummer-friendly flora such as salvia, columbine or trumpet creeper.

ID: tiny; long bill; iridescent, green back; pale underparts; dark tail. *Male:* ruby red throat; black "chin." *Female* and *immature:* fine, dark throat streaking.

Size: *L* 3½–4 in; *W* 4½ in.

Habitat: open, mixed woodlands, wetlands, orchards, meadows, flower gardens and backyards with trees and feeders.

Nesting: on a horizontal tree limb; tiny, deep cup nest of plant down and fibers is held together with spider silk; lichens and leaves are pasted on exterior walls; female incubates 2 tiny, white eggs for 13–16 days; female feeds the young.

Feeding: uses long bill and tongue to probe blooming flowers and sugar-sweetened water from feeders; also eats small insects and spiders.

Voice: most noticeable is the soft buzzing of wings while in flight; also produces a loud *chick* and other high squeaks.

Similar Species: sphinx moths can look similar to hummingbirds at first glance. *Rufous Hummingbird* (p. 337): records increasing rapidly; male has conspicuous, rufous feathers; female has red-spotted throat and reddish flanks; female and juvenile are tricky to identify.

Best Sites: in gardens and at sugarwater feeders; Aullwood Audubon Center; Holden Arboretum.

J F M A M J J A S O N D

BELTED KINGFISHER

Ceryle alcyon

The kingfisher makes shallow dives for small fish from a favored perch over water. As the bird nears the water, it shuts its eyes and snares the prey with a pincerlike snap of its bill. The unfortunate victim is then taken back to the perch, where it is often savagely pounded against an unyielding surface until stunned, at which point the kingfisher gulps it down. • Unusual for the bird world, female Belted Kingfishers are more colorful than males. • The Belted Kingfisher is often first detected by its loud, raucous rattle call, and you are likely to be the reason that the bird sounds off. Belted Kingfishers, which are normally solitary except when nesting, defend hunting territories vigorously year round, and don't hesitate to scold any invader. • These birds are never seen in large numbers but are regular in Ohio wherever suitable hunting waters are found. They are very hardy—individuals will regularly overwinter if open water is available.

ID: bluish upperparts; shaggy crest; blue gray breast band; white "collar"; long, straight bill; short legs; white underwings; small, white patch near eye. *Male:* no "belt." *Female:* rust-colored "belt" is occasionally incomplete.
Size: *L* 11–14 in; *W* 20 in.
Habitat: rivers, large streams, lakes, marshes and beaver ponds, especially near exposed soil banks, gravel pits or bluffs.

Nesting: pair digs earth burrow up to 6 ft long; pair incubates 6–7 white eggs for 22–24 days; both adults feed the young; sometimes uses gravel piles and rarely tree cavities.
Feeding: dives headfirst into water, either from a perch or from hovering flight; eats mostly small fish, aquatic invertebrates and tadpoles.
Voice: fast, repetitive, cackling rattle, like a teacup shaking on a saucer.
Similar Species: none in Ohio.
Best Sites: best populations are found on rivers, creeks and streams.

J F M A M J J A S O N D

RED-HEADED WOODPECKER

Melanerpes erythrocephalus

Adult Red-headed Woodpeckers are such a glorious sight that they inspired Alexander Wilson, author and illustrator of the early-1800s, nine-volume work *American Ornithology,* to become an ornithologist. • Red-headed Woodpeckers are unusual in their food habits, seldom drilling into trees for insects. Instead, they frequently flycatch, making sallies from conspicuous perches. So smitten are they with berries that Audubon wrote in 1842 that "I would not recommend anyone to trust their fruit to the red-heads...a hundred have been shot in a single cherry-tree in one day." We should be so lucky to have such hordes of Red-headed Woodpeckers plundering our crops today—and would hopefully refrain from shooting them. This species is not as plentiful in Ohio as it was in the early 1900s. Its peak numbers coincided with the die-off of once abundant American chestnuts and American elms—the millions of standing dead trees created perfect habitat. Today, the maturation of large forests blanketing southeastern Ohio has largely eliminated Red-headed Woodpecker habitat in that region.

juvenile

ID: bright red head, chin, throat and "bib" bordered with black; black back, wings and tail; white breast, belly, rump, lower back and inner wing patches.
Juvenile: brown head, back, wings and tail; slight brown streaking on white underparts.
Size: *L* 9–9½ in; *W* 17 in.
Habitat: open deciduous woodlands with predominance of oaks and hickories; especially attracted to standing dead trees; perches on telephone poles.
Nesting: male excavates a nest cavity in a dead tree or limb; pair incubates 4–5 white eggs for 12–13 days; both adults feed the young.
Feeding: eats mostly insects, earthworms, spiders, nuts, berries, seeds and fruit; flycatches for insects; hammers dead and decaying wood for grubs; may also eat some young birds and eggs.
Voice: loud series of *kweer* or *kwrring* notes; occasionally a chattering *kerr-r-ruck;* also drums softly in short bursts.
Similar Species: none in Ohio.
Best Sites: frequent in western Ohio woodlots; Goll Woods SNP; Killdeer Plains WA; Killbuck WA.

J F M A M J J A S O N D

RED-BELLIED WOODPECKER

Melanerpes carolinus

One of our most common, conspicuous and easily recognized woodpeckers, the Red-bellied Woodpecker is not at all shy and retiring, but makes its presence known by frequent loud calls. • Red-bellied Woodpeckers occur in woodlots, forests and treed urban areas statewide, but this wasn't always so. Their continuing, northward expansion in recent decades is probably associated with two factors: recovery of eastern forests to a more mature condition and the abundance of bird feeders, which Red-bellied Woodpeckers often visit. • During late winter or early spring courtship time, many homeowners lower their opinions of these handsome birds when the males begin drumming on metal gutters or siding in an effort to produce as much noise as possible to impress the females. • Red-bellied Woodpeckers can be long-lived—banding has documented individuals living for over 20 years. • The Red-bellied Woodpecker is poorly named as its red belly is almost impossible to see in the field.

ID: black-and-white barring on back; white patches on rump and topside base of primaries; reddish tinge on belly. *Male:* red nape extends to forehead. *Female:* red on nape only. *Juvenile:* dark gray crown; streaked breast.
Size: *L* 9–10½ in; *W* 16 in.
Habitat: mature deciduous woodlands; occasionally in wooded residential areas.
Nesting: in a natural cavity or one abandoned by another woodpecker; female selects one of several nest sites excavated by the male; pair incubates 4–5 white eggs for 12–14 days; both adults raise the young.
Feeding: forages in trees, on the ground or occasionally on the wing; eats mostly insects, seeds, nuts and fruit; may also eat tree sap, small amphibians, bird eggs or small fish.
Voice: call is a soft, rolling *churr;* also a variety of *cha* notes; drums in second-long bursts.
Similar Species: *Northern Flicker* (p. 189): yellow underwings; gray crown; brown back with dark barring; black "bib"; large, dark spots on underparts. *Red-headed Woodpecker* (p. 184): all-red head; unbarred, black back and wings; white patch on trailing edge of wing.
Best Sites: any forest, good-sized woodland or backyard feeder; Mohican SF; Lake Katherine SNP; Green Lawn Cemetery.

J F M A M J J A S O N D

185

YELLOW-BELLIED SAPSUCKER

Sphyrapicus varius

Yellow-bellied Sapsuckers have been saddled with a comical name that is often used to poke fun at birders, yet the moniker is quite descriptive. Sapsuckers drill horizontal rows of small sap wells, which are easily recognized as their handiwork. They defend and maintain these well fields, returning frequently to drink the sugary sap. • Over 1000 species of woody plants have been documented as hosts for these sap wells, and the ecological web that results is remarkable. For instance, the Ruby-throated Hummingbird follows sapsuckers and drinks from the wells, and may even time its migration in parts of its range to coincide with peak sapsucker numbers. Several hundred other species use sapsucker wells, including bats, flying squirrels, insects, and other birds such as kinglets and warblers. • Yellow-bellied Sapsuckers are easy to miss because they are shy and retiring, are normally uncommon in migration and are currently listed as endangered in Ohio. There is a small breeding population in the extreme northeastern counties. They are more likely southward in winter.

ID: black "bib"; red forecrown; black-and-white face, back, wings and tail; large, white wing patch; yellow wash on lower breast and belly. *Male:* red "chin." *Female:* white "chin." *Juvenile:* brownish overall, but with large, clearly defined wing patches.
Size: *L* 7–9 in; *W* 16 in.
Habitat: deciduous and mixed forests, especially dry, second-growth woodlands; often attracted to pines in migration.
Nesting: in a cavity; often in a live tree with heart rot; lines the cavity with wood chips; pair incubates 5–6 white eggs for 12–13 days.

Feeding: hammers trees for insects; drills holes in live trees to collect sap and trapped insects; also occasionally flycatches for insects.
Voice: nasal, catlike *meow;* territorial and courtship hammering has a distinctive, 2-speed quality; soft *vee-ooo* when alarmed.
Similar Species: *Red-headed Woodpecker* (p. 184): juvenile lacks white patch on wing. *Downy Woodpecker* (p. 187) and *Hairy Woodpecker* (p. 188): red napes; white backs; lack large, white wing patch and red forecrown.
Best Sites: *In migration:* often returns yearly to favored sap trees such as the Austrian pines around the pond at Green Lawn Cemetery; also Spring Grove Cemetery and Magee Marsh bird trail.

J F M A M J J A S O N D

DOWNY WOODPECKER

Picoides pubescens

Our most widespread and common woodpecker, the Downy uses a wide variety of habitats from mature woods to scruffy fencerows, and regularly visits feeders. • Seemingly always hard at work, Downy Woodpeckers are constantly foraging and devote much time to the excavation of nest holes, which may later be used by a variety of other animals. • Like other members of the woodpecker family, the Downy has evolved many features to help cushion the repeated shocks of a lifetime of hammering, including a strong bill, strong neck muscles, a flexible, reinforced skull and a brain that is tightly packed in its protective cranium. Another feature that the Downy shares with other woodpeckers is its feathered nostrils, which serve to filter out the sawdust it produces when hammering. • The Downy is probably closely related to the western Nuttall's Woodpecker *(P. nuttallii),* with which it is known to hybridize.

ID: clean white belly and back; white barring on black wings; black eye line and crown; short, stubby bill; mostly black tail; white outer tail feathers spotted with black. *Male:* small, red patch on back of head. *Female:* no red patch.
Size: *L* 6–7 in; *W* 12 in.
Habitat: all wooded environments, especially deciduous and mixed forests and areas with tall deciduous shrubs; will even forage in overgrown fields.
Nesting: pair excavates a cavity in a decaying or dying tree trunk or limb and lines it with wood chips; pair incubates 4–5 white eggs for 11–13 days; both adults feed the young.
Feeding: forages on trunks and branches in saplings and shrubs; chips and probes for insect eggs, cocoons, larvae and adults; also eats nuts and seeds; attracted to suet feeders; frequently eats goldenrod gallfly larvae.
Voice: long, unbroken trill; calls are a sharp *pik* or *ki-ki-ki* or a whiny *queek queek;* drums more than the Hairy Woodpecker and at a higher pitch, usually on smaller trees and dead branches.
Similar Species: *Hairy Woodpecker* (p. 188): larger; bill is as long as head is wide; no spots on white outer tail feathers. *Yellow-bellied Sapsucker* (p. 186): large, white wing patch; red forecrown; lacks red nape and clean white back.
Best Sites: all manner of woodlands statewide; readily visits feeders, particularly those with suet.

J F M A M J J A S O N D

HAIRY WOODPECKER

Picoides villosus

Hairy Woodpeckers, while not rare, are outnumbered ten to one by Downy Woodpeckers. At first, these two may be confusing, but Hairy Woodpeckers are noticeably larger with proportionately much bigger bills, and have louder, sharper vocalizations. They also require more extensive, unbroken woodlands with larger trees. • Seventeen subspecies of Hairy Woodpecker have been described across North America from Alaska to Central America. They exhibit Bergman's rule: the northern birds are larger with a gradual decrease in size to the south. • Most woodpeckers have very maneuverable tongues that are, in some cases, more than four times the length of the bill. This is made possible by twin structures that wrap around the perimeter of the skull and store the tongue in much the same way that a measuring tape is stored in its case. The tip of the tongue is sticky with saliva and is finely barbed to help seize wood-boring insects.

ID: pure white belly; white spotting on black wings; black "cheek" and crown; bill is about as long as head is wide; black tail with unspotted, white outer feathers. *Male:* small red patch on back of head. *Female:* no red patch. *Juvenile:* more indistinct patterning, with brown instead of black.
Size: *L* 8–9½ in; *W* 15 in.
Habitat: deciduous and mixed forests; prefers larger tracts with bigger timber.
Nesting: pair excavates a nest site in a live or decaying tree trunk or limb; cavity is lined with wood chips; pair incubates 4–5 white eggs for 12–14 days; both adults feed the young.
Feeding: forages on tree trunks and branches; chips, hammers and probes bark for insect eggs, cocoons, larvae and adults; also eats nuts, fruit and seeds; attracted to feeders with suet, especially in winter.
Voice: loud, sharp call: *peek peek;* long, unbroken trill: *keek-ik-ik-ik-ik;* drums less regularly and at a lower pitch than the Downy Woodpecker, always on tree trunks and large branches.
Similar Species: *Downy Woodpecker* (p. 187): smaller; shorter bill; dark spots on white outer tail feathers. *Yellow-bellied Sapsucker* (p. 186): large, white wing patch; red forecrown; lacks red nape and clean white back.
Best Sites: larger forests, such as Cuyahoga Valley NRA, Burr Oak SP or Mohican SF; at feeders if suitable habitat is nearby.

J F M A M J J A S O N D

NORTHERN FLICKER

Colaptes auratus

A wide-ranging and well-known bird, the Northern Flicker has at least 132 colloquial names. The best known of these was the former "Yellow-shafted Flicker," from the days when it was considered distinct from the western "Red-shafted Flicker." • Flickers are birds of open country with scattered trees and woodlots, and spend much of their time foraging on the ground. Ants are often the object of their pursuit—flickers probably consume more of these insects than any other bird species. One specimen was found to have over 5000 ants in its stomach. • Unfortunately, flicker populations have declined, and the European Starling is thought to be a primary culprit. Flickers' nest holes are suitable for starlings, which often displace the rightful owners. • Like many woodpeckers, the Northern Flicker has zygodactyl feet—each foot has two toes facing forward and two toes pointing backward—that allow the bird to move vertically up and down tree trunks. Stiff tail feathers help to prop up the bird's body while it scales trees and excavates cavities. • Northern Flickers sometimes stage spectacular migrations in spring, with dozens or even hundreds seen in a day along Lake Erie.

ID: brown, barred back and wings; spotted, buff to whitish underparts; black "bib"; yellow underwings and undertail; white rump; long bill; brownish to buff face; gray crown. *Male:* black "mustache" stripe; red nape crescent. *Female:* no "mustache."
Size: *L* 12½–13 in; *W* 20 in.
Habitat: open deciduous woodlands and forest edges, fields and meadows; often enters treed urban neighborhoods.
Nesting: pair excavates a cavity in a dead or dying deciduous tree; will also use a nest box; lines cavity with wood chips; pair incubates 5–8 white eggs for 11–16 days; both adults feed the young.
Feeding: forages on the ground for ants and other terrestrial insects; also eats berries and nuts; probes bark; occasionally flycatches; readily visits feeders for suet, seeds and berries.
Voice: loud, laughing, rapid *kick-kick-kick-kick-kick-kick; woika-woika-woika* issued during courtship.
Similar Species: there are a few records of western "Red-shafted" Northern Flicker subspecies, which has reddish underwings. *Red-bellied Woodpecker* (p. 185): black-and-white pattern on back; more red on head; dark underwings.
Best Sites: *In migration:* common in every county; western L. Erie in vicinity of Ottawa NWR; Gilmore Ponds Preserve; Killdeer Plains WA.

J F M A M J J A S O N D

PILEATED WOODPECKER

Dryocopus pileatus

The spectacular, crow-sized Pileated Woodpecker is quite secretive considering its size, so lucking into one is always a noteworthy experience. Many more Pileateds are heard than seen; their loud, maniacal laughter carries great distances. • The sixth-largest woodpecker in the world, this species has had a checkered history in Ohio. Prior to European settlement, it no doubt occurred throughout the state, as 95 percent of Ohio was blanketed with virgin timber. With the settling and clearing of the state, Pileated Woodpeckers had virtually disappeared by the early 1900s. Fortunately, they have rebounded and are still expanding today as maturing forests have provided suitable habitat—a pair of Pileated Woodpeckers requires about 100 acres of woodland and there must be many large trees present. • Pileateds use their powerful, chisel-like bills to pry for carpenter ants and wood-boring beetles. Evidence of Pileateds is easily detected—their food excavations and nest holes tend to be large and oval-shaped. These cavities are used by many other animals, including Wood Ducks, squirrels, owls and snakes.

ID: predominantly black; white wing linings; flaming red crest; yellow eyes; stout, dark bill; white stripe runs from bill to shoulder; white "chin." *Male:* red "mustache"; red crest extends from forehead. *Female:* no red "mustache"; red crest starts on crown.

Size: *L* 16–19 in; *W* 29 in.

Habitat: extensive tracts of mature deciduous, coniferous or mixed forest; also riparian woodlands or woodlots in suburban and agricultural areas.

Nesting: pair excavates a cavity in a dead or dying tree trunk; excavation can take 3–6 weeks; lines cavity with wood chips; pair incubates 4 white eggs for 15–18 days; both adults feed the young.

Feeding: hammers the base of rotting trees, creating fist-sized or larger holes; eats carpenter ants, wood-boring beetle larvae, berries and nuts.

Voice: loud, fast, laughing *woika-woika-woika-woika;* long series of *kuk* notes; loud, resonant drumming.

Similar Species: none in Ohio.

Best Sites: best areas are extensive forests of southern and eastern Ohio, including all state forests; Cantwell Cliffs SP; Little Beaver Creek SP; Clear Creek MP; Highbanks MP; Lake Katherine SNP; large woodlands in western and northern Ohio, including Goll Woods SNP and Oak Openings MP.

J F M A M J J A S O N D

PASSERINES

Flycatchers   *Shrikes & Vireos*

Jays & Crows   *Larks & Swallows*

Chickadees,   *Kinglets, Bluebirds*
Nuthatches & Wrens   *& Thrushes*

Mimics, Starlings   *Wood-Warblers*
& Waxwings   *& Tanagers*

Sparrows, Grosbeaks   *Blackbirds*
& Buntings   *& Allies*

Finchlike Birds

Passerines are also commonly known as songbirds or perching birds. Although these terms are easier to comprehend, they are not as strictly accurate, because some passerines neither sing nor perch, and many nonpasserines do sing and perch. In a general sense, however, these terms represent passerines adequately: these birds are among the best singers, and they are typically seen perched on a branch or wire.

It is believed that passerines, which all belong to the order Passeriformes, make up the most recent evolutionary group of birds. Theirs is the most numerous of all orders, representing about 43 percent of the bird species in Ohio, and nearly three-fifths of all birds worldwide.

Passerines are grouped together based on the sum total of many morphological and molecular similarities, including such things as the number of tail and flight feathers and reproductive characteristics. All passerines share the same foot shape: three toes face forward and one faces backward, and no passerines have webbed toes. Also, all passerines have a tendon that runs along the back side of the bird's knee and tightens when the bird perches, giving it a firm grip.

Some of our most common and easily identified birds are passerines, such as the Black-capped Chickadee, American Robin and House Sparrow, but the passerines also include some of the most challenging and frustrating birds to identify, until their distinct songs and calls are learned.

191

OLIVE-SIDED FLYCATCHER

Contopus cooperi

No shrinking violet, the Olive-sided Flycatcher was dubbed the "Peregrine of flycatchers" by researcher J.T. Marshall. Often seen sitting conspicuously high on a snag, the Olive-sided Flycatcher engages in "yo-yo flights": the bird sallies forth from a perch, grabs an insect and promptly returns to the same spot. • In the spring, Olive-sided Flycatchers can be heard giving their distinctive, whistled *quick, three beers!* call. • Unfortunately, this species has been declining steadily throughout its range; the greatest numbers now breed in western forests. They are rare migrants here, so detecting more than one in a day would be exceptional. • Olive-sideds return late in spring, generally in the latter half of May to early June with southbound birds arriving by late July. Because this species feeds strictly on flying insects, primarily bees and wasps, this late arrival ensures that food is available. • Olive-sided Flycatchers winter in Central and South America, and have the longest migration of any North American flycatcher.

ID: big head; large bill; dark olive gray "vest"; pale throat and belly; olive gray to olive brown upperparts; white tufts on sides of rump; dark upper mandible; dull yellow orange base to lower mandible; inconspicuous eye ring.
Size: *L* 7–8 in; *W* 13 in.
Habitat: semi-open mixed and coniferous forests near water; prefers burned areas and wetlands.
Nesting: does not nest in Ohio.
Feeding: flycatches insects from a perch, preying mostly on bees, wasps and flying ants.

Voice: *Male:* chipper and lively *quick-three-beers!* with the 2nd note highest in pitch; descending *pip-pip-pip* when excited.
Similar Species: *Other flycatchers* (pp. 193–201): smaller heads and bills; lack prominent olive "vest" and white tufts on sides of tail.
Best Sites: open canopy of large trees with prominent snags for perches; Green Lawn Cemetery; upper end of Hoover Reservoir; Oak Openings MP; Holden Arboretum.

J F M A M J J A S O N D

EASTERN WOOD-PEWEE

Contopus virens

Our most common and widespread woodland flycatcher, the Eastern Wood-Pewee breeds in every Ohio county. It is readily detected by listening for its easily learned *peee-ah-weeee* song, which is repeated throughout the day. • Like their larger and more conspicuous relative the Olive-sided Flycatcher, pewees feed almost exclusively on flying insects, which they capture by sallying out from perches. But pewees are much less conspicuous, normally foraging from shaded, hidden, midcanopy sites. • The Eastern Wood-Pewee has declined significantly over the past several decades, but remains a common species, in part because of its adaptability and its tolerance of forest fragmentation. • The biology of Eastern Wood-Pewees is poorly known, as they construct their nests on the outer reaches of smaller branches high in trees—locations virtually inaccessible to researchers. Behavioral studies on their South American wintering grounds are also difficult owing to the overlapping presence of the Western Wood-Pewee (*C. sordidulus*) and the Tropical Pewee (*C. cinereus*), nonsinging winter birds that are extremely hard to differentiate.

ID: olive gray to olive brown upperparts; 2 narrow, white wing bars; whitish throat; gray breast and sides; whitish or pale yellow belly, flanks and undertail coverts; dark upper mandible; dull, yellow orange base to lower mandible; no eye ring.
Size: *L* 6–6½ in; *W* 10 in.
Habitat: open mixed and deciduous woodlands with a sparse understory, especially woodland openings and edges; larger tracts of riparian forests.
Nesting: in a deciduous tree well away from the trunk; open cup of grass, plant fibers and lichen is bound with spider webs; female incubates 3 darkly blotched, whitish eggs for 12–13 days.
Feeding: flycatches insects from a perch or gleans insects from foliage, especially while hovering.
Voice: clear, slow, plaintive *pee-ah-wee,* with 2nd note lower, followed by a down-slurred *pee-oh,* given with or without intermittent pauses; also gives flat *chip.*
Similar Species: *Olive-sided Flycatcher* (p. 192): larger; white rump tufts; olive gray "vest"; lacks conspicuous white wing bars. *Eastern Phoebe* (p. 199): all-dark bill; lacks conspicuous white wing bars; often pumps its tail. *Empidonax flycatchers* (pp. 194–98): smaller; more conspicuous wing bars; eye rings.
Best Sites: any decent-sized woodlot, forest or treed suburban park.

J F M A M J J A S O N D

YELLOW-BELLIED FLYCATCHER

Empidonax flaviventris

The small flycatchers in the genus *Empidonax* are notoriously difficult to separate visually, but the Yellow-bellied Flycatcher is an exception, at least in spring. Its obvious, yellowish olive underparts make this species more colorful than others in the genus. Even so, the Yellow-bellied is easily overlooked because it returns late, usually mid-May into early June, after the understory shrubs that it frequents have leafed out, making detection difficult. It is also an uncommon bird here, and finding more than two or three in a day is unusual—an exceptional sighting was the 20 found on May 25, 1989, near Toledo. They return by mid-August, and most have passed through by mid-September. • Yellow-bellied Flycatchers breed well north of us, in the extensive Canadian boreal forest. There, they occupy a botanical wonderland of boggy, coniferous woods underlain with a carpet of sphagnum moss and festooned with such stunning plants as goldthread, bunchberry and twinflower. However, birders wishing to visit this species' summer haunts are usually inhibited by the hordes of mosquitoes and black flies.

ID: olive green upperparts; 2 whitish wing bars; yellowish eye ring; white throat; yellow underparts; pale olive breast.
Size: *L* 5–6 in; *W* 8 in.

Habitat: coniferous bogs and fens; shady, spruce and pine forests with a dense shrub understory.

Nesting: does not nest in Ohio.

Feeding: flycatches for insects at low to middle levels of the forest; also gleans vegetation for larval and adult invertebrates while hovering.

Voice: calls include a chipper *pe-wheep, preee, pur-wee* or *killik*. Male: song is a soft *che-luk* or *che-lek,* with lower-pitched 2nd syllable.

Similar Species: *Acadian* (p. 195), *Willow* (p. 197), *Alder* (p. 196) and *Least* (p. 198) *flycatchers:* white eye rings; all but Acadian have browner upperparts; all lack extensive yellow wash from throat to belly; different songs.

Best Sites: Magee Marsh bird trail; along L. Erie in moist areas with thick undergrowth.

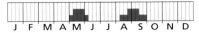

J F M A M J J A S O N D

ACADIAN FLYCATCHER

Empidonax virescens

lycatchers in the genus *Empidonax* are some of the most difficult birds to identify—unless they are singing. The five regularly occurring Ohio species look
similar, but their voices are quite different. The Acadian
Flycatcher has an explosive *peet-sa* song. • Although the
Acadian's song is often heard and easily recognized,
much less is known about its territorial dusk and
dawn songs. These vocalizations are long and elaborate series of soft *chips* and other notes, most often
delivered in the predawn gloom. One bird was
timed singing continuously for 45 minutes.
• For breeding birds, habitat is an excellent
clue to identity. Acadians are denizens of
mature woodlands, often along creeks or
streams, which is a habitat not shared
in Ohio by any other "Empid." • Fall
migrant Acadians, which
are rarely detected, are
our only *Empidonax* to
molt before migrating south, so
a molting bird in fall is almost
certainly this species.

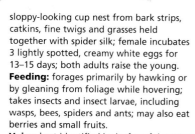

sloppy-looking cup nest from bark strips,
catkins, fine twigs and grasses held
together with spider silk; female incubates
3 lightly spotted, creamy white eggs for
13–15 days; both adults raise the young.
Feeding: forages primarily by hawking or
by gleaning from foliage while hovering;
takes insects and insect larvae, including
wasps, bees, spiders and ants; may also eat
berries and small fruits.
Voice: best identified by its forceful *peet-sa*
song; call is a softer *peet;* may issue a loud
ti-ti-ti-ti-ti during the breeding season.
Similar Species: Other Empidonax *flycatchers* (pp. 194–98): best identified by
voice; not as green above or as white below;
not found breeding in mature woodlands.
Eastern Wood-Pewee (p. 193): larger; less
prominent eye ring; dusky gray breast
divided in the middle; calls are different.
Best Sites: large, mature forests statewide;
Black Hand Gorge SNP; Mohican SF;
Hueston Woods SP.

ID: narrow, yellowish
eye ring; 2 buff to
yellowish wing bars;
large bill has dark
upper mandible and
pinkish yellow lower
mandible; white
throat; faint olive
yellow breast; yellow belly and undertail
coverts; olive green upperparts; very long
primaries.
Size: *L* 5½–6 in; *W* 9 in.
Habitat: fairly mature deciduous woodlands, riparian woodlands and wooded
swamps.
Nesting: up to 20 ft above the ground in a
beech or maple tree; female builds a loose,

J	F	M	A	M	J	J	A	S	O	N	D

195

ALDER FLYCATCHER

Empidonax alnorum

This bird illustrates the fact that ornithologists still have things to learn about North American bird life; the Alder Flycatcher wasn't described as a species until 1973. Prior to that, it was called "Traill's Flycatcher" and was considered part of a complex that included the very similar Willow Flycatcher. • The Alder Flycatcher is not safely separable from the Willow Flycatcher visually, but fortunately their calls aren't too hard to tell apart. Alders give a burry *fee-bee-o*, though the third syllable can be hard to hear; hence the song sometimes sounds like *free beer!* • Alders are much scarcer here since they mostly breed to the north of us and local populations are generally found only in northwestern and northeastern Ohio. Migrants are not often detected owing to the difficulty of identifying nonsinging birds. • South of the northern tier of counties, "Traill's" types are likely to be Willow Flycatchers.

ID: olive brown upperparts; 2 dull white to buff wing bars; faint, whitish eye ring; dark upper mandible; orange lower mandible; long tail; white throat; pale olive breast; pale yellowish belly.
Size: *L* 5½–6 in; *W* 8½ in.
Habitat: wet, shrubby thickets, often dominated by alder, willow and dogwood, bordering lakes or streams or in wetlands.
Nesting: in a dense bush or shrub, usually less than 3 ft above the ground; small cup nest is loosely woven of grass and other plant materials; female incubates 3–4 darkly spotted, white eggs for 12–14 days; both adults feed the young.

Feeding: flycatches from a perch for beetles, bees, wasps and other flying insects; also eats berries and occasionally seeds.
Voice: snappy *fee-bee-o* or *free beer;* call is a dry, flat *pit* or *pip.*
Similar Species: *Other* Empidonax *flycatchers* (pp. 194–98): smaller; upperparts not as brown; longer wings; different eye ring; best identified by voice.
Best Sites: small numbers in wet, shrubby thickets in northern Ohio. *Breeding:* Irwin Prairie SNP; Gott Fen SNP.

| J | F | M | A | M | J | J | A | S | O | N | D |

WILLOW FLYCATCHER

Empidonax traillii

The Willow Flycatcher is our most common *Empidonax* of open habitats, breeding in every county. It is easily found and identified by its explosive *fitz-bew!* song, heard in a variety of dry to wet successional habitats dominated by brushy thickets. It often gives its call note, a soft, liquid *whit*, which is distinctly different from the very similar-looking Alder Flycatcher's flat, dry *pip* or *pit* call notes. • Although the Willow Flycatcher certainly benefited from the clearing of Ohio's forests and is much more widespread today than prior to European settlement, another less desirable species, the Brown-headed Cowbird, has also profited from these changes and is now abundant. An Ohio study found that an amazing 80 percent of Willow Flycatcher nests were parasitized by cowbirds. • There are currently four distinct sub-species of Willow Flycatcher, two of which breed in Ohio—subspecies *traillii* and *campestris*.

ID: olive brown upperparts; 2 whitish wing bars; no eye ring; white throat; yellowish belly; pale olive breast.
Size: *L* 5½–6 in; *W* 8½ in.

Habitat: shrubby successional areas in old fields, along stream and wetland margins, and reclaimed strip mines.

Nesting: in a dense shrub, usually 3–7 ft above the ground; female builds an open cup nest with grass, bark strips and plant fibers and lines it with down; female incubates 3–4 brown-spotted, whitish to pale buff eggs for 12–15 days.

Feeding: flycatches insects; also gleans insects from vegetation, usually while hovering.

Voice: call is a quick *whit. Male:* song is a quick, sneezy *fitz-bew* that drops off at the end, issued up to 30 times per minute.

Similar Species: *Eastern Wood-Pewee* (p. 193): larger; lacks eye ring and conspicuous wing bars. *Alder Flycatcher* (p. 196): song is *fee-bee-o;* usually found in wetter areas. *Least Flycatcher* (p. 198): song is a clear *che-bek;* bolder white eye ring; greener upperparts; pale gray white underparts. *Acadian Flycatcher* (p. 195): song is a forceful *peet-sa;* yellowish eye ring; greener upperparts; yellower underparts. *Yellow-bellied Flycatcher* (p. 194): song is a liquid *che-lek;* yellowish eye ring; greener upperparts; yellower underparts.

Best Sites: brushy habitats statewide; reclaimed strip mines such as Woodbury WA; Gilmore Ponds Preserve; Ottawa NWR; Cowan Lake SP.

J F M A M J J A S O N D

LEAST FLYCATCHER

Empidonax minimus

The genus name *Empidonax* means "King of the Gnats," an appropriate name for these small flycatchers that feed largely on tiny insects. • With its drab, undistinguished plumage features, the Least Flycatcher is quite difficult to distinguish. Fortunately, its *che-bek* call is easily learned and diagnostic. At least in spring and summer, this songbird vocalizes more or less constantly, delivering up to 60 songs a minute. • This is our most common migrant *Empidonax* flycatcher; about 300 individuals were detected at Green Lawn Cemetery on May 14, 1981. • Least Flycatchers are much scarcer as breeding birds, but numbers are increasing. Most numerous in the extreme northeastern corner of the state, they should be watched for in shrubby, successional habitats throughout northern Ohio. • These territorial birds are quite aggressive when nesting, driving off virtually all small and medium-sized species of birds. This aggression is thought to account for the low rate of parasitism by Brown-headed Cowbirds.

ID: grayish olive upperparts; 2 bold, white wing bars; bold, white eye ring; fairly long, narrow tail; mostly dark bill has yellow orange lower base; white throat; gray breast; gray white to yellowish belly and undertail coverts.

Size: *L* 4½–6 in; *W* 7½ in.

Habitat: open deciduous or mixed woodlands; forest openings and edges; often in second-growth woodlands; breeders prefer brushy thickets, often in or around wetlands or water.

Nesting: in a small tree or shrub, often against the trunk; female builds a small cup nest with plant fibers and bark and lines it with fine materials; female incubates 4 creamy white eggs for 13–15 days; both adults feed the young.

Feeding: flycatches insects and gleans trees and shrubs for insects while hovering; may also eat some fruit and seeds.

Voice: constantly repeated, dry *che-bek che-bek*.

Similar Species: *Alder Flycatcher* (p. 196): song is *fee-bee-o;* faint eye ring; usually found in wetter areas. *Willow Flycatcher* (p. 197): song is an explosive *fitz-bew;* greener upperparts; yellower underparts; lacks eye ring. *Acadian Flycatcher* (p. 195): song is a forceful *peet-sa;* yellowish eye ring; greener upperparts; yellower underparts. *Yellow-bellied Flycatcher* (p. 194): song is a liquid *che-lek;* yellowish eye ring; greener upperparts; yellower underparts.

Best Sites: *Spring migration:* Green Lawn Cemetery; Magee Marsh bird trail; Sheldon Marsh SNP. *Fall migration:* common, but more difficult to identify as is less apt to vocalize.

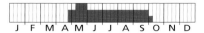

J F M A M J J A S O N D

EASTERN PHOEBE

Sayornis phoebe

Our hardiest flycatchers, Eastern Phoebes return by early March and routinely overwinter in southern Ohio, at least in mild years. Although rather drab, they are easily recognized by their frequent, emphatic *fee-bee* song and their chronic tail pumping. • Eastern Phoebes are far more numerous today than prior to European settlement, when breeding birds would have been confined to cliff faces. Although some birds still nest on cliff ledges, particularly in the Hocking Hills, phoebes have adapted well to human-made structures and now commonly nest on the eaves of buildings, in culverts and under bridges. They are common breeders in southern and eastern Ohio, but are much less common in the heavily agricultural western and northwestern regions. • At favored nest sites, phoebes will renovate the nest and use it year after year. • Eastern Phoebes are frequently found around water, in part because more insects are found near water in cold weather. Phoebes don't hesitate to turn to fruit and seeds for sustenance during cold snaps.

breeding

ID: gray brown upperparts; white underparts with gray wash on breast and sides; belly may be washed with yellow in fall; no eye ring; no obvious wing bars; all-black bill; dark legs; frequently pumps its tail.

Size: *L* 6½–7 in; *W* 10½ in.

Habitat: open deciduous woodlands and forest edges and clearings; usually near water.

Nesting: under the ledge of a building, culvert, bridge or other structure; rarely on a cliff; cup-shaped mud nest is lined with moss, grass, fur and feathers; female incubates 4–5 sparsely spotted, white eggs for 14–16 days; both adults feed the young.

Feeding: flycatches flies, wasps, flying beetles, grasshoppers, mayflies and other insects; occasionally plucks aquatic invertebrates and small fish from the water's surface; will eat fruit and seeds in winter.

Voice: hearty, snappy *fee-bee,* delivered frequently; call is a sharp *chip.*

Similar Species: *Eastern Wood-Pewee* (p. 193): pale wing bars; bicolored bill; does not pump its tail. *Olive-sided Flycatcher* (p. 192): larger; dark "vest"; white, fluffy patches border rump. Empidonax *flycatchers* (pp. 194–98): most have eye ring and conspicuous wing bars.

Best Sites: Hocking Hills, Cantwell Cliffs SP; Conkle's Hollow SNP; nesting pairs at Green Lawn Cemetery (on Howald mausoleum on north side of pond); Blackhand Gorge SNP (in abandoned railroad tunnel on north side of Licking R.).

J F M A M J J A S O N D

GREAT CRESTED FLYCATCHER

Myiarchus crinitus

The Great Crested Flycatcher is one of our largest and showiest flycatchers, but it can be surprisingly easy to overlook as it forages high in the dense canopies of tall trees—until it calls, that is, when it unleashes a very loud *wheee-eep!* that carries great distances. Tracking down the singer will reveal a stout, large-billed and long-tailed flycatcher bedecked in hues of cinnamon, yellow, gray and brown. • Great Crested Flycatchers are probably more common today than prior to European settlement, as they are generally not inhabitants of mature, unbroken forests; rather, they prefer semi-open woodlots, younger woods with scattered large trees and large cemeteries with numerous big trees. • These are the only eastern North American flycatchers that nest in cavities. They are known to nest in every Ohio county and are fairly common statewide wherever suitable habitat is available, but are most frequent in southern and eastern Ohio. • Great Crested Flycatchers can be quite long-lived—the record, documented by banding, was of a bird over 14 years old.

ID: bright yellow belly and undertail coverts; gray throat and upper breast; reddish brown tail; peaked, "crested" head; dark, olive brown upperparts; heavy, black bill.

Size: *L* 8–9 in; *W* 13 in.

Habitat: deciduous and mixed woodlands and forests, usually near openings or edges.

Nesting: in a tree cavity, nest box or other artificial cavity; nest is lined with grass, bark strips and feathers; often hangs a shed snakeskin or plastic wrap at the entrance hole; female incubates 5 creamy white to pale buff eggs, marked with lavender, olive and brown, for 13–15 days.

Feeding: usually in the upper branches of deciduous trees, where it flycatches for flying insects; may also glean caterpillars and occasionally fruit.

Voice: loud, whistled *wheep!* and a rolling *prrrrreet!*

Similar Species: distinctive. *Ash-throated Flycatcher:* vagrant in adjoining states, not yet seen in Ohio; more washed-out colors with pale yellow belly. Any *Myiarchus* flycatcher seen after mid-October should be carefully scrutinized.

Best Sites: woodlots, treed parklands and forest edges statewide; Edge of Appalachia Preserve; Scioto Trail SF; Oak Openings MP.

J F M A M J J A S O N D

EASTERN KINGBIRD

Tyrannus tyrannus

astern Kingbirds are aptly named—both their genus and species names mean "tyrant"—and when on their nesting territories, both sexes will attack and drive off any perceived threat that invades their turf, regardless of size. Crows, hawks, squirrels and even Bald Eagles will be relentlessly harassed, dive-bombed and even pecked. Occasionally this strategy backfires, as was the case when a kingbird dared to alight on a flying Red-tailed Hawk's back, the better to jab the hapless raptor, and was picked off from above by an American Kestrel! • Kingbirds become highly social in migration, forming large flocks that are conspicuous by their diurnal migrations. On August 9, 1988, 147 birds were recorded in Portage County; however, that pales in comparison to a Florida flock seen on August 28, 1964, that was estimated at 500,000 birds. • Eastern Kingbirds are found in every Ohio county, where they inhabit open-country fields with scattered trees, thickets and fencerows.

ID: dark gray to black upperparts; white underparts; white-tipped tail; black bill; small head crest; thin, orange red crown (rarely seen); black legs; no eye ring.

Size: *L* 8½ in; *W* 15 in.

Habitat: rural fields with scattered trees or hedgerows, clearings in fragmented forests, open roadsides, burned areas and near human settlements.

Nesting: on a horizontal tree or shrub limb; also on a standing stump or an upturned tree root; pair builds a cup nest of weeds, twigs and grass and lines it with root fibers, fine grass and fur; female incubates 3–4 darkly blotched, white to pinkish white eggs for 14–18 days.

Feeding: flycatches aerial insects; infrequently eats berries.

Voice: call is a quick, loud, chattering *kit-kit-kitter-kitter;* also a buzzy *dzee-dzee-dzee.*

Similar Species: *Western Kingbird* (p. 338): accidental; yellow belly and undertail coverts; grayer back; white-edged outer tail feathers. *Tree Swallow* (p. 214): iridescent, dark blue back; more streamlined body; smaller bill; lacks white tip on tail. *Olive-sided Flycatcher* (p. 192) and *Eastern Wood-Pewee* (p. 193): all-white underparts; lack white tip on tail.

Best Sites: Gilmore Ponds Preserve; Killdeer Plains WA; La Due Reservoir.

NORTHERN SHRIKE

Lanius excubitor

Often seen perched atop a small tree or shrub, the Northern Shrike looks refined and peaceful as it casually surveys its winter domain. Nothing could be further from the truth; this fascinating songbird is one of the most vicious predators in the avian world. *Lanius* means "butcher," and the shrike is often called "Butcher Bird" with good reason. The Northern Shrike is highly carnivorous, its winter prey being small rodents and other songbirds. Fearless and known to attack birds up to the size of Blue Jays and Rock Pigeons, the shrike normally dispatches its victims by using its powerful bill to sever the cervical vertebrae. The mouse or bird is often taken to a favored thorn, where it is unceremoniously impaled—like a butcher hanging meat on a hook—so the shrike can leisurely shred it apart. • Northern Shrikes breed in the arctic tundra, and are relatively rare visitors to Ohio, with only a handful of reports most years. Occasionally, as with arctic birds of prey, they stage southward irruptions. Most sightings are from the tier of counties bordering Lake Erie, but the species occurs occasionally southward into Holmes County and Killdeer Plains Wildlife Area.

ID: black tail and wings; pale gray upperparts; finely barred, light underparts; black "mask" does not extend above hooked bill. *Immature:* faint "mask"; light brown upperparts; brown or gray barring on underparts. *In flight:* white wing patches; white-edged tail.

Size: *L* 10 in; *W* 14½ in.

Habitat: open country, including fields, shrubby areas, forest clearings and roadsides.

Nesting: does not nest in Ohio.

Feeding: swoops down on prey from a perch or chases prey through the air; eats small birds, shrews and mice; may impale prey on a thorn or barb for later consumption.

Voice: usually silent; infrequently gives a long, grating laugh: *raa-raa-raa-raa.*

Similar Species: generally unmistakable. *Loggerhead Shrike* (p. 338): thicker "mask" extends above bill onto forehead; shorter, thinner bill with less prominent hook; lacks faint barring on underparts. *Northern Mockingbird* (p. 241): slimmer; longer tail; much thinner bill; lacks black "mask."

Best Sites: Killdeer Plains WA; Magee Marsh WA; Maumee Bay SP; Mentor Headlands area.

J F M A M J J A S O N D

WHITE-EYED VIREO

Vireo griseus

The White-eyed Vireo is best found by listening for its song, a loud, emphatic *pick up the beer, check!* Partial to skulking in thick undergrowth, this species can be frustratingly difficult to view in spite of its persistent calls from just inside the wall of shrubbery. • White-eyed Vireos have increased greatly in Ohio in the last 100 years, as they are birds of scruffy successional growth, not mature woodlands. They are most numerous in southern and eastern regions and scarcest in western and northeastern areas of the state. • Young clear-cuts, reverting fields and overgrown pastures provide good habitat for White-eyed Vireos and for a suite of species that includes the Blue-winged Warbler, Yellow-breasted Chat and Prairie Warbler. However, because such habitat is often considered "weedy" and transitional, little thought is given to its protection or management. • Brown-headed Cowbirds parasitize White-eyed Vireos heavily; perhaps 50 percent of all nests are affected.

ID: yellow "spectacles"; olive gray upperparts; white underparts; yellow sides and flanks; 2 whitish wing bars; dark wings and tail; pale eyes.

Size: *L* 5 in; *W* 7½ in.

Habitat: dense shrubby undergrowth and thickets, overgrown fields, young second-growth woodlands, woodland clearings, reclaimed strip mines and woodlot edges.

Nesting: in a deciduous shrub or small tree; cup nest hangs from a horizontal fork; pair incubates 4 lightly speckled, white eggs for 13–15 days; both adults feed the young.

Feeding: gleans insects from branches and foliage during very active foraging; often hovers while gleaning.

Voice: loud, snappy, song of 3–9 notes, usually beginning and ending with "chick": *chick-ticha-wheeyou, pick up the beer, check!;* occasionally mimics other birds such as Whip-poor-will.

Similar Species: *Bell's Vireo* (p. 204): rare; shares same habitat; dark eyes; white eye ring; faint yellowish throat; thicker bill; much duller underparts. *Other vireos* (pp. 205–09): do not normally share the same habitat.

Best Sites: Clear Creek MP; Crown City WA; Edge of Appalachia Preserve; Indian Creek WA.

J F M A M J J A S O N D

BELL'S VIREO
Vireo bellii

B ell's Vireo is one of our rarest breeding birds—only a few pairs nest in Ohio annually. Its habitat preference is similar to that of the more common White-eyed Vireo, and sometimes the two occur in close proximity. The first Ohio Bell's Vireo was not recorded until 1962, reflecting this species' recent eastward expansion. Now, a few are reported each year, but the number of breeding pairs seems to be dwindling. • Bell's Vireo is devilishly difficult to view, as it sings and forages from the densest of thickets and is loathe to reveal itself. Fortunately, its loud, sputtering song helps locate it. • A relatively new threat to our few breeders is the invasion of exotic shrubs such as bush honeysuckle and glossy buckthorn. The dominance of these non-natives likely reduces the viability of Bell's Vireo habitat. • Virtually all Ohio breeding records correlate with the distribution of former prairie tracts, suggesting that this species has long been present in very small numbers but has been overlooked.

ID: gray to green upperparts; white to yellow underparts; 2 white wing bars (upper bar is usually faint if present); whitish "eyebrow"; whitish eye ring or lores or both. *Eastern (Midwest) form:* olive green upperparts; yellow underparts.
Size: L 4½–5 in; W 7 in.
Habitat: brushy fields and second-growth scrub; almost all known nesting sites are in areas of original prairie.
Nesting: small hanging cup of woven vegetation is suspended from a horizontal shrub branch, usually within 6 ft of the ground; pair incubates 3–5 darkly dotted, white eggs for about 14 days; both adults feed the young.

Feeding: insects are gleaned from foliage and caught by hawking or hovering; also may eat spiders and very few berries.
Voice: song is a rapid, nonmusical series of harsh notes: *cheedle cheedle cheedle cheew, cheedle cheedle cheedle cheee,* increasing in volume and often ending with an upward or downward inflection on the last note.
Similar Species: *Ruby-crowned Kinglet* (p. 231): 2 distinct, white wing bars; bold, broken eye ring; sings 3-part, descending phrases. *White-eyed Vireo* (p. 203): immature is stockier, with more prominent "spectacles," and lacks thin, dark eye line.
Best Sites: Buck Creek SP; nested at Resthaven WA; also nested at Irwin Prairie SNP, but glossy buckthorn infestations may have eliminated it there.

J F M A M J J A S O N D

YELLOW-THROATED VIREO

Vireo flavifrons

The canopy-dwelling Yellow-throated Vireo is Ohio's flashiest member of the genus, sporting a brilliant yellow throat and bright yellow "spectacles." • Birders often have to work for a look at this species, because it feeds slowly and deliberately high in the dense leafy tops of mature trees. Although somewhat similar in habitat requirements to the much more common Red-eyed Vireo, the Yellow-throated Vireo prefers more fragmented woodlands, particularly where tall, scattered red oaks and white oaks are found. • As butterflies and moths *(lepidoptera)* constitute a major part of its diet, spraying for the invasive gypsy moth may have a detrimental impact on the Yellow-throated Vireo. If the chemicals used are nonselective and also kill other native species of *lepidopterans*, a major food source for the Yellow-throated Vireo will be eliminated. • Yellow-throated Vireos are most closely related to Blue-headed Vireos and may occasionally hybridize with them.

ID: bright yellow "spectacles," chin, throat and breast; olive upperparts, except for gray rump and dark wings and tail; 2 white wing bars; white belly and undertail coverts.
Size: *L* 5½ in; *W* 9½ in.
Habitat: mature deciduous woodlands with minimal understory.
Nesting: in a deciduous tree; pair builds a hanging cup nest in the fork of a horizontal branch; pair incubates 4 darkly spotted, creamy white to pinkish eggs for 14–15 days; each parent takes on guardianship of half the fledged young.
Feeding: forages by inspecting branches and foliage in the upper canopy; eats mostly insects; also eats seasonally available berries.

Voice: song is a slowly repeated series of hoarse phrases with long pauses in between: *ahweeo, eeoway, away;* has distinctive raspy or burry quality; calls include a throaty *heh heh heh.*
Similar Species: easily distinguished from other vireos. *Pine Warbler* (p. 261): olive yellow rump; thinner bill; faint, darkish streaking along sides; yellow belly; faint "spectacles." *Yellow-breasted Chat* (p. 281): larger and stockier; white "spectacles"; lacks wing bars; generally occupies different habitat and gives different calls.
Best Sites: found in suitable habitat statewide; probably nests in every county; woods around upper end of Alum Creek Reservoir; Mohican SF; Oak Openings MP.

J F M A M J J A S O N D

205

BLUE-HEADED VIREO

Vireo solitarius

In 1997, the American Ornithologist's Union (AOU) split the former "Solitary Vireo" into three species: the western Cassin's Vireo *(V. cassinii)* and Plumbeous Vireo *(V. plumbeus)* and the eastern Blue-headed Vireo. • The first of our vireos to return in spring, Blue-headed Vireo breeders are on territory by early to mid-April. They are very localized nesters that require extensive hemlock gorges, though some birds have recently begun using older pine plantings in the Oak Openings and in northeast Ohio. Blue-headed Vireos are part of a suite of birds that, in Ohio, are obligate hemlock species. These are northern species with disjunct populations here. A well-developed hemlock community may harbor Blue-headed Vireos, Hermit Thrushes and Black-throated Green, Canada and Magnolia warblers—all rare hemlock breeders. Continued northward expansion into Ohio of the woolly adelgid, an insect that infests and kills hemlocks, could be devastating to Ohio's hemlock birds.

ID: white "spectacles"; blue gray head; 2 white wing bars; olive green upperparts; white underparts; yellow sides and flanks; yellow highlights on dark wings and tail; stout bill; dark legs.
Size: *L* 5–6 in; *W* 9½ in.
Habitat: primarily hemlock gorges; occasionally pine plantations; migrants use all manner of woodlands.
Nesting: in a horizontal fork in a coniferous tree or tall shrub; hanging, basketlike cup nest is made of grass, roots, plant down, spider silk and cocoons; pair incubates 3–5 lightly spotted, whitish eggs for 12–14 days.
Feeding: gleans branches for insects; frequently hovers to pluck insects from vegetation.

Voice: *churr* call. *Male:* slow, purposeful, slurred, robinlike notes with moderate pauses in between: *chu-wee, taweeto, toowip, cheerio, teeyay;* song has shorter, higher-pitched phrases than other vireos.
Similar Species: *White-eyed Vireo* (p. 203): yellow "spectacles"; light-colored eyes. *Yellow-throated Vireo* (p. 205): yellow "spectacles" and throat.
Best Sites: migrants are widespread; breeders are localized. *Summer:* Clear Creek MP; Conkle's Hollow SNP; Mohican SF; Stebbins Gulch within Holden Arboretum.

J F M A M J J A S O N D

WARBLING VIREO

Vireo gilvus

The Warbling Vireo is common in Ohio and breeds in every county. You might never know it, though, because this species is quite hard to detect as it sluggishly forages high in the leafy canopies of tall trees. Fortunately, it is an incessant singer, giving its pleasing, warbling song throughout even the hottest summer days. Even when the general location of the songster has been pinpointed, actually spotting the bird can take a long time and then the visual reward is not that great. The Warbling Vireo is the drabbest of the vireos, and is quite a contrast to the flashy Baltimore Oriole, a common associate. • The Warbling Vireo is strongly associated with the eastern cottonwood tree, and thus is most commonly found along streams and lakes. • This bird is the "motor mouth" of the bird world—a single bird might sing its song 4000 times a day! It will even sing while sitting on the nest, incubating eggs. One mnemonic translation of its song is *if I sees you, I will seize you and I'll squeeze you till you squirt.*

breeding

Size: *L* 5–5½ in; *W* 8½ in.
Habitat: primarily along lakes and streams with tall cottonwoods, sometimes in parks and cemeteries if large, scattered trees are present.
Nesting: in a horizontal fork in a deciduous tree or shrub; hanging, basketlike cup nest is made of grass, roots, plant down, spider silk and a few feathers; pair incubates 4 darkly speckled, white eggs for 12–14 days.

ID: partial, dark eye line borders white "eyebrow"; olive gray upperparts; greenish flanks; white to pale gray underparts; gray crown; no wing bars.

Feeding: gleans foliage for insects, occasionally while hovering.
Voice: song is a musical warble of slurred whistles; call is a harsh nasal *eeah.*
Similar Species: *Philadelphia Vireo* (p. 208): yellow breast, sides and flanks; full, dark eye line borders white "eyebrow." *Red-eyed Vireo* (p. 209): black eye line extends to bill; blue gray crown; red eyes. *Tennessee Warbler* (p. 248): blue gray crown and nape; olive green back; slimmer bill.
Best Sites: riparian areas or lakeshores; Charles Mills L.; Cowan Lake SP; Grand Lake St. Marys; cottonwoods around parking lots at Magee Marsh bird trail.

J F M A M J J A S O N D

PHILADELPHIA VIREO

Vireo philadelphicus

The Philadelphia Vireo and the Blue-headed, Yellow-throated and Red-eyed vireos sing classic vireo songs of short, somewhat robinlike phrases interspersed with pauses. Initially, these species can be hard to tell apart by song, but with a bit of practice the latter three separate easily. However, the Philadelphia Vireo, which nests far to the north of Ohio and only occurs here as an uncommon migrant, sounds quite similar to the Red-eyed Vireo, even to experienced birders. • Partly owing to confusion with Red-eyed Vireo songs, but also because of the relative inaccessibility of its northern, early successional, deciduous woodland habitats, the Philadelphia Vireo remains one of North America's least studied and most poorly known birds. • Although named for the city in which the first specimen was collected by John Cassin in 1851, the Philadelphia Vireo only occurs as a fairly rare migrant there.

breeding

ID: gray "cap"; full, dark eye line bordered by bold, white "eyebrow"; dark, olive green upperparts; pale yellow breast, sides and flanks; white belly (underparts may be completely yellow in fall); robust bill; pale eyes.
Size: *L* 4½–5 in; *W* 8 in.
Habitat: occurs in a variety of woodland types, but prefers foraging at lower elevations than most other vireos, often feeding in woodland understory vegetation.
Nesting: does not nest in Ohio.
Feeding: gleans vegetation for insects; frequently hovers to glean food from foliage.

Voice: song is very similar to that of Red-eyed Vireo, but is often slightly slower, higher pitched and not as variable: *look-up way-up tree-top see-me;* call also similar to Red-eyed Vireo's softly nasal call, but is often broken into distinct, multiple syllables.
Similar Species: *Warbling Vireo* (p. 207): yellow on flanks only in fall bird; less distinct eye line; smoky gray, rather than greenish, upperparts. *Red-eyed Vireo* (p. 209): black-bordered, blue gray "cap"; red eyes; lacks yellow breast.
Best Sites: *Spring migration:* Green Lawn Cemetery; Magee Marsh bird trail; Mentor Headlands.

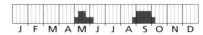

J F M A M J J A S O N D

RED-EYED VIREO

Vireo olivaceus

Prior to European settlement, when Ohio was 95 percent forested, the Red-eyed Vireo was probably the most abundant bird of any species breeding in the state, discounting the extinct Passenger Pigeon *(Ectopistes migratorius)*. Today, it is still likely our most numerous neotropical nester, found in every county and wherever woodlands occur. A pair requires only about one acre of forest for their territory, so densities can be quite high—a daily tally exceeding 100 singing males can easily be detected in the extensive woodlands of southeastern Ohio. • Hearing a Red-eyed Vireo is not difficult; this bird can sing incessantly all day, even in the heat of midsummer, and the male delivers about 40 phrases per minute. Theoretically, a bird could sing over 30,000 phrases a day. Because of this nonstop chatter, someone—apparently with low regard for the clergy—long ago dubbed the Red-eyed Vireo "Preacher Bird." • Visual confirmation can take some effort, as this bird sluggishly forages high in the leafy canopy.

breeding

ID: dark eye line; white "eyebrow"; black-bordered, blue gray crown; olive green upperparts; olive "cheek"; white to pale gray underparts; may have yellow wash on sides, flanks and undertail coverts, especially in fall; red eyes; no wing bars.

Size: *L* 6 in; *W* 10 in.

Habitat: deciduous woodlands with a shrubby understory; large shade trees in urban parks and neighborhoods.

Nesting: in a horizontal fork in a deciduous tree or shrub; hanging, basketlike cup nest is made of grass, roots, spider silk and cocoons; female incubates 4 darkly spotted, white eggs for 11–14 days.

Feeding: gleans foliage for insects, especially caterpillars; often hovers; also eats berries.

Voice: call is a short, scolding *neeah*. *Male:* song is a continuous, variable, robinlike run of quick, short phrases, with distinct pauses in between: *Look-up, way-up, tree-top, see-me, here-I-am!*

Similar Species: *Philadelphia Vireo* (p. 208): yellow breast; lacks black border to blue gray "cap"; song is very similar but slightly higher pitched. *Warbling Vireo* (p. 207): duller overall; dusky eye line does not extend to bill; lacks black border on gray "cap." *Tennessee Warbler* (p. 248): blue gray "cap" and nape; olive green back; slimmer bill.

Best Sites: Shawnee SF; Cuyahoga Valley NRA; Mohican SF.

J F M A M J J A S O N D

BLUE JAY

Cyanocitta cristata

Blue Jays are a common, widely recognized species throughout Ohio, but are not always popular owing to their aggressiveness at feeders and penchant for nest robbing. Nevertheless, this is one of our most fascinating birds and possibly one of the most important ecologically. Blue Jays are great consumers of tree nuts—an individual might harvest several thousand acorns, beech nuts and the like in a single fall! The jay caches most of them by burying them in the soil, but a large percentage is lost or forgotten. Thus, the Blue Jay is an important vector for tree dispersal; some authorities have gone so far as to suggest that jays are responsible for the rapid northward expansion of oaks following the last glaciations. • Although many Blue Jays winter here, some are highly migratory and the spring migration along Lake Erie is one of our most fantastic birding spectacles. Huge numbers queue up along the shoreline in May. They fly in flocks along the shore, often disappearing to the west, then return heading east, then back again. In spite of their bravado, Blue Jays are afraid of flying over the lake, and eventually circle around it by following the shoreline.

ID: blue crest; black "necklace"; blue upperparts; white underparts; white bar and flecking on wings; dark bars and white corners on blue tail; black bill.

Size: *L* 11–12½ in; *W* 16 in.

Habitat: mixed deciduous forests, agricultural areas, scrubby fields and towns.

Nesting: in the crotch of a tree or tall shrub; pair builds a bulky stick nest and incubates 4–5 greenish, buff or pale blue eggs, spotted with gray and brown, for 16–18 days; birds are quiet and inconspicuous during breeding season.

Feeding: forages on the ground and among vegetation for nuts, berries, eggs, nestlings and birdseed; also eats insects and carrion.

Voice: noisy, screaming *jay-jay-jay;* nasal *queedle queedle queedle-queedle* sounds like a muted trumpet; often imitates various sounds, including calls of the Red-tailed Hawk and Red-shouldered Hawk.

Similar Species: none in Ohio.

Best Sites: almost anywhere; spectacular migrations in early to mid-May at places such as Magee Marsh bird trail and Maumee Bay SP.

J F M A M J J A S O N D

AMERICAN CROW

Corvus brachyrhynchos

I n recent years, the wary American Crow has become a much more common sight in urban neighborhoods. This shift in habitat reflects this bird's intelligence: the crow, which is legal game and hunted year round, has learned that not only is it safe from guns in town, but that an abundant supply of food is present. Cities provide a ready source of garbage, roadkills and open lawns with grubs and earthworms for this omnivorous bird. • Few nests are seen in neighborhoods, though a nest is probably almost always within view if you know where to look. • The American Crow is a common resident in all counties, and it occurs in large numbers as a migrant. Migrating flocks are most evident along Lake Erie in early spring—an estimated 10,000 were observed passing along the lakeshore at Cleveland in 1931, and over 2000 were seen at Mentor Headlands on March 11, 1990. Recently, West Nile virus is thought to have decreased numbers. • American Crows can form enormous winter roosts, sometimes numbering into the tens of thousands.

Habitat: ubiquitous.

Nesting: in a conifer, often a red cedar, or in any deciduous tree where the bulky nest can be hidden; large stick-and-branch nest is lined with fur and soft plant materials; female incubates 4–6 darkly blotched, gray green to blue green eggs, for about 18 days.

Feeding: opportunistic; feeds on carrion, small vertebrates, other birds' eggs and

ID: glossy, purple black plumage; square-shaped tail; black bill and legs; slim, sleek head and throat.

Size: *L* 17–21 in; *W* 3 ft.

nestlings, berries, seeds, invertebrates and human food waste.

Voice: distinctive, far-carrying, repeated *caw-caw-caw*.

Similar Species: *Common Raven:* accidental; much larger; heavier bill; wedge-shaped tail; call is a deep croak. *Fish Crow:* not yet found in Ohio, but could turn up along Ohio R.; call is a brief, very nasal *cah-ah*.

Best Sites: large movements in early spring along L. Erie; large roosts sometimes form in winter; Mansfield, Columbus and Springfield have hosted impressive congregations.

J F M A M J J A S O N D

HORNED LARK

Eremophila alpestris

Horned Larks are birds of open, barren, largely bird-free habitats, where they escape detection as their cryptic plumage blends with the ground. They are surprisingly common, however, nesting in every county and reaching peak abundance in heavily farmed regions. They are less common in heavily forested southeastern Ohio. • Stopping along a road bisecting large fields in early spring and listening for the rapid, musical, tinkling song will almost always reveal Horned Larks. The black tail, which contrasts with the pale back, is a good field mark. • Horned Larks nest quite early, usually incubating eggs by mid-March. This species is most evident in winter, when large flocks in the thousands can form and are often seen along roadsides, together with Lapland Longspurs and Snow Buntings. • The wide-ranging Horned Lark is divided into 21 subspecies; Ohio nesters are the prairie subspecies, *E. a. praticola.* Two other subspecies visit in winter, the nominate form *alpestris* and *hoyti*. The brownish back color varies among subspecies and generally correlates with the color of local soils.

ID: *Male:* small black "horns" (can be hard to see); black line under eye extends from bill to "cheek"; light yellow to white face; dull brown upperparts; black breast band; dark tail with white outer tail feathers; pale throat. *Female:* less distinctively patterned; duller plumage overall.
Size: *L* 7 in; *W* 12 in.
Habitat: open areas, including pastures, croplands, sparsely vegetated fields, weedy meadows and airfields.
Nesting: on the ground; in a shallow scrape lined with grass, plant fibers and roots; female chooses the nest site and incubates 3–4 brown-blotched, pale gray to greenish white eggs for 10–12 days.

Feeding: gleans the ground for seeds; feeds insects to its young during the breeding season.
Voice: song is a long series of tinkling, twittered whistles; call is a high weak *see-to*, often heard from birds flying overhead.
Similar Species: *Sparrows* (pp. 284–304), *Lapland Longspur* (p. 303) and *American Pipit* (p. 244): all lack distinctive facial pattern, "horns" and solid black breast band; Horned Larks also walk, while sparrows hop.
Best Sites: listen for distinctive song in agricultural country. *Spring migration:* large numbers at Maumee Bay SP in March; also nearby roads through farmland. *Winter:* big flocks forage along State Route 294 between Harpster and Little Sandusky, near Killdeer Plains WA.

J F M A M J J A S O N D

PURPLE MARTIN

Progne subis

For centuries, humans have been enamored by Purple Martins, North America's largest swallows, and have tried to entice these birds to nest nearby. Native Americans used to put hollowed-out gourds around their villages to lure martins; today nest boxes are commonly used by enthusiasts to attract them. In fact, probably all Purple Martins nesting east of the Mississippi River are doing so in artificial boxes. Nesting in "wild" sites—tree cavities—largely ceased by 1900. • Purple Martins prefer wide open country, almost always with a body of water nearby. By placing appropriate nest boxes, there is a good chance that martins will eventually establish residence. Care must be taken to prevent House Sparrows and European Starlings from commandeering nest boxes; these introduced species will readily displace martins. • Purple Martins form enormous roosts in early autumn before migrating to their wintering grounds in South America—over 700,000 birds were tallied on a 12-acre island in South Carolina in 1996. Closer to home in the 1960s and 1970s, up to 30,000 martins roosted at the former Ohio Penitentiary site in Columbus. Presumably, they were there by their own choice.

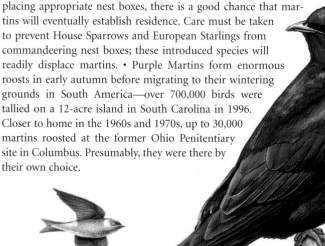

ID: glossy, dark blue body; slightly forked tail; pointed wings; small bill. *Male:* dark underparts. *Female:* sooty gray underparts.
Size: *L* 7–8 in; *W* 18 in.
Habitat: open habitats, often near water, including marshes, agricultural land and around lakes and reservoirs.
Nesting: communal; usually in an apartment-style birdhouse; nest materials include feathers, grass, mud and vegetation; female incubates 4–5 white eggs for 15–18 days.
Feeding: mostly while in flight; eats flies, ants, bugs, dragonflies and mosquitoes;

may also walk on the ground, taking insects and rarely berries.
Voice: rich, liquid, robinlike *pew-pew,* often heard in flight.
Similar Species: *Other swallows* (pp. 214–18): smaller; lighter plumage. *European Starling* (p. 243): longer bill (yellow in summer); lacks forked tail; faster wingbeats and flight.
Best Sites: migrants can appear almost anywhere; breeding birds are best sought around nest boxes; Magee Marsh WA; many landowners in Amish areas around Holmes Co. have established thriving colonies.

| J | F | M | A | M | J | J | A | S | O | N | D |

TREE SWALLOW

Tachycineta bicolor

After the Barn Swallow, this species is generally our most frequently encountered swallow, and certainly the hardiest. A few Tree Swallows routinely return to Ohio by late February, and they become common by late March. • It isn't uncommon to see large flocks that can number into the thousands in early spring and again in late summer and fall. Because this species requires extensive open country, almost always with a body of water close at hand, Tree Swallows are far more numerous today than prior to European settlement when Ohio was heavily forested. • The primary factor that restricts breeding populations is the availability of nest cavities. Tree Swallows prefer woodpecker holes in trees; competition for such nest sites is fierce and is probably what drives much of this bird's ecology. For instance, their very early arrival in spring is likely geared toward giving them an edge over other species that use cavities but don't return as early. • This species readily adapts to artificial nest boxes, and the proliferation of bluebird and Wood Duck boxes has greatly increased breeding numbers. Tree Swallows breed at least locally in every Ohio county and are common migrants statewide.

ID: iridescent, dark blue or green head and upperparts; white underparts; no white on "cheek"; dark rump; small bill; long, pointed wings; shallowly forked tail. *Female:* slightly duller. *Immature:* brown upperparts; white underparts.
Size: *L* 5½ in; *W* 14½ in.
Habitat: open areas, normally with water nearby; marshes, lakeshores, agricultural country and even urban areas with ponds and nest boxes.
Nesting: in a tree cavity or nest box lined with weeds, grass and feathers; female incubates 4–6 white eggs for up to 19 days.

Feeding: catches flies, midges, mosquitoes, beetles and ants on the wing; also takes stoneflies, mayflies and caddisflies over water; can subsist on berries and seeds during cold snaps.
Voice: song is a liquid, chattering twitter; alarm call is a metallic, buzzy *klweet*.
Similar Species: *Violet-green Swallow:* accidental; white face and sides of rump. *Northern Rough-winged Swallow* (p. 215): similar to dull female and immature Tree Swallow, but has less demarcation between white underparts and dark upperparts.
Best Sites: nesters and migrants abundant around marshes at Magee Marsh WA and Ottawa NWR; breeders along bluebird nest box trails.

J F M A M J J A S O N D

NORTHERN ROUGH-WINGED SWALLOW

Stelgidopteryx serripennis

Northern Rough-winged Swallows are associated with riparian habitats and are most often seen hawking for insects low over rivers and streams. • Favored nest sites are unused burrows of Bank Swallows, Belted Kingfishers and small mammals, usually in eroded banks buffering a stream. Increasingly, Rough-wingeds use drainage pipes and tiles for nest sites, and will sometimes nest far from water in crevices of cliffs and road cuts. • Northern Rough-winged Swallows closely resemble Bank Swallows both in appearance and habitat. In fact, their discovery was serendipitous; John James Audubon shot a few in 1819 thinking them to be Bank Swallows, and it wasn't until the birds were in hand did he realize that this drab swallow was another species. • Unlike our other swallows, Northern Rough-wingeds do not normally collect in large flocks in late summer and fall, but at least a few are often present in big, mixed, postbreeding swallow congregations. These flocks provide a good opportunity to study swallow flight characteristics; with practice the various species can be told from afar by differing flight styles. • This species is very widespread, nesting in every county, but is generally not abundant.

ID: brown upperparts; light brownish gray underparts; small bill; dark "cheek"; dark rump. *In flight:* long, pointed wings; notched tail.

Size: *L* 5½ in; *W* 14 in.

Habitat: foraging birds may be seen anywhere over open habitats, but are most often observed coursing low over streams and rivers, where they usually nest.

Nesting: occasionally in small colonies; at the end of a burrow lined with leaves and dry grass; sometimes reuses a kingfisher burrow, rodent burrow or other land crevice; mostly the female incubates 4–8 white eggs for 12–16 days.

Feeding: catches flying insects on the wing; occasionally eats insects from the ground; drinks on the wing.

Voice: often emits a raspy *chirt, chirt* in flight.

Similar Species: *Bank Swallow* (p. 216): smaller; dark breast band; more fluttery, butterfly-like flight. *Tree Swallow* (p. 214): dull female and immature are whiter below with cleaner demarcation between dark upperparts and white underparts.

Best Sites: along streams; observe from bridge over Licking R. at Blackhand Gorge SNP; mixed flocks in autumn migration at Killdeer Plains WA and Hebron Fish Hatchery.

215

BANK SWALLOW

Riparia riparia

One of the world's most widely distributed swallows, the Bank Swallow ranges throughout North America and across the Old World, where it is known as "Sand Martin." • Inextricably intertwined with the ecology of streams and rivers, the vast majority of Bank Swallows nest in soft, eroding banks and bluffs along waterways. Recently, they have also learned to use sand and gravel piles within quarries as nesting sites. Bank Swallows nest colonially in burrows excavated in soft substrates, and colonies can number into the hundreds. Colonies are often short-lived owing to constant erosion and habitat disturbance. • Our smallest swallow is an excellent digger—nest burrows are usually about 2 feet deep. Birds use their wings, feet and bill to excavate nests, and generally build new ones each year. • Increasing development within watersheds may have an impact on populations: hardening of watersheds creates more abrupt flooding and erosion, which is good for creating nesting habitat, but might cause increased inundation of active colonies. • Bank Swallows form huge flocks in August after nesting. Perhaps the largest on record—with a staggering 1 million birds—was at Cedar Point National Wildlife Refuge on August 8, 1931.

ID: brown breast band is diagnostic but hard to see in flight; brown upperparts; light underparts; long, pointed wings; shallowly forked tail; white throat; dark "cheek"; small legs. *In flight:* thin wings; small head; erratic, butterfly-like flight interspersed with very short glides.
Size: *L* 5½ in; *W* 13 in.
Habitat: steep banks, lakeshore bluffs and sand or gravel pits.
Nesting: colonial; pair excavates or sometimes reuses a long burrow in a steep earthen bank or cliff; end of burrow is lined with grass, rootlets, weeds, straw and feathers; pair incubates 4–5 white eggs for 14–16 days.

Feeding: catches flying insects; drinks on the wing.
Voice: twittering chatter: *speed-zeet speed-zeet.*
Similar Species: *Northern Rough-winged Swallow* (p. 215): no dark breast band. *Tree Swallow* (p. 214): iridescent, dark bluish to greenish upperparts; no dark breast band. *Cliff Swallow* (p. 217): blue gray head and wings; buff forehead and rump; no dark breast band.
Best Sites: larger streams such as Scioto R. or Great Miami R.; colonies on steep bluffs along upper end of Alum Creek; large flocks in August at Hebron Fish Hatchery and Killdeer Plains WA.

J F M A M J J A S O N D

CLIFF SWALLOW

Petrochelidon pyrrhonota

Cliff Swallows, the legendary swallows of Capistrano, were once restricted to cliffs throughout the western part of the country. Ohio has no suitable natural cliffs, but after European settlement and the opening up of the vast eastern deciduous forest created open country conducive to foraging swallows, Cliff Swallows began an eastward expansion. The first colony was discovered in Cincinnati in 1820, and the species rapidly expanded throughout the state. • By the 1960s, populations were in steep decline. The primary factor was the introduction and rapid increase of House Sparrows that readily usurped the domelike mud nests of Cliff Swallows, which were usually built on barns and other structures. In recent decades, Cliff Swallows have largely switched habitats, and now nest on dams and bridges over water. Populations have steadily increased and Cliff Swallows are once again becoming a common sight. • Small flocks of up to 150 birds can form in late summer. In recent years, flocks of over 2000 birds have been seen in Pickaway County.

ID: orangy rump; buff forehead; blue gray head and wings; rusty "cheek," nape and throat; buff breast; white belly; spotted undertail coverts; nearly square tail.

Size: *L* 5½ in; *W* 13½ in.

Habitat: steep banks, bridges and buildings, usually near watercourses; forages over water, fields and marshes.

Nesting: colonial, occasionally hundreds of nests; under bridges, on cliffs and buildings; pair builds a gourd-shaped mud nest with a small opening near the bottom; pair incubates 4–5 brown-spotted, white to pinkish eggs for 14–16 days.

Feeding: catches flying insects in midair; occasionally eats berries; drinks on the wing.

Voice: twittering chatter: *churrr-churrr;* also an alarm call: *nyew.*

Similar Species: *Barn Swallow* (p. 218): deeply forked tail; dark rump; usually has rust-colored underparts and forehead. *Other swallows* (pp. 213–18): lack buff forehead and rump patch. *Cave Swallow:* southern species, increasingly vagrant in northern U.S. and should someday occur here; orangy forehead; pale throat; buffy flanks.

Best Sites: watch all bridges and dams for colonies; most reservoirs in central eastern and northeastern Ohio; increasing elsewhere; big colony at State Route 32 bridge over Scioto R. in Pike County; also bridges and dam at Hoover Reservoir.

J F M A M J J A S O N D

BARN SWALLOW

Hirundo rustica

The Barn Swallow is the most abundant and wide-ranging swallow on the globe; consequently, it is probably the most widely recognized swallow as it is easily identified and nests in close proximity to humans. • This swallow attaches its cup-shaped mud nest to the eaves of buildings, beams in barns and under bridges. Historically, Barn Swallows would have nested on natural cliff faces and rock outcrops, but the switch to human-made structures was made at the time of early European settlement. It is now almost unheard of for these birds to nest in natural sites. • Its accessibility has made the Barn Swallow one of the best-studied species, and many interesting facets of its biology have been uncovered. For example, tail length and shape correlate with vigor, and individuals with longer tails are more desirable mates, better withstand parasites, have higher reproductive success and live longer on average. • The Barn Swallow was the catalyst for bird conservation. Prompted by overharvesting of Barn Swallows, George Bird Grinnell wrote an 1886 editorial in *Forest and Stream* blasting excessive hunting for the millinery trade. This article led to formation of the first Audubon chapter.

ID: long, deeply forked tail; rust-colored throat and forehead; blue black upperparts; rust- to buff-colored underparts; long, pointed wings.

Size: *L* 7 in; *W* 15 in.

Habitat: open rural and urban areas where bridges, culverts and buildings are found.

Nesting: singly or in small, loose colonies; on a vertical or horizontal building structure under a suitable overhang, on a bridge or in a culvert; half or full cup nest is made of mud and grass or straw; pair incubates 4–7 brown-spotted, white eggs for 13–17 days.

Feeding: catches flying insects on the wing.

Voice: continuous twittering chatter: *zip-zip-zip;* also *kvick-kvick.*

Similar Species: *Cliff Swallow* (p. 217): nearly square tail; buff rump and forehead; light-colored underparts.

Best Sites: easily found and ubiquitous inhabitant of open country, particularly around farms; occurs in large numbers in late summer, frequently in association with other swallows.

CAROLINA CHICKADEE

Poecile carolinensis

A chickadee is easy enough to recognize, but there is an identification issue in Ohio. The Carolina Chickadee is *the* chickadee of the southern two-thirds of Ohio, while the very similar Black-capped Chickadee occupies the northern third. Only in irruption winters, which occur every decade or so, do Black-cappeds regularly appear south into Carolina territory. The two species have a narrow zone of overlap extending approximately from Paulding County to Columbiana County. Within this zone, they not only hybridize but may also imitate each other's songs, making identification problematic. • Carolina Chickadees are attracted to backyards with feeders, and can sometimes be induced to nest if appropriate boxes are placed and there are adequate shrubs and trees in the vicinity. • In winter, chickadees form small, mixed flocks with titmice, nuthatches, creepers and other species. • If a Northern Saw-whet Owl or Eastern Screech-Owl is encountered, angry chickadees will often mob it; birders can sometimes find owls by searching for the source of the chickadees' wrath. However, when a much more dangerous Sharp-shinned Hawk or Cooper's Hawk appears, the chickadees usually adopt the "Sleeking Posture," compressing their feathers and sitting tight.

ID: black "cap" and "bib"; white "cheek"; gray upperparts; white underparts; buffy flanks.
Size: *L* 5¾ in; *W* 7½ in.
Habitat: deciduous and mixed woods, riparian woodlands, wooded neighborhoods and parklands.
Nesting: pair excavates or enlarges the interior of a natural tree cavity; may also use a nest box or secondary woodpecker cavity; cavity is lined with soft plant material and animal hair; female incubates 5–8 finely speckled, white eggs for 11–14 days; both parents raise the young.
Feeding: gleans insects, seeds and berries from vegetation; may hawk for insects or glean while hanging upside down on branches or while hovering; may visit seed and suet feeders.

Voice: whistled song has 4 clear notes: *fee-bee fee-bay;* note that Black-capped Chickadee can sing Carolina's song and vice versa; call is faster, higher and less hoarse version of Black-capped's *chick-a-dee-dee-dee.*
Similar Species: *Black-capped Chickadee* (p. 220): very similar; lower edge of black "bib" not as defined; secondaries and wing coverts have broad, hockey–stick-shaped, white edgings; clear, entirely white "cheek" patch.
Best Sites: in suitable woodlands throughout range; also in parks that put out feeders, such as Blendon Woods MP.

J F M A M J J A S O N D

BLACK-CAPPED CHICKADEE

Poecile atricapillus

Because the tiny Black-capped Chickadee is normally nonmigratory and winters in areas of frigid temperatures, it has evolved coping mechanisms to aid in survival. During bitterly cold nights, it can enter what is essentially a regulated hypothermia, in which the body temperature drops significantly to conserve energy until feeding can resume. • This wide-ranging bird is found across North America north to Alaska and Hudson Bay, and is probably more closely related to the western Mountain Chickadee *(P. gambeli)* than to the Carolina Chickadee. • Black-capped Chickadees frequently cache seeds and apparently possess phenomenal memories as they can relocate these stores up to a month later. • Like Carolina Chickadees, Black-cappeds are readily lured to yards by feeders, where their amusing antics and interesting behavior can be easily studied. • Common in northern Ohio, the Black-capped generally only encroaches south into the range of the Carolina Chickadee during winter irruption years. The two species are frequently confused, so care should be taken when identifying and reporting the Black-capped south of its normal range.

ID: black "cap" and "bib"; white "cheek"; gray back and wings; white underparts; light buff sides and flanks; dark legs; conspicuous white edging on wing feathers.
Size: *L* 5–6 in; *W* 8 in.
Habitat: deciduous and mixed forests, woodlots, riparian woodlands, wooded urban parks and backyards with bird feeders.
Nesting: excavates a cavity in a soft, rotting stump or tree; cavity is lined with fur, feathers, moss, grass and cocoons; female incubates 6–8 finely speckled, white eggs for 12–13 days.

Feeding: gleans vegetation, branches and the ground for small insects and spiders; visits backyard feeders; also eats conifer seeds and invertebrate eggs.
Voice: call is a husky, rather slow *chick-a-dee-dee-dee;* song is a slow, whistled *swee-tee* or *fee-bee.*
Similar Species: *Carolina Chickadee* (p. 219): call is faster and higher-pitched; neater edge on black "bib"; less prominent white edgings on wing feathers.
Best Sites: easy to find in appropriate habitat; Oak Openings MP; Findley SP; interesting area is Killdeer Plains WA where both Black-capped and Carolina species have been reported.

J F M A M J J A S O N D

TUFTED TITMOUSE

Baeolophus bicolor

The Tufted Titmouse is familiar to almost everyone who feeds birds and its loud, whistled *peter, peter, peter* call is a frequently heard woodland sound. It wasn't always this way; the Tufted Titmouse has been expanding its range northward over the past century, though it was already well established in every Ohio county by 1900. It's thought that maturation of forests that were cut over soon after European settlement, steady climatic warming and the increasing prevalence of bird feeders have allowed this expansion. • The Tufted Titmouse covets sunflower seeds and is interesting to watch at feeders. Often, it holds a seed with its feet and, using its bill like a jackhammer, pounds away until the husk falls off. Often the bird flies away with the seed. In this case, the titmouse is probably creating a cache—it hoards food by hiding it in the bark of trees and then returns later for its pickings. • Tufted Titmice nest in tree cavities, either natural ones or those created by woodpeckers. Twenty-three other native bird species use cavities and need dead or dying trees in which to excavate them; people who practice wood-land management need to recognize the value of standing dead timber and allow some of these trees to remain.

ID: gray crest and upperparts; black forehead; white underparts; buffy flanks.
Size: *L* 6–6½ in; *W* 10 in.
Habitat: deciduous woodlands, groves and suburban parks with large, mature trees.
Nesting: in a natural cavity or woodpecker cavity lined with soft vegetation and animal hair; female fed by the male; female incubates 5–6 finely dotted, white eggs for 12–14 days; both adults and occasionally a "helper" raise the young.

Feeding: forages on the ground and in trees, often hanging upside down like a chickadee; eats insects supplemented with seeds, nuts and fruits; will eat seeds and suet from feeders.
Voice: noisy, scolding call, similar to that of a chickadee; song is a whistled *peter peter* or *peter peter peter*.
Similar Species: none in Ohio.
Best Sites: easily located in woodlands throughout Ohio, but absent from L. Erie islands; often visits feeders.

J F M A M J J A S O N D

RED-BREASTED NUTHATCH

Sitta canadensis

Red-breasted Nuthatches are irregular in Ohio and can be virtually impossible to find in lean years. However, they stage cyclical southward invasions into our state every two to three years, sometimes appearing as early as August. In invasion winters they seem to be everywhere, as was evidenced by a one-day count of 154 in the 1998 Toledo-area Christmas bird count. These irruptions are triggered by food shortages in northern boreal forests, not by weather conditions. • Breeding Red-breasted Nuthatches are strictly associated with conifers, and even the wintering birds that visit Ohio occupy areas with pine, spruce and hemlock if available. Very small numbers breed here annually, mainly in northeastern Ohio, but more birds stay and breed following large irruptions, occasionally even in southern counties. • The Red-breasted Nuthatch is the only one of the four North American nuthatches that regularly stages long distance movements, and birds can sometimes stray far from the beaten path. In 1989, one nuthatch crossed the Atlantic and ended up in Norfolk, England. This tiny, 10-gram bundle of feathers thrilled the Brits who were fortunate enough to see it.

ID: rusty underparts; gray blue upperparts; white "eyebrow"; black eye line; black "cap"; straight bill; short tail; white "cheek." *Male:* deeper rust on breast; black crown. *Female:* light red wash on breast; dark gray crown.

Size: *L* 4½ in; *W* 8½ in.

Habitat: *Breeding:* hemlock gorges and pine stands, spruce-fir and pine plantations. *In migration* and *winter:* around conifers, but also in mixed woodlands, especially those near bird feeders.

Nesting: excavates a cavity or uses abandoned woodpecker nest; usually smears entrance with pitch; nest of bark shreds,

grass and fur; female incubates 5–6 brown-spotted, white eggs for about 12 days.

Feeding: forages down trees while probing under loose bark for larval and adult invertebrates; eats pine and spruce seeds in winter; often seen at feeders.

Voice: call is a slow, continually repeated, nasal *yank-yank-yank* or *rah-rah-rah-rah*, often likened to a tin horn; also a short *tsip*.

Similar Species: *White-breasted Nuthatch* (p. 223): larger; lacks black eye line and rusty red underparts.

Best Sites: conifer stands; ridge-top groves of Virginia pine around Lake Katherine SNP; cemeteries with large spruce and pines, such as Woodlawn, Green Lawn and Spring Grove cemeteries; often visits backyard feeders.

J F M A M J J A S O N D

WHITE-BREASTED NUTHATCH

Sitta carolinensis

Nuthatches are instantly recognizable by their peculiar habit of creeping headfirst down tree trunks. In the world of bark ecology, this gives them a distinct advantage because they occupy a niche not used by the other nine species of Ohio breeding birds that are specifically adapted for tree trunk foraging. Woodpeckers, Brown Creepers and Black-and-white Warblers all travel upward as they glean the bark for insects; by traveling headfirst down the tree, nuthatches gain a unique visual perspective that can help them spot prey that would otherwise be missed. • The odd common name "nuthatch" stems from this bird's habit of lodging large nuts in crevices, then using its bill to hammer the nut until it splits open. • White-breasted Nuthatches are inveterate food cachers, storing a great many seeds and nuts in bark and tree crevices, but almost always returning within hours to retrieve their plunder. • It is thought that pairs remain bonded for life, and observation of nuthatches reveals that the male and female stay in close proximity throughout the year. • This bird is our most common nuthatch; it breeds in every county and is a frequent feeder visitor.

ID: white underparts; white face; gray blue back; rusty undertail coverts; short tail; straight bill; short legs. *Male:* black "cap." *Female:* dark gray "cap."

Size: *L* 5½–6 in; *W* 11 in.

Habitat: deciduous forests, woodlots, parks and backyards with large trees.

Nesting: in a natural cavity or an abandoned woodpecker nest in a large deciduous tree; female lines the cavity with bark, grass, fur and feathers; female incubates 5–8 white eggs, spotted with reddish brown, for 12–14 days.

Feeding: forages headfirst down trees in search of larval and adult invertebrates; also eats many nuts and seeds; regularly visits feeders.

Voice: song is a frequently repeated *werwerwerwerwer;* calls include *ha-ha-ha ha-ha-ha, ank ank* and *ip.*

Similar Species: *Red-breasted Nuthatch* (p. 222): smaller; black eye line; rusty red underparts.

Best Sites: any forest or decent-sized woodland; very common at backyard feeders.

J F M A M J J A S O N D

BROWN CREEPER

Certhia americana

The Brown Creeper's world is inextricably intertwined with the branches and trunks of trees. In fact, this bird resembles nothing so much as a piece of bark with legs, scurrying spirally up and around tree trunks as it carefully inspects the surface for insects. After systematically working one tree from bottom to top, the creeper drops like a detached leaf to the base of another and resumes the never-ending hunt for prey. • If danger is detected, Brown Creepers freeze and flatten against the tree, becoming virtually undetectable, so well does their cryptic plumage blend in. • The best way to gain an idea of the true frequency of creepers is to learn their often-emitted call, a high-pitched, sibilant *tseee*. These birds can be surprisingly abundant if observers are clued in to their presence: a 1980 Toledo-area Christmas bird count recorded 111 in one day. Brown Creepers are generally rare nesters but appear to be increasing, primarily in northeastern Ohio and occasionally elsewhere. • The unfortunate demise of the American elm owing to Dutch elm disease benefited the Brown Creeper by creating an abundance of dead, standing timber with exfoliating bark—ideal habitat for a bird that hides its nest under flaps of loose bark.

ID: brown back is heavily streaked with buffy white; white "eyebrow"; white underparts; downcurved bill; long, pointed tail feathers; rusty rump.

Size: *L* 5–5½ in; *W* 7½ in.

Habitat: mature deciduous and mixed forests and woodlands, especially in wet areas with large, dead trees and riparian corridors with dead American elms.

Nesting: under loose bark; nest of grass and conifer needles is woven together with spider silk; female incubates 5–6 whitish eggs, dotted with reddish brown, for 14–17 days.

Feeding: creeps up tree trunks and large limbs, probing loose bark for adult and larval invertebrates.

Voice: song is a faint, high-pitched *trees-trees-trees see the trees;* call is a high-pitched *tseee.*

Similar Species: unique; no other Ohio bark forager has similar plumage or body structure.

Best Sites: woodlands statewide from fall through spring.

J	F	M	A	M	J	J	A	S	O	N	D

CAROLINA WREN

Thryothorus ludovicianus

The Carolina Wren is a persistent year-round singer; its loud, ringing song is often heard, even during winter months. Principally a southerner, it has been expanding northward over the past decade, owing to climatic warming and overall habitat changes that have created more of the open habitats that this species favors. This bird is common throughout the southern two-thirds of Ohio, but becomes local and less common to the north. • The status of these wrens in northern latitude regions like Ohio is often tenuous as they are vulnerable during cold winters with extended, deep snowfalls. The brutal winter of 1977–78 reduced Ohio's Carolina Wren population by an estimated 90 percent; it took two decades for them to completely rebound. • Our largest wren, the Carolina Wren is especially fond of spiders. It is often seen skulking about sheds, barns and old homes, seeking arachnid prey. • Carolina Wrens hide their nests in cavities, overhangs or dense vegetation, and often pick odd spots such as discarded boots, hanging flowerpots, mailboxes or even glove compartments in abandoned cars.

ID: long, prominent, white "eyebrow"; rusty brown upperparts; rich buff underparts; white throat; slightly downcurved bill.
Size: *L* 5 in; *W* 7 in.

Habitat: generally an edge species; scruffy woodlands and ravines, thickets, brush piles, woodland edges, overgrown yards and unkempt farms.

Nesting: dense vegetation, overhangs or cavities; both adults fill the cavity with twigs and vegetation and line it with finer materials; nest cup may be domed and may include a snakeskin; female incubates 4–5 brown-blotched, white eggs for 12–16 days; both adults feed the young.

Feeding: usually forages in pairs on the ground and among vegetation; eats mostly insects and other invertebrates; also takes berries, fruits and seeds; will visit feeders for peanuts and suet.

Voice: song is a loud, repetitious *tea-kettle tea-kettle tea-kettle* that may be heard at any time of day or year; loud buzzing chatter; scolding call is a harsh *zwee, zwee*.

Similar Species: *Other wrens* (pp. 226–29): smaller; lack bright rusty plumage, long tail and bold, white eye line. *Bewick's Wren* (p. 339): rare; white-tipped corners to tail; habitually flicks tail from side to side; gray sides of neck; generally lacks strong buffy tones below.

Best Sites: Clear Creek MP; Burr Oak SP; Mohican SF; Lake Katherine SNP.

J F M A M J J A S O N D

HOUSE WREN

Troglodytes aedon

Because shrubby suburban landscapes mimic the woodland clearings and edges that the House Wren occupies naturally, this species is a regular associate of people and one of our best-known birds. • House Wrens begin returning in mid-April and are widely established by early May, the males being quite conspicuous as they constantly sing their loud, bubbly songs. The males with the most potential nest sites in their territory apparently have the best chance of being chosen by the females. House Wrens strictly use cavities for nesting—the scientific name *Troglodytes* means "cave dweller"—and they can easily be lured to nest in appropriate boxes. They are quite aggressive toward other species nesting within their territories, and are known to enter other nests and puncture the eggs. Some authorities feel that competition from House Wrens is an important factor in virtually eliminating the Bewick's Wren from Ohio and this region. • House Wrens range from Canada all the way to southern South America—the broadest longitudinal distribution of any New World passerine.

ID: brown upperparts; fine, dark barring on upperwings and lower back; faint, pale "eyebrow" and eye ring; short tail is finely barred with black and held upraised; whitish throat; whitish to buff underparts; faintly barred flanks.
Size: *L* 4½–5 in; *W* 6 in.
Habitat: thickets and shrubby openings in or at the edges of deciduous or mixed woodlands; often in shrubs and thickets near buildings.
Nesting: in a natural cavity or abandoned woodpecker nest; also in an artificial cavity; nest of sticks and grass is lined with feathers, fur and other soft materials; female incubates 6–8 heavily marked, white eggs for 13–15 days.
Feeding: gleans the ground and vegetation for insects, especially beetles, caterpillars, grasshoppers and spiders.
Voice: call is a harsh, scolding rattle; song is a smooth, running, bubbly warble: *tsi-tsi-tsi-tsi oodle-oodle-oodle-oodle*, lasting about 2–3 seconds.
Similar Species: *Carolina Wren* (p. 225): larger and much brighter rusty coloration; bold, white eye line. *Winter Wren* (p. 227): smaller; darker overall; much shorter, stubby tail; prominent, dark barring on flanks.
Best Sites: well-vegetated neighborhoods, parks and woodland edges throughout the state.

J F M A M J J A S O N D

WINTER WREN

Troglodytes troglodytes

Perhaps the most amazing aspect of the lilliputian Winter Wren is its song. Many birders, upon exposure to its vocal barrage, are incredulous that so much sound can emanate from something so small. Lasting up to 10 seconds, the song is an impossibly complicated, ethereal jumble of sweet, gushing notes; any person's life should truly be enriched by exposure to this wondrous melody. • Occasionally, Winter Wrens are heard singing in spring migration when they can be common, but the best chance of hearing one is on its breeding grounds. Unfortunately, the few breeders in Ohio are confined to cool, shaded hemlock gorges in northeastern Ohio and Mohican State Forest. • This is the only North American wren that occurs outside the Americas. • An almost spherical, chocolate brown ball of feathers, the Winter Wren is one of our smallest birds, measuring but 4 inches and weighing in at 9 grams.

ID: very short, stubby, upraised tail; fine, pale buff "eyebrow"; dark brown upperparts; lighter brown underparts; prominent, dark barring on flanks.

Size: *L* 4 in; *W* 5½ in.

Habitat: a ground forager, most often found near exposed roots along streams, and in brush piles and around the bases of trees in woodlands; nests locally in cool, shaded hemlock ravines.

Nesting: in a natural cavity, under bark or upturned tree roots; bulky nest made of twigs, moss, grass and fur; male frequently builds up to 6 "dummy" nests prior to egg laying; female incubates 5–7 sparsely speckled, white eggs for 14–16 days.

Feeding: forages on the ground and in trees for beetles, wood-boring insects and other invertebrates.

Voice: song is a warbled, tinkling series of quick trills and twitters, often more than 8 seconds long and repeated with undiminished enthusiasm; call is a sharp *jip-jip*.

Similar Species: *House Wren* (p. 226): tail is longer than leg; less conspicuous barring on flanks; paler overall. *Carolina Wren* (p. 225): long, bold, white "eyebrow"; much larger; long tail.

Best Sites: migration hotspots such as Magee Marsh bird trail, Mentor Headlands and Green Lawn Cemetery; also among exposed tree roots along ravines and streams.

J F M A M J J A S O N D

SEDGE WREN

Cistothorus platensis

The Sedge Wren is one of our most enigmatic species from an ecological perspective. Birds often do not appear on territory until July or August and display no real fidelity to specific sites from year to year. This may be because their habitats of open grasslands, sedge stands and hayfields, often in moist soils, are ephemeral. These habitats quickly grow into vegetation types no longer suitable for breeding Sedge Wrens, or are mowed or otherwise altered for agricultural purposes. It is also known that Great Plains populations—from Wisconsin to Saskatchewan, Canada—breed in May and June, whereas many eastern populations such as ours often breed much later. It's possible that there is a postbreeding dispersal of western birds to the east, where they nest again when midwestern habitats are more conducive to successful breeding. • Learning to recognize the rapid, staccato series of *chip* notes comprising their song is key—Sedge Wrens lurk in dense grasses and can be quite hard to spot. They are often loosely colonial, and several territorial singing males can be found in a small area.

ID: short, narrow tail, often upraised; faint, pale "eyebrow"; dark crown and back faintly streaked with white; barred wing coverts; whitish underparts with buff orange sides, flanks and undertail coverts.

Size: *L* 4–4½ in; *W* 5½ in.

Habitat: wet sedge meadows; wet grassy fields; sometimes in drier hayfields and grasslands.

Nesting: usually less than 3 ft from the ground; well-built globe nest with a side entrance is woven from sedges and grasses; female incubates 4–8 unmarked, white eggs for about 14 days.

Feeding: forages low in dense vegetation, where it picks and probes for adult and larval insects and spiders; occasionally catches flying insects.

Voice: song is a few short, staccato notes followed by a rattling trill: *chap-chap-chap-chap, chap, churr-r-r-r-r;* call is a sharp, staccato *chat* or *chep.*

Similar Species: *Marsh Wren* (p. 229): broad, conspicuous, white "eyebrow"; prominent, white streaking on black back; unstreaked crown; prefers cattail marshes.

Best Sites: very irregular; may turn up in odd places; Killdeer Plains WA; Voice of America property in Butler Co.; large marshes at Magee Marsh WA, particularly where Canada bluejoint grass occurs.

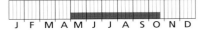

J F M A M J J A S O N D

MARSH WREN

Cistothorus palustris

An unmistakable, mechanical gurgling trill, somewhat like an old-fashioned sewing machine, coming from the marshes heralds the presence of the Marsh Wren, a species more likely to be heard than seen. However, like all wrens, this species has a bit of a busybody streak, so by making loud squeaking or pishing sounds you can encourage these birds to investigate and perhaps offer glimpses of themselves. • In the early 1900s, Marsh Wrens were known from half of Ohio's 88 counties, and populations were often quite large. Today, the only real stronghold is the western Lake Erie marshes. This population reduction is associated with the estimated loss of 92 percent of Ohio's wetlands; relatively few good breeding sites remain. Newer detrimental factors are the increased invasion of marshes by non-native plants such as giant reed and purple loosestrife, as well as marsh management practices that reduce stands of broad-leaved cattail, the obligate host plant for Marsh Wrens. • Eastern and western U.S. populations are probably in the process of evolutionary divergence and differ dramatically in vocalizations and subtly in plumage. They may prove to be distinct species.

ID: white "chin" and belly; white to light brown upperparts; black triangle on upper back is streaked with white; bold, white "eyebrow"; unstreaked brown crown; long, thin, downcurved bill.
Size: *L* 5 in; *W* 6 in.
Habitat: large cattail and bulrush marshes interspersed with open water; occasionally tall grass–sedge marshes.
Nesting: in a marsh among cattails or tall emergent vegetation; globelike nest is woven from cattails, bulrushes, weeds and grass and lined with cattail down; female incubates 4–6 heavily marked, white to pale brown eggs for 12–16 days.

Feeding: gleans vegetation and flycatches for adult aquatic invertebrates, especially dragonflies and damselflies.
Voice: rapid, rattling, staccato warble; call is a harsh *chek*.
Similar Species: *Sedge Wren* (p. 228): smaller and paler; streaked crown, less prominent white streaks on back.
Best Sites: large cattail stands; Spring Valley WA; Springville Marsh WA; large marshes in and around Magee Marsh WA; small numbers overwinter some years.

J F M A M J J A S O N D

GOLDEN-CROWNED KINGLET

Regulus satrapa

The Golden-crowned Kinglet has the distinction of being our smallest passerine, scarcely bigger than a Ruby-throated Hummingbird. • On October 7, 1954, famed Ohio ornithologist Milton Trautman estimated that up to 50,000 of these tiny birds were present on Lake Erie's South Bass Island; it is still possible to record hundreds in a day. • Behavioral traits that can help identify kinglets at a distance are their habit of being energetic little masses of perpetual motion, flitting about constantly, and their chronic wing-flicking. • Golden-crowned Kinglets are worth a close look; the males in particular are resplendent with their brilliant, golden and orange crown stripes. It's helpful to become familiar with this species' faint, high-pitched *tsee-tsee-tsee* call; many more kinglets will be heard than seen. Making pishing or squeaking sounds almost always lures these bantams into close range. • In recent decades in Ohio, Golden-crowned Kinglets have become rare and local nesters. They can, however, be very common in migration. • Hazards can be unique to one so tiny—kinglets have been known to perish after becoming ensnared by the prickly burs of the widespread common burdock plant.

ID: olive back; darker wings and tail; light underparts; dark "cheek"; 2 white wing bars; black eye line; white "eyebrow"; black border around crown. *Male:* reddish orange crown. *Female:* yellow crown.
Size: *L* 4 in; *W* 7 in.
Habitat: all types of woodlands, but displays strong affinity for conifers when available.
Nesting: usually in a spruce or conifer; hanging nest is made of moss, lichens, twigs and leaves; female incubates 8–9 tiny, whitish to pale buff eggs, spotted with gray and brown, for 14–15 days.

Feeding: gleans and hovers among the forest canopy for insects, berries and occasionally sap.
Voice: song is a faint, high-pitched, accelerating *tsee-tsee-tsee-tsee, why do you shilly-shally?;* call is a very high-pitched *tsee tsee tsee.*
Similar Species: *Ruby-crowned Kinglet* (p. 231): bold, broken, white eye ring; lacks strongly striped crown.
Best Sites: easily found in woodlands statewide in migration, and often in winter, at least milder ones; large numbers can occur, particularly along L. Erie.

J F M A M J J A S O N D

RUBY-CROWNED KINGLET

Regulus calendula

One of the treats of spring migration is the surprisingly loud, rollicking, complex warbling song of the Ruby-crowned Kinglet, always preceded by a few clear whistles that sound like pressurized steam escaping a teakettle. • At first glance, Ruby-crowned Kinglets might be mistaken for dull warblers or *Empidonax* flycatchers; however, their tiny size, very active foraging behavior and constant wing-flicking easily identify them. The male's scarlet crest is rarely seen; it's usually raised only when the bird is singing or agitated. • As is the Golden-crowned Kinglet, this species is a very common migrant but is not nearly so hardy, and wintering birds are rare. The primary factor determining northern limits of winter distribution is average temperature, and Ohio is too cold on average for Ruby-crowned Kinglets to survive. • Nesting well north of Ohio in the boreal forest, Ruby-crowned Kinglets have an interesting breeding ecology. Females lay up to 12 eggs, the largest clutches of any comparably sized North American passerine, and the total clutch might weigh over 7 grams—matching the weight of the bird!

ID: bold, broken eye ring; 2 bold, white wing bars; olive green upperparts; dark wings and tail; whitish to yellowish underparts; short tail; flicks its wings. *Male:* small, red crown is usually hidden. *Female:* lacks red crown.
Size: *L* 4 in; *W* 7½ in.
Habitat: all types of woodlands; unlike the Golden-crowned Kinglet, has no strong affinity for conifers.
Nesting: does not nest in Ohio.

Feeding: gleans for insects and spiders; hovers more than the Golden-crowned; also eats seeds and berries.
Voice: *Male:* song is an accelerating and rising *tea-tea-tea-tew-tew-tew look-at-Me, look-at-Me, look-at-Me;* call is a low, dry *che-dit.*
Similar Species: *Golden-crowned Kinglet* (p. 230): dark "cheek"; black border around crown; male has orange crown with yellow border; female has yellow crown.
Best Sites: *In migration:* woodlands statewide.

J F M A M J J A S O N D

231

BLUE-GRAY GNATCATCHER

Polioptila caerulea

Blue-gray Gnatcatchers are fidgety, fussy little birds that possess one of the least-distinguished vocal arrays of all our birds; they emit continuous, scratchy, banjolike twangs and nasal *szeee* notes as they flit restlessly about, gleaning insects from branches and leaves. • Gnatcatchers are one of the earliest of our highly migratory, insectivorous songbirds to return in spring. Nesting commences soon after arrival, with both sexes working together to create a tiny, cuplike nest containing numerous lichens and cemented together with spider webs. It's possible to observe nest building, as construction takes place prior to the trees leafing out, but nests are often 50 to 60 feet up, which makes up-close studies difficult. • In spite of their tiny dimensions, Blue-gray Gnatcatchers are quite bold and regularly mob potential predators such as Eastern Screech-Owls and Cooper's Hawks. Their agitated calls usually recruit other species to join the action, and the frenzy can often lead birders to see an interesting raptor that might otherwise go undetected.

♂

breeding

Nesting: on a branch halfway to the trunk; cup nest of plant fibers and bark chips is decorated with lichens and lined with fine vegetation, hair and feathers; female incubates 3–5 brown-dotted, pale bluish white eggs for 11–15 days; male feeds the female and young.

Feeding: moves up and down through foliage, flicking its tail constantly, possibly to flush prey into view; eats small insects and spiders.

Voice: calls are thin and high-pitched: single *see* notes or a short series of "mewing" or chattering notes; can mimic several species.

Similar Species: *Golden-crowned Kinglet* (p. 230) and *Ruby-crowned Kinglet* (p. 231): olive green overall; short tails; wing bars.

Best Sites: mature woodlands, particularly along rivers and lakes, such as Blackhand Gorge SNP and Sheldon Marsh SNP; also Little Beaver Creek SP.

ID: blue gray upperparts; long, thin tail; white eye ring; pale gray underparts; black upper tail with white outer tail feathers; dark legs. *Breeding male:* black forehead. *Juvenile:* similar to adult; pale bill; brown-washed upperparts.

Size: *L* 4½ in; *W* 6 in.

Habitat: deciduous woodlands with at least some large trees, particularly along streams, ponds, lakes and swamps; also neighborhoods and parks with mature shade trees.

J F M A M J J A S O N D

EASTERN BLUEBIRD

Sialia sialis

The Eastern Bluebird is one of the most popular and widely recognized Ohio birds. "Bluebird trails" have been created that consist of nest boxes placed and maintained by hundreds of enthusiasts, and organizations devoted to bluebirds have even sprung up. As a result of such efforts, it's likely that the Eastern Bluebird is more common in Ohio than ever. It's little wonder people want this bird around, for not only is the male bluebird colored in one of the most richly vivid shades of blue imaginable, but it also delivers a most pleasing, warbling song. • Eastern Bluebirds are highly migratory, with small flocks appearing early in spring and larger numbers in February or March. Occasionally, however, this intrepidness gets them into trouble—late-hitting, savage winter storms in the spring of 1895 virtually eliminated Eastern Bluebirds from the entire Great Lakes region! • Far more bluebirds will be detected flying overhead if you are alert to their soft *teeu* or *tu-a-wee* call, given often while on the wing. • Many Eastern Bluebirds overwinter, particularly in mild years, switching from their largely insectivorous diet to one dominated by berries and fruits.

ID: chestnut red "chin," throat, breast and sides; white belly and undertail coverts; dark bill and legs. *Male:* deep blue upperparts. *Female:* thin, white eye ring; gray brown head and back tinged with blue; blue wings and tail; paler chestnut on underparts.

Size: *L* 7 in; *W* 13 in.

Habitat: cropland fencelines, meadows, fallow and abandoned fields, pastures, forest clearings and edges, golf courses, large lawns and cemeteries.

Nesting: in an abandoned woodpecker cavity, natural cavity or nest box; female builds a loose cup of grass, weed stems and small twigs; mostly the female incubates 4–5 pale blue eggs for 13–16 days.

Feeding: swoops from a perch to pursue flying insects; also forages on the ground for invertebrates.

Voice: song is a rich, warbling *turr, turr-lee, turr-lee;* call is a soft whistled *teeu* or *tu-a-wee.*

Similar Species: not likely to be confused with any other Ohio bird.

Best Sites: nest boxes or other cavities in open country, parks, on golf courses and farms and along some state highways.

J F M A M J J A S O N D

VEERY

Catharus fuscescens

Thrushes in the genus *Catharus* are rather drab, which helps them to blend with the dappled shade of their woodland understory habitats. The Veery is the brightest of the group—seen in good light it is quite striking with its rich, tawny rufous overtones. • This bird's odd name is probably imitative of its call note, a soft, whistled *veeerr*. Its song is exceptionally beautiful and more than adequate compensation for its somewhat plain looks. The Veery sings a rapidly cascading series of sweet, ethereal notes that have an odd, haunting resonance as if sung within a pipe. • Veerys can be numerous in spring migration—a few dozen or more might be encountered on an exceptional day. They are also locally common nesters in the northern third of Ohio, gradually scarcer to the south, and prefer younger successional woods in wet or poorly drained sites. • An excellent place to observe our three breeding woodland thrushes is Clear Creek Metropark. Several pairs of Veeries breed in the damp, young woods buffering Clear Creek, the rare (as a nester) Hermit Thrush occupies a few of the dense, hemlock-choked side ravines, and Wood Thrushes commonly nest along the rich, forested slopes above the creek.

ID: reddish brown or tawny upperparts; very thin, grayish eye ring; faintly streaked, buff throat and upper breast; light underparts; gray flanks and face patch; pinkish legs.

Size: *L* 7 in; *W* 12 in.

Habitat: cool, moist deciduous and mixed forests and woodlands with a dense understory of shrubs and ferns; often in disturbed woodlands.

Nesting: on the ground or in a shrub; female builds a bulky nest of leaves, weeds, bark strips and rootlets; female incubates 3–4 pale greenish blue eggs for 10–15 days.

Feeding: gleans the ground and lower vegetation for invertebrates and berries.

Voice: *Male:* song is a fluty, descending *da-vee-ur, vee-ur, vee-ur, veer, veer, veer;* call is a high, whistled *veeerr.*

Similar Species: *Wood Thrush* (p. 238): much heavier breast spotting; stronger eye ring. *Other spot-breasted thrushes* (pp. 235–38): heavier breast spotting; stronger eye ring; no bright rufous back.

Best Sites: migrant traps such as Mentor Headlands, Magee Marsh bird trail and Green Lawn and Spring Grove cemeteries; breeders at Clear Creek MP, Oak Openings MP, Goll Woods SNP and Mohican SF.

GRAY-CHEEKED THRUSH

Catharus minimus

The least common of our spot-breasted thrushes, Gray-cheeked Thrushes pass through Ohio in fewer numbers than their close relative, the Swainson's Thrush. • As with other species in the genus *Catharus,* the Gray-cheeked Thrush has a distinctive call note: a surprisingly loud, nasal *jeeer*. Tuning into this sound will reveal many more birds than would otherwise be detected. Because they call regularly while passing over at night, these thrushes can be identified as they wing overhead on the way to their boreal forest breeding grounds. Under peak conditions, over 100 birds have been heard flying over in an evening. However, its beautiful, ethereal song is rarely heard in Ohio, in part because it sings very infrequently in migration. • Seeing more than a few Gray-cheeked Thrushes on the ground during daylight hours would be exceptional. Their shy nature makes locating them difficult—they are quick to flit away into dense cover as soon as they detect an intruder. • Patiently searching through the May or September hordes of Swainson's Thrushes will eventually turn up a Gray-cheeked. The key feature to focus on is the facial pattern—Gray-cheeked Thrushes lack the bright, buffy spectacles of the Swainson's Thrush.

ID: gray brown upperparts; gray face; inconspicuous eye ring may not be visible; heavily spotted breast; light underparts; brownish gray flanks.

Size: *L* 7–8 in; *W* 13 in.

Habitat: variety of forested areas, parks and wooded cemeteries.

Nesting: does not nest in Ohio.

Feeding: hops along the ground, picking up insects and other invertebrates; may also feed on berries during migration.

Voice: typically thrushlike in tone, ending with a clear, usually 3-part whistle, with the middle note higher pitched: *wee-o, wee-a, titi wheeee;* call is a downslurred *jeeer*.

Similar Species: *Swainson's Thrush* (p. 236): prominent, buffy eye ring and "cheek." *Hermit Thrush* (p. 237): reddish tail, often raised and lowered; olive brown upperparts; lacks gray "cheek." *Veery* (p. 234): much brighter reddish brown upperparts; very light breast spotting.

Best Sites: woodlands, parks and cemeteries; L. Erie coastal woodlands such as Mentor Headlands and Magee Marsh bird trail.

SWAINSON'S THRUSH

Catharus ustulatus

Swainson's Thrush is our most common, migrant spot-breasted thrush. There have been exceptional days when several hundred were detected in spring by a single observer, though more typical peaks would be several dozen. This pales in comparison with the numbers that can be heard passing overhead at night under good migratory conditions. Observers familiar with the oft-emitted, easily learned flight call—a soft musical *peep* likened to the sound made by spring peeper frogs—have recorded over 1000 winging by in an evening. • Swainson's Thrush can be a late migrant, sometimes lingering well into June, and there are a few records of summering nonbreeders. This has occasionally prompted speculation about breeding, but Swainson's Thrush nests well to the north of Ohio and is intimately associated with boreal spruce-fir forests, a habitat not found here. • This species has a song unique among the *Catharus* thrushes. The rapid, ethereal, flutelike notes spiral upward, rather than descending like the Veery and Gray-cheeked Thrush.

ID: gray brown upperparts; conspicuous buff eye ring; buff wash on "cheek" and upper breast; spots arranged in streaks on throat and breast; white belly and undertail coverts; brownish gray flanks.
Size: *L* 7 in; *W* 11½ in.
Habitat: all manner of woodlands, even relatively open parklands in cemeteries.
Nesting: does not nest in Ohio.
Feeding: forages much higher in trees than do other thrushes; sometimes flycatches;

gleans vegetation and forages on the ground for invertebrates; also eats berries.
Voice: song is a slow, rolling, rising spiral: *Oh, Aurelia will-ya, will-ya will-yeee;* call is a sharp but pleasant *wick* or *prit;* flight call is a soft musical *peep.*
Similar Species: *Gray-cheeked Thrush* (p. 235): gray "cheek"; lacks conspicuous buffy spectacles. *Hermit Thrush* (p. 237): reddish tail (often raised and lowered) and rump; grayish brown upperparts; darker breast spotting on whiter breast. *Veery* (p. 234): upperparts are more reddish; faint breast streaking; lacks bold eye ring.
Best Sites: woodlots along L. Erie.; woodlands, parks and cemeteries; listen for flight calls on quiet May or September evenings to discover large numbers passing overhead.

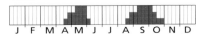

HERMIT THRUSH

Catharus guttatus

The thrush family is known for its beautiful singers, and the Hermit Thrush may have the most mellifluous voice of all. Its haunting tones inspired 19th-century naturalist John Burroughs to write "…as the hermit's evening hymn goes up from the deep solitude below me, I experience that serene exaltation of sentiment of which music, literature and religion are but faint types and symbols."
• Hermit Thrushes are common migrants, arriving and largely moving on before the other spot-breasted thrushes begin to appear. They are also quite hardy, being the only *Catharus* thrushes that winter in North America. Quite secretive and hard to detect, wintering birds are more numerous than realized, and can be expected in much of southern Ohio.
• Hermit Thrushes have an affinity for the fruit of sumac shrubs; making squeaking or pishing sounds around this habitat will often lure one into view.
• Since it rarely sings in migration, the best place to enjoy this spectacular songster is at dusk or dawn in the cool hemlock gorges of Hocking County, where most of the bigger ravines harbor a pair or two.

ID: reddish brown tail (often raised and lowered) and rump; grayish brown upperparts; black-spotted throat and breast; light underparts; gray flanks; thin, whitish eye ring.
Size: *L* 7 in; *W* 11½ in.
Habitat: breeds in hemlock gorges; migrants are found in woodlands, parks and cemeteries; wintering birds are found around berry-producing shrubs such as sumac, also honeysuckle.
Nesting: usually on the ground; occasionally in a small tree or shrub; female builds a bulky cup nest of grass, twigs, moss, ferns and bark strips; female incubates 4 pale blue to greenish blue eggs for 11–13 days.

Feeding: forages on the ground and gleans vegetation for insects and other invertebrates; also eats berries.
Voice: song is a series of beautiful, flutelike notes, both rising and falling in pitch; a small questioning note may precede the song; calls include a faint *chuck* and a fluty *treee*.
Similar Species: *Other spot-breasted thrushes* (pp. 234–38): do not have strongly reddish tail, which is often raised and lowered, contrasting with brownish olive back.
Best Sites: nesting areas, such as Old Man's Cave SP, Cantwell Cliffs SP, Conkle's Hollow SNP, Mohican SF.

J F M A M J J A S O N D

WOOD THRUSH

Hylocichla mustelina

A poster child for declining neotropical birds, the Wood Thrush's populations have been steadily decreasing overall owing to habitat loss on its North American breeding grounds and its Central American wintering range. A more insidious factor is increased Brown-headed Cowbird parasitism; cowbirds have expanded dramatically as forests have been fragmented, and Wood Thrushes are heavily victimized. • Fortunately, the Wood Thrush is common and widespread in Ohio, nesting in every county. It will nest in relatively small woodlots, but reaches peak abundance in our larger forests, where several birds might be heard singing simultaneously. Favored habitat is dominated by large trees that form a closed canopy, a well-developed secondary shrub layer and a thick litter of decaying leaves. • The beautiful, flutelike song is unmistakable and one of the richest sounds in Ohio woods. The Wood Thrush has split windpipes in its syrinx (vocal organ) that allows for the delivery of two notes together. This creates the odd harmonic effect that renders such a unique quality to its song.

Feeding: forages on the ground and gleans vegetation for insects and other invertebrates; also eats berries.

Voice: *Male:* flutelike phrases of 3–6 notes, with each note at a different pitch and followed by a trill: *Will you live with me? Way up high in a tree, I'll come right down and...seeee!;* calls include a *pit pit* and *bweebeebeep.*

ID: plump body; large, black spots on white breast, sides and flanks; bold, white eye ring; rusty head and back; brown wings, rump and tail.

Size: *L* 8 in; *W* 13 in.

Habitat: older second-growth to mature, preferably undisturbed, deciduous woodlands.

Nesting: low in the fork of a deciduous tree; female builds a bulky cup nest of grass, twigs, moss, weeds, bark strips and mud; nest is lined with softer materials; female incubates 3–4 pale greenish blue eggs for 13–14 days.

Similar Species: *Ovenbird* (p. 271): much smaller and browner; black-and-russet crown stripes; streaky spots on underparts. *Other thrushes* (pp. 233–39): smaller spots on underparts; most have colored wash on sides and flanks; all lack bold, white eye ring and rusty "cap" and back.

Best Sites: Shawnee SF; Burr Oak SP; Cuyahoga Valley NRS; Oak Openings MP.

J F M A M J J A S O N D

AMERICAN ROBIN

Turdus migratorius

Even nonbirders can easily identify the American Robin, which is probably our most common songbird. Robins are birds of open country and adapt easily to human-made habitats. The best way to realize how abundant they are is to listen to the spring and summer dawn chorus, when most birds are in full voice—American Robins generally greatly outnumber all other singers. • American Robins have always wintered in Ohio, but have greatly increased in recent years as year-round inhabitants. In cold weather, they become largely frugivorous; the increased numbers of ornamental fruit-bearing plants such as hawthorn and serviceberry, plus the proliferation of bush honeysuckles and other escaped and established exotic shrubs, provide a ready source of food that has increased the number of overwintering robins. • Our largest and most wide-ranging North American thrush, the American Robin is split into seven subspecies. Birders should watch for exceptionally dark individuals in winter, which may be *T. m. nigrideus*, a boreal forest subspecies of eastern Canada that can occur in Ohio.

ID: gray brown back; dark head; white throat streaked with black; white under-tail coverts; incomplete, white eye ring; black-tipped, yellow bill. *Male:* deep brick red breast; black head. *Female:* light red orange breast; dark gray head. *Immature:* heavily spotted breast.

Size: *L* 10 in; *W* 17 in.

Habitat: residential lawns and gardens, pastures, urban parks, forests, woodlands and farmland.

Nesting: in a tree or shrub, sometimes on a building or other structure; sturdy cup nest of grass, moss and loose bark is cemented with mud; female incubates

4 light blue eggs for 11–16 days; may raise up to 3 broods per year.

Feeding: forages on the ground and among vegetation for larval and adult insects, earthworms, other invertebrates and berries.

Voice: song is a series of rich, liquid warbles delivered rapidly with pauses between: *cheerily cheer-up cheerio;* call is a rapid *tut-tut-tut.* It's useful to master this song, as there is a suite of other species that give "robinlike" songs.

Similar Species: unmistakable. *Varied Thrush* (p. 339): very rare; dark breast band; orange wing bars and eye stripe.

Best Sites: urban lawns and gardens.

J F M A M J J A S O N D

GRAY CATBIRD

Dumetella carolinensis

Of our three mimics—the other two being the Northern Mockingbird and the Brown Thrasher—the Gray Catbird is easily the most numerous. You might not know it if you aren't familiar with catbird vocalizations, as this is a bird of dense shrubby thickets and is a master of concealment. Curious by nature, it can often be enticed to show itself by making squeaking or pishing sounds. • Named for one of their many calls, which vaguely resembles a cat's meow, Gray Catbirds are the least accomplished of the mimics, seldom delivering sounds that are identifiable as another species. Rather, they issue a nonstop, rapid-fire series of squeaky notes in haphazard order, often jumbling along in a disjointed cacophony all day long. • As the successional habitats in which Gray Catbirds live are prime areas for Brown-headed Cowbirds, catbird nests can be heavily parasitized. However, catbirds are one of only a dozen or so songbirds that can recognize cowbird eggs. Consequently, Gray Catbirds invariably shove an offending egg from the nest immediately upon detection. • Although catbirds are highly migratory and winter far south of Ohio, a few attempt to overwinter every year, particularly in the southern one-third of the state.

ID: dark gray overall; black "cap"; long tail may be dark gray to black; chestnut under-tail coverts; black eyes, bill and legs.
Size: *L* 8½–9 in; *W* 11 in.

Habitat: dense thickets, brambles, shrubby or brushy areas and hedgerows, often near water.

Nesting: in a dense shrub or thicket; bulky cup nest is loosely built with twigs, leaves and grass and is lined with finer materials;

female incubates 4 greenish blue eggs for 12–15 days.
Feeding: forages on the ground and in vegetation for ants, beetles, grasshoppers, caterpillars, moths and spiders; also eats berries and visits feeders.
Voice: calls include a catlike *meoow* and a harsh *check-check;* song is a variety of warbles, squeaks and poorly mimicked phrases repeated only once and often interspersed with a *mew* call.
Similar Species: relatively unmistakable.
Best Sites: dense shrubby vegetation, particularly if water is nearby; very common along the Magee Marsh WA bird trail, around C.J. Brown and Hoover reservoirs and at Pymatuning SP.

J F M A M J J A S O N D

NORTHERN MOCKINGBIRD

Mimus polyglottos

Northern Mockingbirds are highly skilled mimics, imitating many bird species so well that even the best birders can be fooled. These birds don't restrict themselves to imitating other birds—their vocal repertoire can encompass nearly 200 sounds, everything from human whistling to truck backup beeps and sirens. • Northern Mockingbirds are extroverts that deliver their seemingly unending imitations from prominent locations. Bachelors can keep you awake with their loud singing, which sometimes continues throughout the night—and it can be especially irksome when they select a chimney for a perch and funnel the sound directly into the house. • Northern Mockingbirds are known to aggressively drive off invaders, including Red-tailed Hawks and even cats! • Ohio is on the northern edge of this southern bird's range; the mockingbird is common only in the southern two-thirds of Ohio, and becomes infrequent north of that, though the species is expanding northward. • The scientific name *polyglottos* means "many tongues."

ID: gray upperparts; dark wings; 2 thin, white wing bars; long, dark tail with white outer tail feathers; light gray underparts. *Immature:* paler overall; spotted breast. *In flight:* large white patch at base of black primaries.
Size: *L* 10 in; *W* 14 in.
Habitat: hedges, suburban gardens and orchard margins with an abundance of available fruit; hedgerows of multiflora roses are especially important in winter.
Nesting: often in a small shrub or small tree; cup nest of twigs, grass, fur and leaves; female incubates 3–4 brown-blotched, bluish gray to greenish eggs for 12–13 days.
Feeding: gleans vegetation and forages on the ground for beetles, ants, wasps and grasshoppers; also eats berries and wild fruit; visits feeders for suet and raisins.

Voice: song is a medley of mimicked phrases, with the phrases often repeated 3 times or more; calls include a harsh *chair* and *chewk*.
Similar Species: *Northern Shrike* (p. 202) and *Loggerhead Shrike* (p. 338): much rarer; black "mask"; thick, stubby bill; blacker wings; shorter tail. *Gray Catbird* (p. 240): much squeakier, softer song generally lacking in recognizable imitations. *Brown Thrasher* (p. 242): sings notes in doublets, pausing between phrases.
Best Sites: east end of Cleveland Lakefront Park; Green Lawn Cemetery; Crown City WA.

J F M A M J J A S O N D

241

BROWN THRASHER

Toxostoma rufum

This is the only one of our three mimics in decline; in sheer numbers it may be the scarcest of the Ohio mimics and the least likely to be found in urban areas. Still, it is fairly common in drier, more open landscapes with scattered brush, treelines and small woodlots. • Normally, the Brown Thrasher is much more retiring than its relative, the Northern Mockingbird. But, with over 1100 documented sounds, its vocal repertoire is even more expansive, though without as many obvious imitations. • The genus name *Toxostoma*, which means "bow beak," refers to the long, downcurved bill typical of all eight North American thrasher species. • Brown Thrashers are short-range migrants that winter in the southern U.S. and will only rarely winter in Ohio, mostly in the southern one-third of the state. They are most conspicuous here upon arrival in April when males often sing from treetops. The Brown Thrasher is the only thrasher species that breeds east of the Mississippi River.

ID: reddish brown upperparts; pale underparts with heavy, brown streaking; long, downcurved bill; orange yellow eyes; long, rufous tail; 2 white wing bars.

Size: *L* 11½ in; *W* 13 in.

Habitat: dense shrubs and thickets, overgrown pastures (especially those with hawthorns), woodland edges and brushy areas; generally drier sites; rarely close to human habitation.

Nesting: usually in a low shrub; often on the ground; cup nest of grass, twigs and leaves is lined with fine vegetation; pair incubates 4 brown-spotted, bluish white to pale blue eggs for 11–14 days.

Feeding: gleans the ground and vegetation for larval and adult invertebrates; occasionally tosses leaves aside and uncovers food with its bill; also eats seeds and berries.

Voice: sings a large variety of phrases, with each phrase usually repeated twice: *dig-it dig-it, hoe-it hoe-it, pull-it-up, pull-it-up;* calls include a loud crackling note, a harsh *shuck,* a soft *churr* and a whistled, 3-note *pit-cher-ee.*

Similar Species: *Thrushes* (pp. 233–39): occur in vastly different habitats; much different behaviorally; lack long tail and downcurved bill.

Best Sites: reclaimed strip mine grasslands, such as The Wilds, Tri-Valley WA and Egypt Valley WA; a few pairs in overgrown fields of Ottawa NWR and at Oak Openings MP.

J F M A M J J A S O N D

EUROPEAN STARLING

Sturnus vulgaris

The European Starling was introduced to North America in 1890 and 1891, when about 60 individuals were released into New York's Central Park as part of the local Shakespeare society's plan to introduce all the birds mentioned in their favorite author's writings. What a mistake that was! From this meager beginning, starlings have overrun virtually the entire North American continent.
• Probably the worst ecological impact from this species is the damage done to our native cavity-nesting birds. European Starlings also nest in cavities and outcompete many of the natives, causing significant population reductions in some cases.
• European Starlings may be the most numerous birds of any species in Ohio. They sometimes form massive winter roosts numbering into the tens of thousands. Starlings can be surprisingly long-lived; the North American record for a wild bird is 17 years, 8 months. • Perhaps the most noteworthy aspect of European Starlings is their voice. While we commonly hear their grating clatters, rattles and screams, they are accomplished mimics capable of a vast array of sounds. Starlings can readily imitate many other birds, mechanical sounds of all sorts and even human speech!

breeding

ID: short, squared tail; dark eyes. *Breeding:* iridescent, blackish plumage; yellow bill. *Nonbreeding:* blackish wings; feather tips are heavily spotted with white and buff. *Juvenile:* gray brown plumage; brown bill. *In flight:* pointed, triangular wings.
Size: *L* 8½ in; *W* 16 in.
Habitat: agricultural areas, townsites, woodland and forest edges, landfills and roadsides.
Nesting: in a tree, nest box or other artificial cavity; nest of grass, twigs and straw; mostly the female incubates 4–6 bluish to greenish white eggs for 12–14 days.
Feeding: very diverse diet includes many invertebrates, berries, seeds and human food waste; forages mostly on the ground.
Voice: variety of whistles, squeaks and gurgles; imitates other birds throughout the year.
Similar Species: none in Ohio.
Best Sites: widespread; fast-food restaurant parking lots.

J F M A M J J A S O N D

AMERICAN PIPIT

Anthus rubescens

American Pipits are inconspicuous in virtually every respect, and can easily be missed. They are ground feeders that are most attracted to mudflats and low, wet swales in large expanses of short grass and barren agricultural fields, where their cryptic plumage renders them nearly invisible. Note that, like many habitual ground-feeding birds, pipits walk rather than hop; this method of locomotion instantly rules out a number of other species with which they might be confused. • Only when one is attuned to their high-pitched *pip-it* call note, often given in flight, does the true extent of American Pipit migration become apparent. There have been spring days along Lake Erie when in excess of 1000 birds have passed overhead. • If pipits are detected on the ground, it pays to check around carefully—almost invariably far more birds will be present than are evident at first glance. • American Pipits nest in high arctic tundra and in similar alpine areas across North America, Siberia and Greenland. They are common migrants in Ohio but are less numerous in the unglaciated regions. Small numbers sometimes overwinter.

breeding

ID: faintly streaked, gray brown upperparts; lightly streaked "necklace" on upper breast; streaked sides and flanks; dark legs; dark tail with white outer tail feathers; buff-colored underparts; slim bill and body.

Size: *L* 6–7 in; *W* 10½ in.

Habitat: muddy agricultural fields, beaches, pastures and the shores of wetlands, lakes and rivers.

Nesting: does not nest in Ohio.

Feeding: gleans the ground and vegetation for terrestrial and freshwater invertebrates and seeds.

Voice: the only vocalization likely to be heard here is its familiar flight call: *pip-it pip-it.*

Similar Species: *Horned Lark* (p. 212): black "horns," "bib" and facial markings. *Lapland Longspur* (p. 303): heavier bill; boldly streaked flanks; strong facial markings and "bib." *Sprague's Pipit:* accidental; streaked back and crown; paler legs; stouter bill.

Best Sites: Killdeer Plains WA; Big Island WA; Charlie's Pond; large numbers along L. Erie, particularly at spots such as Mentor Headlands and Maumee Bay SP.; more likely southward in winter.

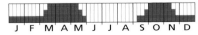

J F M A M J J A S O N D

CEDAR WAXWING

Bombycilla cedrorum

Cedar Waxwings appear impeccably regal in their ornate plumage, with wings and tail that appear to have been dipped in wax. They also appear to possess manners beyond most birds—gregarious and quite fond of berries, waxwings are known to pass fruit from bird to bird. • Highly evolved socially, Cedar Waxwings don't appear to establish territories and are almost always encountered in groups. This behavior is likely related to their rather nomadic wanderings, necessitated by their diet of cyclical crops of fruit that are best exploited en masse and in a short time frame. Nomadic winter flocks can number into the hundreds or thousands, appearing wherever fruit trees are plentiful. • The increase in fruit-bearing trees and shrubs used in landscaping has benefited waxwings. A particular favorite—which also occurs in the wild—is the serviceberry. A mob of Cedar Waxwings can plunder the fruit from one of these trees in less than a day. • Breeding waxwings are strongly associated with riparian habitats and are frequently seen flycatching from tall trees.

female incubates 3–5 pale gray to bluish gray eggs for 12–16 days.

Feeding: catches flying insects on the wing or gleans vegetation; eats large amounts of berries and wild fruit, especially in fall and winter.

Voice: faint, high-pitched, trilled whistle: *tseee-tseee-tseee*. A "hearing test" bird: people who lose the upper ranges of hearing cannot hear this species.

Similar Species: *Bohemian Waxwing* (p. 339): very rare; larger; chestnut undertail coverts; small, white, red and yellow markings on wings; juvenile has chestnut undertail coverts and white wing patches.

Best Sites: wherever berry bushes are found, though visits may be brief. *Summer:* larger streams and lakes where most birds breed; Blackhand Gorge SNP; Sheldon Marsh SNP; Maumee River Rapids; Newell's Run.

ID: cinnamon crest; brown upperparts; black "mask"; yellow wash on belly; gray rump; yellow terminal tail band; white undertail coverts; small, red "drops" on wings. *Juvenile:* no "mask"; streaked underparts; gray brown body.

Size: *L* 7 in; *W* 12 in.

Habitat: wooded urban parks and gardens, forest edges, and in particular riparian woodlands during summer.

Nesting: in a coniferous or deciduous tree or shrub; cup nest of twigs, grass, moss and lichens is often lined with fine grass;

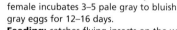

J F M A M J J A S O N D

BLUE-WINGED WARBLER

Vermivora pinus

Prior to European settlement, the Blue-winged Warbler was a southerner, only expanding northward into Ohio after the original forests were cleared. This range expansion has had a detrimental impact on the Golden-winged Warbler, a related species that has virtually disappeared as an Ohio breeder. • One of our showiest and most easily identified warblers, the Blue-winged Warbler is easy to locate by listening for its loud *beee-bzzz* song. Not as prone to skulking in dense growth as some other birds of brushy, successional habitats, this species can usually be spotted with a bit of perseverance. Sometimes it pays to find the singer since both the "Lawrence's Warbler" and "Brewster's Warbler"—hybrids formed by crossbreeding with the Golden-winged Warbler—can sing the Blue-winged's song. • One of the most interesting hybrid wood-warblers was the so-called "Cincinnati Warbler," first collected in its namesake city in 1880 and again in Michigan in 1948. This bird turned out to be a hybrid of the Blue-winged Warbler and the Kentucky Warbler.

breeding

ID: bright yellow head and underparts, except for white to yellowish undertail coverts; olive yellow upperparts; bluish gray wings and tail; black eye line; thin, dark bill; 2 white wing bars; bold, white tail spots on underside of tail.
Size: *L* 4½–5 in; *W* 7½ in.
Habitat: second-growth woodlands; locust thickets in reclaimed strip mines; shrubby, overgrown fields, pastures and woodland edges and openings.
Nesting: on or near the ground, concealed by vegetation; female builds a narrow, inverted, cone-shaped nest of grass, leaves and bark strips and lines it with soft materials; female incubates 5 brown-spotted, white eggs for 10–11 days.
Feeding: gleans insects and spiders from the lower branches of trees and shrubs.
Voice: buzzy, 2-note song: *beee-bzzz.*
Similar Species: Other *Vermivora warblers* (pp. 247–50): lack bright yellow plumage with contrasting bluish gray wings and obvious wing bars. *Prothonotary Warbler* (p. 269): lacks wing bars and eye line; occurs in very different habitat.
Best Sites: overgrown brushy fields; reclaimed strip mines such as Crown City WA and Egypt Valley WA; Clear Creek MP; Oak Openings MP.

J F M A M J J A S O N D

GOLDEN-WINGED WARBLER

Vermivora chrysoptera

The Golden-winged Warbler together with the genetically very similar Blue-winged Warbler may comprise a "superspecies" of two color morphs. Such a relationship is supported by the relatively frequent fertile hybrids that are produced when the two meet. The "Brewster's Warbler" hybrid is the product of pure individuals of both species, whereas the much rarer "Lawrence's Warbler" results from mating with at least one hybrid. • The Golden-winged Warbler is the more northerly breeder of the two, but rapid northward expansion of the more dominant Blue-winged is a factor that has greatly reduced U.S. populations of the Golden-winged. • The Golden-winged was never widespread in Ohio, reaching its peak in the 1930s in the Oak Openings region, where several dozen pairs nested. Today, there are probably no successfully breeding Golden-winged Warblers in Ohio, though the odd unmated individual sometimes turns up. • When Blue-winged Warblers invade Golden-winged populations, it takes only a few decades for the latter species to be completely eliminated.

breeding

ID: yellow forecrown and wing patch; dark "chin," throat and white-bordered "mask" over eye; bluish gray upperparts and flanks; white undersides; white spots on underside of tail. *Female* and *immature:* duller overall; gray throat and "mask."
Size: *L* 4½–5 in; *W* 7½ in.
Habitat: moist shrubby fields, woodland edges and early-succession forest clearings.
Nesting: on the ground, concealed by vegetation; female builds an open cup nest of grasses, leaves and grapevine bark and lines it with softer materials; female incubates 5 sparsely marked, pinkish to creamy white eggs for 9–11 days.
Feeding: gleans insects and spiders from tree and shrub canopies.
Voice: call is a sweet *chip*; buzzy song begins with a higher note: *zee-bz-bz-bz;* note that hybrids and seemingly pure Blue-wingeds may also sing this song.
Similar Species: yellow crown and wings and chickadee-like head pattern are distinctive.
Best Sites: rare; Magee Marsh bird trail; Green Lawn Cemetery; Sheldon Marsh SNP.

J F M A M J J A S O N D

TENNESSEE WARBLER

Vermivora peregrina

The Tennessee Warbler and the Orange-crowned Warbler have the dullest plumages of the *Vermivora* warblers, but the Tennessee compensates for its lack of visual pizzazz with a loud, staccato song that is easily learned. During this bird's relatively brief May passage, it can seem like this tune plays from all manner of wooded areas, including shade trees in suburbia. • This warbler's common name is misleading as the species actually nests far to the north in Canada's boreal forests. However, the scientific name is spot on—*Vermivora* means "worm eater" (the Tennessee eats a lot of wormlike spruce budworms), and *peregrina* means "wanderer." These small birds migrate to Central and South America to winter. • Tennessee Warbler populations can fluctuate dramatically from year to year, as their primary prey on the breeding grounds—the spruce budworm— is prone to cyclical fluctuations. Nevertheless, this bird is typically one of our more common migrant warblers.

breeding

ID: *Breeding male:* blue gray "cap"; olive green back, wings and tail edgings; white "eyebrow"; black eye line; clean white underparts; thin bill. *Breeding female:* yellow wash on breast and "eyebrow"; olive gray "cap." *Nonbreeding:* olive yellow upperparts; yellow "eyebrow"; yellowish underparts except for white undertail coverts; male may have white belly.
Size: *L* 4½–5 in; *W* 8 in.
Habitat: woodlands or areas with tall shrubs.

Nesting: does not nest in Ohio.
Feeding: gleans foliage and buds for small insects, caterpillars and other invertebrates; also eats berries; rarely visits suet feeders.
Voice: male's song is a loud, sharp, accelerating *ticka-ticka-ticka swit-swit-swit-swit chew-chew-chew-chew-chew*; call is a sweet *chip.*
Similar Species: *Warbling Vireo* (p. 207): stouter; thicker bill; much less green on upperparts. *Philadelphia Vireo* (p. 208): stouter; thicker bill; yellow breast and sides. *Orange-crowned Warbler* (p. 249): noticeably yellowish undertail coverts; lacks white "eyebrow" and blue gray head.
Best Sites: almost all woodlands in late May.

ORANGE-CROWNED WARBLER

Vermivora celata

The Orange-crowned Warbler's common name is not its best descriptor: the male's dull orange crown patch is almost never visible in the field, though highly agitated birds will reveal these feathers. • In marked contrast to most of our wood-warblers, the Orange-crowned Warbler is quite drab in plumage. This somber plumage actually renders this species distinctive once it is recognized as a warbler. • These warblers tend to be spotted at eye level, as they favor low thickets and tangles, and generally shun treetops. • Ohio birders covet sightings of the Orange-crowned because this species is much more common in the western U.S. and is one of the rarer warblers to pass through our state. Spotting more than a few Orange-crowned Warblers in a season is a noteworthy phenomenon. • Orange-crowned Warblers can be quite hardy, and there are several winter records.

ID: olive yellow to olive gray body; faintly streaked underparts; bright yellow undertail coverts; thin, faint, dark eye line; bright yellow "eyebrow" and broken eye ring; thin bill; faint orange crown patch is rarely seen.
Size: *L* 5 in; *W* 7 in.
Habitat: woodlands or areas with tall shrubs.
Nesting: does not nest in Ohio.
Feeding: gleans foliage for invertebrates, berries, nectar and sap; often hover-gleans.
Voice: song is a faint trill that breaks downward halfway through; call is a clear, sharp distinctive *chip*.

Similar Species: *Tennessee Warbler* (p. 248): blue gray head; dark eye line; bold, white "eyebrow"; brighter green upperparts and whiter underparts, including undertail coverts; lacks broken eye ring. *Ruby-crowned Kinglet* (p. 231): smaller; constantly flicks wings; white wing bars.
Best Sites: unpredictable and difficult to find; brushy areas at migrant traps, such as near a pond or bridge at Green Lawn Cemetery, Magee Marsh bird trail and wooded edges buffering Calamus Swamp.

J F M A M J J A S O N D

NASHVILLE WARBLER

Vermivora ruficapilla

The Nashville Warbler has a most unusual distribution, with two widely separated summer populations: one in eastern North America and the other in the West. These populations are believed to have been isolated thousands of years ago when a single core population was split during continental glaciation. • One of our more common migrant warblers, several dozen Nashville Warblers might be detected on a good day in early May. Fifty-seven were banded in one day at a Lake Erie site on May 9, 1994. • Nashville Warblers are extremely rare as breeders, with only three documented records from the far northeastern corner of the state. However, some breeders may be missed because Nashville Warbler breeding habitat can be rather uninviting. They should be watched for in dense shrub zones surrounding bogs, fens and other wetlands.

Nesting: on the ground under a fern, sapling or shrubby cover; female builds a cup nest of grass, bark strips, ferns and moss and lines it with conifer needles, fur and fine grasses; female incubates 4–5 brown-spotted, white eggs for 11–12 days.

Feeding: gleans foliage for insects such as caterpillars, flies and aphids.

Voice: song begins with a thin, high-pitched *see-it see-it see-it see-it,* followed by a trilling *ti-ti-ti-ti-ti;* call is a metallic *chink.*

Similar Species: *Common Yellowthroat* (p. 277) and *Wilson's Warbler* (p. 279): all-yellow underparts; females lack grayish head and bold, white eye ring. *Connecticut Warbler* (p. 275) and *Mourning Warbler* (p. 276): females have grayish to brownish "hood" and yellow between legs.

ID: bold, white eye ring; yellow green upperparts; yellow underparts; white between legs. *Male:* blue gray head; may show small, chestnut red crown (normally hard to see). *Female* and *immature:* duller overall; light eye ring; olive gray head; blue gray nape.

Size: *L* 4½–5 in; *W* 7½ in.

Habitat: *In migration:* almost any type of woody cover, often at lower levels. *Breeding:* low thickets and shrub zones, often in and around wetlands; sometimes in clearings in hemlock gorges.

Best Sites: easily found statewide in first half of May; almost any type of wooded area.

J F M A M J J A S O N D

NORTHERN PARULA

Parula americana

One of the earlier returning warblers, Northern Parulas can be heard singing on territory by mid-April. It is vital to learn the song, as these little sprites are treetop specialists. Fortunately, the typical song is quite distinctive—a rapid, rising buzz with an emphatically sharp ending note. Alternate songs are quite different and can be reminiscent of a Golden-winged Warbler. • Although migrants are not common, they do appear statewide. On the other hand, breeders are largely confined to southern Ohio and are usually found in riparian woods with peak numbers occurring in hemlock gorges. Northern Parulas are expanding northward, and nesters are now sporadically encountered into central Ohio. • "Sutton's Warbler" is a hybrid of the Northern Parula and the Yellow-throated Warbler that could eventually occur in Ohio, likely in the southeast. • Parula means "little titmouse," a rather odd, yet distinctive, name for our smallest warbler.

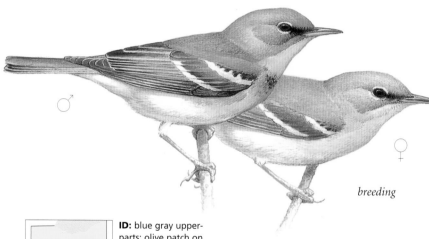

breeding

ID: blue gray upperparts; olive patch on back; 2 bold, white wing bars; bold, white eye ring is broken by black eye line; yellow "chin," throat and breast; white belly and flanks. *Male:* 1 black and 1 orange breast band.
Size: *L* 4½ in; *W* 7 in.
Habitat: hemlock gorges and mature wooded riparian corridors.
Nesting: often in a hemlock; female weaves a small, nest into hanging strands of tree lichens; may add lichens to a dense cluster of conifer boughs; pair incubates 4–5 brown-marked, whitish eggs for 12–14 days.
Feeding: forages for insects and other invertebrates by hovering, gleaning or hawking; feeds from the tips of branches and occasionally on the ground.

Voice: typical song is a rising, buzzy trill ending with an abrupt, lower-pitched *zip*.
Similar Species: fairly distinctive. *Cerulean Warbler* (p. 266): streaking on breast and sides; lacks white eye ring. *Blue-winged Warbler* (p. 246): yellow underparts. *Yellow-rumped Warbler* (p. 257): yellow rump and crown; lacks yellow throat. *Yellow-throated Warbler* (p. 260): heavy, black streaking along sides.
Best Sites: *In migration:* Magee Marsh bird trail. *Breeding:* Clear Creek MP; streamside habitat in Shawnee SF and Zaleski SF; streams with older growth riparian forests in southern Ohio.

J F M A M J J A S O N D

251

YELLOW WARBLER

Dendroica petechia

The Yellow Warbler is one of our most common and widespread breeding warblers, occurring statewide in wet, shrubby thickets, particularly where willows are found. This species returns in mid-April and enlivens wetlands with its lively, loud, clear *sweet-sweet-sweet I'm so sweet* song. Spectacular numbers can occur during peak May migration in the western Lake Erie marshes—on a good day of birding you may detect several hundred! • Because Yellow Warblers occur in the same areas as Brown-headed Cowbirds, they are heavily parasitized by that species. However, Yellow Warblers are one of the few species that can recognize cowbird eggs, and these warblers will sometimes simply build a new nest on top of the old one that contains the cowbird eggs. Occasionally, cowbirds strike repeatedly—once a stack of five warbler nests was found! • Yellow Warblers have the widest range of any North American *Dendroica* warbler, and have been divided into three groups and 43 subspecies.

breeding

ID: rather stocky, bright yellow body; thick, black bill; relatively short tail; black eyes; yellow highlights in dark olive tail and wings. *Breeding male:* red breast streaks.
Size: *L* 5 in; *W* 8 in.
Habitat: favors willow thickets; marshes, riparian borders, edges of ponds, brushy fields and other wetland habitats.
Nesting: in a deciduous tree or small shrub; female builds a compact cup nest of grass, weeds and shredded bark and lines it with plant down and fur; female incubates 4–5 brown-spotted, greenish white eggs for 11–12 days.
Feeding: gleans foliage and vegetation for invertebrates, especially caterpillars,

inchworms, beetles, aphids and cankerworms; occasionally hover-gleans.
Voice: song is a fast, frequently repeated *sweet-sweet-sweet I'm so sweet;* call is a loud *chip;* flight call is a clear *zzip.*
Similar Species: distinctive; the only *Dendroica* with yellow tail spots. *Orange-crowned Warbler* (p. 249): darker olive plumage; lacks reddish breast streaks. *Wilson's Warbler* (p. 279): shorter, darker tail; male has black "cap"; female has darker crown and upperparts. *Common Yellowthroat* (p. 277): darker face and upperparts; female lacks yellow highlights in wings.
Best Sites: almost anywhere that wet thickets occur; biggest numbers are found in western L. Erie marshes.

J F M A M J J A S O N D

CHESTNUT-SIDED WARBLER

Dendroica pensylvanica

Chestnut-sided Warblers are among our most attractive neotropical migrants, and can be quite common in migration. A good spring morning of birding might produce a few dozen of these visual treats with their crisp white underparts boldly striped with rich chestnut and their bright yellow "caps." John James Audubon, who spent four decades wandering throughout eastern North America in the early 1800s, wasn't so fortunate—he saw but one Chestnut-sided Warbler in his life! • Chestnut-sided Warblers are birds of successional habitats—young, scruffy forests and edges such as arise several years after logging. In Audubon's time, forests were mostly primeval and unbroken with little Chestnut-sided Warbler habitat. Today, forests are constantly being harvested; consequently, Chestnut-sided Warbler populations continue to increase. • This species has become an increasingly common Ohio nester, primarily in northeastern Ohio, but has nested or at least summered sporadically to southern Ohio.

breeding

ID: *Breeding:* chestnut brown sides; white underparts; yellow "cap"; black legs; yellowish wing bars; black "mask." *Breeding male:* bold colors. *Female:* washed-out colors; dark streaking on yellow "cap." *Nonbreeding:* yellow green crown, nape and back; white eye ring; gray face and sides; white underparts. *Immature:* similar to nonbreeding, but with brighter yellow wing bars.

Size: *L* 5 in; *W* 8 in.

Habitat: shrubby, second-growth deciduous woodlands and brushy abandoned fields and orchards, especially in areas regenerating after logging or fire.

Nesting: low in a shrub or sapling; small cup nest is made of bark strips, grass, roots and weed fibers and lined with fine grasses, plant down and fur; female incubates 4 brown-marked, whitish eggs for 11–12 days.

Feeding: gleans trees and shrubs at midlevel for insects.

Voice: loud, clear song: *so pleased, pleased, pleased to MEET-CHA!;* musical *chip* call.

Similar Species: generally unmistakable. *Bay-breasted Warbler* (p. 264): black face; dark chestnut hindcrown, upper breast and sides; buff belly and undertail coverts; white wing bars. *Yellow Warbler* (p. 252): songs can be very similar.

Best Sites: *In migration:* anywhere there is decent woody cover; large numbers at Sheldon Marsh SNP and Magee Marsh WA. *Summer:* numbers increasing, particularly northward.

J F M A M J J A S O N D

MAGNOLIA WARBLER

Dendroica magnolia

The Magnolia Warbler might be described as the "field mark" bird. Breeding males have it all: wing bars, tail spots, "necklace," streaked underparts, eye line and "mask." These features plus their bright, lemon yellow underparts conspire to create a creature that inspires gasps of admiration. • In migration, the Magnolia is one of our most common warblers and is easy to find. In summer, however, it is a rare breeder in the larger, undisturbed hemlock gorges of Hocking and Lake counties, and the Mohican State Forest. Here it is far removed from the core breeding population and is one of a suite of rare Ohio hemlock nesters that includes the Blue-headed Vireo, Black-throated Green Warbler, Canada Warbler and Hermit Thrush.

breeding

ID: *Breeding male:* yellow underparts with bold, black streaks; black "mask"; white "eyebrow"; blue gray crown; dark upperparts; white wing bars often blend into larger patch. *Female* and *nonbreeding male:* duller overall; pale "mask"; 2 distinct, white wing bars; streaked olive back. *In flight:* yellow rump; white tail patches.

Size: *L* 4½–5 in; *W* 7½ in.

Habitat: *In migration:* all types of trees and woody habitat. *Breeding:* almost exclusively confined to hemlock gorges; rarely pine or spruce plantings.

Nesting: in a conifer; loose cup nest is made of grass, twigs and weeds and lined with rootlets; female incubates 4 variably marked, white eggs for 11–13 days.

Feeding: gleans vegetation and buds; occasionally flycatches for beetles, flies, wasps, caterpillars and other insects; sometimes eats berries.

Voice: song is a quick, rising *pretty pretty lady* or *wheata wheata wheet-zu;* call is a nasal *tzek.*

Similar Species: *Prairie Warbler* (p. 262): immature lacks grayish chest band, white tail band and yellow rump. *Canada Warbler* (p. 280): lacks white wing bars and tail spots. *Kirtland's Warbler* (p. 340): similar to young spring female Magnolia except Kirtland's has different tail pattern, noticeably larger size and constantly bobs its tail.

Best Sites: *Breeding:* Clear Creek MP; Conkle's Hollow SNP; Mohican SF; Lake Co. metropark sites such as Hell Hollow MP and Penitentiary Glen MP. *In migration:* any good wooded area; landscaped residential areas.

J F M A M J J A S O N D

CAPE MAY WARBLER

Dendroica tigrina

Cape May Warblers are closely associated with the boreal spruce-fir forest, where they specialize in feeding on spruce budworms. In years when this insect's population explodes, Cape May Warbler numbers increase; conversely, in lean years numbers plummet. These fluctuations mean that in some years it's hard to find this beautiful warbler in migration in Ohio. Canadian spraying campaigns to control budworms may also lead to decreases in Cape May populations.
• Cape May Warbler males are stunning in their breeding colors; *tigrina* means "striped like a tiger" and refers to the black streaks on the male's yellow underparts.
• John James Audubon named two birds collected in 1811 in Kentucky "Carbonated Warblers"; these were thought at one time to be hybrids of the Cape May Warbler and the Blackpoll Warbler. It is now thought that they were most likely first-year male Cape Mays. • The Cape May is surprisingly hardy—there are at least 10 early winter Ohio records, and one bird overwintered near Cincinnati in 1980.

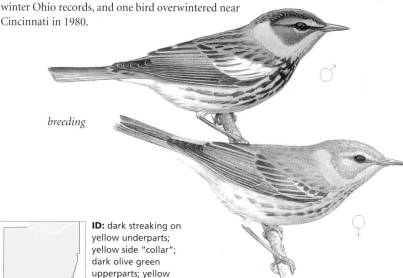

breeding

ID: dark streaking on yellow underparts; yellow side "collar"; dark olive green upperparts; yellow rump; clean white undertail coverts. *Breeding male:* chestnut brown "cheek" on yellow face; dark crown; large, white wing patch. *Female:* paler overall; 2 faint, thin, white wing bars; grayish "cheek" and crown.
Size: *L* 5 in; *W* 8 in.
Habitat: mature coniferous and mixed forests, especially in dense, old-growth stands of white spruce and balsam fir; in migration, prefers conifers but will use various other woody plants.
Nesting: does not nest in Ohio.
Feeding: gleans treetop branches and foliage for flies, beetles, moths, wasps and other insects; occasionally hover-gleans.

Voice: song is a very high-pitched, weak *see see see see;* call is a very high-pitched *tsee.*
Similar Species: generally distinctive. *Yellow-rumped Warbler* (p. 257): similar to fall female Cape Mays, but has obvious yellow rump patch and browner upperparts. *Pine Warbler* (p. 261): female is larger with browner upperparts and more obvious wing bars.
Best Sites: widespread, but has an affinity for tall conifers, especially in Spring Grove, Green Lawn and Woodlawn cemeteries.

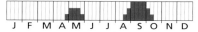

J F M A M J J A S O N D

255

BLACK-THROATED BLUE WARBLER

Dendroica caerulescens

The Black-throated Blue Warbler is a good example of sexual dimorphism—in fact, John James Audubon initially thought the female Black-throated Blue was a different species! • These warblers are normally uncommon migrants in Ohio, so spotting more than four or five in a day is unusual. • Oftentimes, the easily learned, low, buzzy *zee zee zee zreeeee* song is the first clue that a Black-throated Blue is present. The male will often be quite tame and approachable. • There are two historic nesting records from a boggy, northeastern Ohio woodland that was largely destroyed; however, territorial males occasionally summer in suitable hemlock gorge habitats, and the species may breed in the region again someday. Populations are on the rise overall as recovery of forests produces more habitat. Black-throated Blue Warblers require woodlands that are at least 50 years old, where they occupy dense, understory shrubs.

breeding

ID: *Male:* black face, throat, upper breast and sides; dark blue upperparts; clean white underparts and wing patch. *Female:* olive brown upperparts; unmarked buff underparts; faint, white "eyebrow"; small buff to whitish wing patch is not always visible.

Size: *L* 5–5½ in; *W* 7½ in.

Habitat: *Breeding:* mature hemlock gorges; boggy shrub zones of wetlands. *In migration:* young, scruffy woodlands; dense shrub zones of woods.

Nesting: in a dense shrub or sapling, usually within 3 ft of the ground; female builds an open cup nest of weeds, bark strips and spider webs lined with moss, hair and pine needles; female incubates 4 brown-blotched, creamy white eggs for 12–13 days.

Feeding: thoroughly gleans understory for caterpillars, moths, spiders and other insects; occasionally eats berries and seeds.

Voice: song is a slow, wheezy *zee zee zee zreeeee*, sometimes described as *I am soo lay-zee*, rising slowly throughout; call is a short *tip*.

Similar Species: male is unmistakable. *Orange-crowned* (p. 249), *Tennessee* (p. 248), female *Cerulean* (p. 266) or *Pine* (p. 261) *warblers:* wing bars; lack diagnostic white wing mark and very brownish tones of female Black-throated Blue Warbler.

Best Sites: usually in woodland understory shrubs in migration; Magee Marsh WA bird trail; Springville Marsh SNP; Dawes Arboretum.

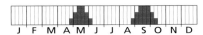

J F M A M J J A S O N D

YELLOW-RUMPED WARBLER

Dendroica coronata

Yellow-rumped Warblers, or "Butter-butts," occur in great numbers and are easily our most common migrant warblers. An estimated 7500 were seen along the Magee Marsh Wildlife Area bird trail on October 5, 1985. • The Yellow-rumped Warbler is quite hardy, and is the only warbler to regularly overwinter in Ohio—while numbers vary from year to year, there are always some birds to be found, particularly in the southern part of the state. In colder months, Yellow-rumpeds switch largely to a diet of berries. • This species was formerly known as the "Myrtle Warbler," while its western counterpart was called "Audubon's Warbler." While currently considered subspecies of the Yellow-rumped Warbler, the two forms are quite distinctive and may be split into separate species again. The "Myrtle Warbler" has a white throat; the "Audubon's Warbler" has a yellow throat and is very rare in Ohio.

breeding

ID: yellow fore-shoulder patches and rump; white underparts with dark streaking; faint, white wing bars; thin "eyebrow." *Male:* yellow crown; blue gray upperparts with black streaking; black "cheek" and breast band. *Female:* gray brown upperparts with dark streaking.
Size: *L* 5–6 in; *W* 9 in.
Habitat: occurs in almost every habitat type including woodlands, marshes, thickets, old fields, cedar stands and ornamental landscaping.
Nesting: does not nest in Ohio.
Feeding: hawks and hovers for beetles, flies, wasps, caterpillars, moths and other insects; also gleans vegetation; often eats berries.
Voice: *Male:* song is a tinkling trill, often given in 2-note phrases that rise or fall at the end (varies among individuals); call is a sharp, distinctive *chip* or *check*.
Similar Species: *Magnolia Warbler* (p. 254): yellow underparts; yellow throat; bold, white "eyebrow"; white patches on tail; lacks yellow crown. *Cape May Warbler* (p. 255): heavily streaked yellow throat, breast and sides; lacks yellow crown.
Best Sites: *In migration:* virtually anywhere. *Winter:* red cedar groves; areas with lots of poison ivy; thickets along stream bottoms.

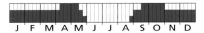

J	F	M	A	M	J	J	A	S	O	N	D

BLACK-THROATED GREEN WARBLER

Dendroica virens

Offering one of the easier songs to learn, Black-throated Green Warblers sing a lazy, wheezy *zee-zee-zee zu zee* persistently throughout the day. An avian motormouth, one bird was documented singing over 450 songs in one hour. • This common migrant appears a bit in advance of most warblers. Some years, they can be exceptionally abundant—400 were present along the bird trail at Magee Marsh Wildlife Area on May 7, 1983. • The localized breeding populations are almost exclusively confined to hemlock gorges. In Ohio, hemlock forests occur on the steep, shaded slopes of small streams, which have a cooler micro-climate and are reminiscent of the northern boreal forest. Hemlock communities support numerous rare and unusual plants and animals, including a group of birds normally found breeding well to the north of Ohio.

breeding

ID: yellow face; may show faint dusky "cheek" or eye line; black upper breast band; streaking along sides; olive crown, back and rump; dark wings and tail; 2 bold, white wing bars; white lower breast, belly and undertail coverts. *Male:* black throat. *Female:* yellow throat; thinner wing bars.

Size: *L* 4½–5 in; *W* 7½ in.

Habitat: *Breeding:* local; almost always confined to hemlock gorges; occasionally nests in large stands of pines. *In migration:* all types of woodlands.

Nesting: usually in a conifer; compact cup nest of grass, weeds, twigs, bark, lichens and spider silk is lined with moss, fur, feathers and plant fibers; female incubates 4–5 brown-speckled, creamy white to gray eggs for about 12 days.

Feeding: gleans vegetation and buds for beetles, flies, wasps, caterpillars and other insects; sometimes takes berries; frequently hover-gleans.

Voice: wheezy *zee-zee-zee zu zee* or *zoo zee zoo zoo zee;* call is a fairly soft *tick.*

Similar Species: very distinctive. *Townsend's Warbler:* 2 records; western vagrant; yellow breast; dark "cheek" patch bordered with yellow.

Best Sites: *Breeding:* Conkle's Hollow SNP; Mohican SF; Stebbins Gulch within Holden Arboretum (access restricted except for guided walks). *In migration:* woodlands statewide.

J F M A M J J A S O N D

BLACKBURNIAN WARBLER

Dendroica fusca

If a nonbirder were to stumble onto a Blackburnian Warbler at close range, we'd likely have a new convert to birding. The flaming orange throat of this species is dazzling, and its overall appearance rivals just about anything to be found in the tropics. • One of our rarest breeding warblers, the Blackburnian nests very sporadically in hemlock forests in Hocking County, Mohican State Forest and a few northeastern Ohio locales. It is a relatively common spring migrant, though less so in fall. • The Blackburnian is one of a suite of *Dendroica* warblers that breed in the northern coniferous forests. It differentiates itself ecologically by occupying a treetop niche, where it specializes in foraging along the outer branches. It continues that behavior in migration, and is almost always a neck-straining, top-of-the-tree bird.

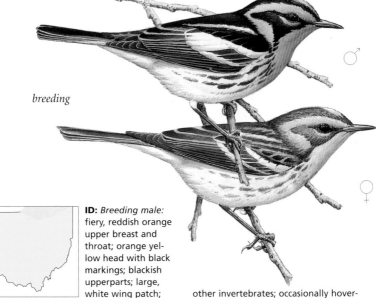

breeding

ID: *Breeding male:* fiery, reddish orange upper breast and throat; orange yellow head with black markings; blackish upperparts; large, white wing patch; yellowish to whitish underparts; dark streaking on sides and flanks. *Female:* paler version of male; more yellowish upper breast and throat.
Size: *L* 5–5½ in; *W* 8½ in.
Habitat: *Breeding:* very rare, localized nester in hemlock gorges. *In migration:* almost anywhere, but favors larger, mature trees and older-growth forests.
Nesting: high in a mature conifer; female builds a cup nest of bark, twigs and plant fibers and lines it with conifer needles, moss and fur; female incubates 3–5 brown-blotched, white to greenish white eggs for about 13 days.
Feeding: forages on uppermost branches, gleaning budworms, flies, beetles and other invertebrates; occasionally hover-gleans.
Voice: song is a soft, faint, high-pitched *ptoo-too-too-too tititi zeee* or *see-me see-me see-me see-me;* call is a short *tick.*
Similar Species: distinctive. *Cerulean Warbler* (p. 266): female looks similar to 1st-year Blackburnian female, but has lighter "cheek" patch, shorter tail and lacks pale streaks on back.
Best Sites: Green Lawn, Woodlawn and Spring Grove cemeteries (especially in larger oaks and spruces); Shawnee SF (old-growth oak-dominated ridge tops); Magee Marsh bird trail.

J F M A M J J A S O N D

YELLOW-THROATED WARBLER

Dendroica dominica

The hardy Yellow-throated Warbler's early April arrival is heralded by its delicate song—a series of descending, sweet notes, percolating down from high in streamside sycamore trees. In fact, Yellow-throated Warblers were once called "Sycamore Warblers." Ohio birds are closely associated with sycamores, though they will very rarely occupy upland forests of mixed oak and pine. • Common statewide at the time of the earliest bird surveys in the 1800s, Yellow-throated Warbler numbers plummeted until their range had contracted to very limited areas in extreme southern Ohio by the early 1900s. This reduction was likely caused by the intense deforestation of riparian corridors. Fortunately, this handsome warbler has reclaimed much of its former range and continues to expand northward. • Yellow-throated Warblers and Black-and-white Warblers share a peculiar foraging behavior, often crawling along branches in the manner of a Brown Creeper.

breeding

ID: yellow throat and upper breast; triangular, black "mask"; black "forehead"; bold, white "eyebrow" and "ear" patch; white underparts with black streaking on sides; 2 white wing bars; bluish gray upperparts.

Size: *L* 5–5½ in; *W* 8 in.

Habitat: primary habitat is mature bottomland riparian forests with numerous sycamores; much less frequently occupies upland oak-pine woodlands; accidental in winter, when birds sometimes visit feeders.

Nesting: in a deciduous or pine tree; female builds a cup nest of fine grasses, bark shreds and plant fibers and lines it with plant down and feathers; female incubates 4 variably marked, pale greenish or grayish white eggs for about 12 days.

Feeding: primarily insectivorous; gleans insects from tree trunks and foliage by creeping along tree surfaces; often flycatches insects in midair; wintering birds may eat suet from feeders.

Voice: boisterous song is a series of downslurred whistles with a final rising note: *tee-ew tee-ew tee-ew tew-wee;* call is a loud *churp;* learn the song as this is a treetop specialist and not easily detected visually.

Similar Species: distinctive. *Blackburnian Warbler* (p. 259): adult female and 1st-year male lack gray upperparts, conspicuous, white eye line and white spot on side of neck.

Best Sites: rivers bordered with large sycamores; most streams in southern two-thirds of state have breeding pairs; plentiful along Ohio R. tributary streams in southern Ohio.

J F M A M J J A S O N D

PINE WARBLER

Dendroica pinus

Pine Warblers are intimately associated with pines. Because the range of our few native species of pine is quite restricted in Ohio, so is the breeding distribution of Pine Warblers. They are most common along ridges in unglaciated southeastern Ohio, where stands of Virginia and pitch pine occur, as does the shortleaf pine in the southernmost counties. Elsewhere, Pine Warblers rarely occupy large, mature plantations of red pines and white pines, such as are found in the Oak Openings and Mohican State Forest. • This is our first breeding warbler to set up territory—males are often heard singing by early March. Pine Warblers also regularly overwinter in very small numbers. • The Pine Warbler is one of a suite of birds that are "trill-singers," a similar-sounding group that includes the Chipping Sparrow and Worm-eating Warbler.

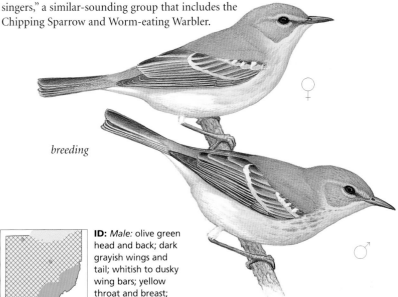

breeding

ID: *Male:* olive green head and back; dark grayish wings and tail; whitish to dusky wing bars; yellow throat and breast; faded, dark streaking or dusky wash on sides of breast; white undertail coverts and belly; faint, yellow, broken eye ring; a little duller in fall. *Female:* duller, especially in fall.
Size: *L* 5–5½ in; *W* 8½ in.
Habitat: *In migration:* mixed and deciduous woodlands. *Breeding:* open, mature pine woodlands and rarely mature pine plantations.
Nesting: toward the end of a pine limb; female builds a deep, open cup nest of twigs, bark, weeds, grasses, pine needles and spider webs and lines it with feathers; pair incubates 3–5 brown-speckled, whitish eggs for about 10 days.
Feeding: eats mostly insects, berries and seeds; often creeps along branches; may hang upside down on branch tips.

Voice: song is a short, musical trill; call note is a sweet *chip*.
Similar Species: *Prairie Warbler* (p. 262): distinctive, dark facial stripes; darker streaking on sides; yellowish wing bars. *Bay-breasted Warbler* (p. 264) and *Blackpoll Warbler* (p. 265): nonbreeding and immature birds have dark streaking on head or back or both; long, thin, yellow "eyebrows." *Yellow-throated Vireo* (p. 205): bright yellow "spectacles"; gray rump; lacks streaking on sides.
Best Sites: best sought in breeding areas where they nest; ridge tops in Shawnee, Hocking, Tar Hollow and Zaleski state forests; Lake Katherine SNP.

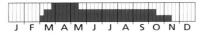

J F M A M J J A S O N D

PRAIRIE WARBLER

Dendroica discolor

The Prairie Warbler's easily learned song is a rapidly ascending, buzzy trill that escalates up the scale in a most distinctive manner. Male Prairie Warblers only sing for a short period—from late April to early July—but when they do, the unmistakable song, combined with its delivery from an exposed perch, makes these birds fairly easy to spot. • The Prairie Warbler is not a bird of the prairies, but of successional habitats such as regenerating clear-cuts and brushy, old fields. Peak numbers are found in the open red cedar glades that occur on the limestone soils of southwestern Ohio, from Adams County west to the Cincinnati area. • Coal strip-mine reclamation sites, largely covered with open grassland dotted with scruffy thickets of black locust, autumn-olive and other brushy plants, often support numerous Prairie Warblers.

♂

breeding

♀

ID: *Male:* bright yellow face and underparts, except for white under-tail coverts; dark "cheek" stripe and eye line; black streaking on sides; olive gray upperparts; inconspicuous chestnut streaks on back; 2 faint, yellowish wing bars. *Female* and *immature:* similar to male, but duller.
Size: *L* 4½–5 in; *W* 7 in.
Habitat: dry, open, scrubby fields; red cedar thickets; reclaimed strip mines; also woodland edges and young pine plantations with deciduous scrub.
Nesting: often in crotch of sapling, less than 10 ft from ground; small cuplike nest of grasses, bark, pine needles and thistle down, often bonded with spider webs;

female incubates 3–5 brown-spotted, pale gray eggs for about 12 days.
Feeding: gleans, hover-gleans and occasionally hawks for prey; mainly insectivorous; also eats berries and tree sap exposed by sapsuckers; nestlings favor caterpillars.
Voice: buzzy song is an ascending series of *zee* notes; call is sweet *chip*.
Similar Species: distinctive. *Pine Warbler* (p. 261): larger; bigger bill; lacks dark spot on side of throat, obvious dark line below eye and chestnut streaking on back. *Magnolia Warbler* (p. 254): fall bird resembles immature Prairie, but has white wing bars, gray neck band and white bands on tail.
Best Sites: cedar glades of southeastern Ohio, including Chaparral Prairie SNP and vicinity and Indian Creek WA; reclaimed strip mines such as Crown City WA and Egypt Valley WA; young clear-cut areas within Shawnee SF.

J F M A M J J A S O N D

PALM WARBLER

Dendroica palmarum

Palm Warblers breed farther north than almost any other warbler, occupying a distinctly untropical habitat of cold, tamarack-ringed bogs and open conifer forests. However, the first specimen was collected on wintering grounds on Hispaniola, and palm trees were undoubtedly in the scene. • Arriving in mid-April, Palm Warbler numbers peak in early May and are most numerous in Lake Erie shoreline habitats, where they gather before flying across the water. • Fall migration generally brings fewer birds, and like many warblers, these have molted into drabber plumage. Still, Palms are distinctive in that they forage extensively on the ground, and emphatically and constantly pump their tails. • The two well-marked subspecies—Western and Eastern—are separated by breeding ranges. Almost all Ohio migrants are of the duller Western group; however, there are a number of Eastern records. Eastern birds appear in late fall, and winter records would likely be this subspecies. They are more colorful, with yellow underparts and brighter brown upperparts.

breeding

ID: chestnut brown "cap" (may be inconspicuous in fall); yellow "eyebrow"; yellow throat and undertail coverts; yellow or white breast and belly; dark streaking on breast and sides; olive brown upperparts; may show dull, yellowish rump; frequently bobs its tail.
Size: *L* 4–5½ in; *W* 8 in.
Habitat: all types of woodlands, but most often in open, scruffy habitats, old fields, thickets, marshes and even open lawns.
Nesting: does not nest in Ohio.
Feeding: generally forages on the ground, or low in bushes or small trees; gleans the ground and vegetation for wide variety of insects and berries while perched or hovering; occasionally hawks for insects; may take some seeds.
Voice: song is a weak, buzzy trill with a quick finish; call is a sharp *sup* or *check*.
Similar Species: spring birds distinctive. *Yellow-rumped Warbler* (p. 257): similar to duller fall Palm Warbler, but has bright yellow rump and does not wag tail.
Best Sites: *In migration:* statewide, but largest numbers along L. Erie; Mentor Headlands; causeway road through Sheldon Marsh SNP; grassy areas along parking lots at Magee Marsh bird trail.

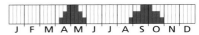

BAY-BREASTED WARBLER

Dendroica castanea

The boreal forest that stretches across northern North America, from Alaska to Newfoundland, is dominated by conifers, primarily spruce and fir, and is home to a small suite of neotropical warblers, including the Bay-breasted Warbler. These birds are primary predators of the spruce budworm, an insect that experiences periodic population outbreaks. Bay-breasted Warbler numbers fluctuate with spruce budworm populations, and this affects the Bay-breasted numbers that Ohio observers see in migration in a given year. • Spring males are a stunning sight, bedecked in rich chestnut tones, and are absolutely unmistakable. The same cannot be said of their song, which is a very high-pitched series of lispy notes, beyond the range of hearing for some observers. • Bay-breasted Warblers are most closely related to the Blackpoll Warbler, and there are three known hybrids.

breeding

Habitat: a broad range of trees, shrubs and woodlands; do not particularly favor conifers as on breeding grounds.

Nesting: does not nest in Ohio.

Feeding: usually forages at midlevel of trees; gleans vegetation and branches for a variety of insects, caterpillars and adult invertebrates.

Voice: song is an extremely high-pitched *seeee-seese-seese-seee;* call is a high *see.*

Similar Species: spring males unmistakable. *Blackpoll Warbler* (p. 265): resembles fall females and immatures, but normally has streaks on flanks and breast, is grayer on sides of neck and usually has yellow legs; lacks buffy tones on flanks.

Best Sites: L. Erie shoreline habitats such as Mentor Headlands, Sheldon Marsh SNP and Magee Marsh WA bird trail.

ID: *Breeding male:* black face and "chin"; chestnut crown, throat, sides and flanks; creamy yellow belly, under-tail coverts and patch on side of neck; 2 white wing bars. *Breeding female:* paler overall; dusky face; whitish to creamy underparts and neck patch; faint chestnut "cap"; rusty wash on sides and flanks. *Nonbreeding:* yellow olive head and back; dark streaking on crown and back; whiter underparts. *Immature:* resembles nonbreeding adult, but has less prominent streaking on upperparts; lacks chestnut sides and flanks.

Size: *L* 5–6 in; *W* 9 in.

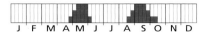

BLACKPOLL WARBLER

Dendroica striata

The Blackpoll Warbler is the greatest warbler migrant: weighing less than a wet teabag, eastern migrants are known to fly south over the Atlantic, leaving land at Cape Cod and flying for 88 nonstop hours to the northern coast of Venezuela. In a single year, a Blackpoll Warbler may fly up to 15,000 miles! In migration, they adjust their flying altitude—sometimes flying at altitudes of up to 20,000 feet—to take advantage of shifting prevailing winds to reach their destination. • On their northern breeding grounds, Blackpolls are similar to the Bay-breasted Warbler in habitat preferences, but occupy an even broader range, from Alaska to Newfoundland. They are common migrants through Ohio, and many more will be detected if you are familiar with their high-pitched song: a series of rapid notes that sounds like steam being intermittently released from a kettle.

♀

breeding

♂

ID: 2 white wing bars; black streaking on white underparts; white undertail coverts. *Breeding male:* black "cap" and "chin" stripe; white "cheek"; black-streaked, olive gray upperparts; white underparts; orange legs. *Breeding female:* streaked, yellow olive head and back; small, dark eye line; pale "eyebrow." *Nonbreeding:* olive yellow head, back, rump, breast and sides; yellow "eyebrow"; dark legs. *Immature:* paler streaking.
Size: *L* 5–5½ in; *W* 9 in.
Habitat: coniferous and mixed scrub; open coniferous growth on dry fens and bogs; backsides of ridged riverbanks and sparsely vegetated beach ridges; occurs in a wide range of trees and shrubs in migration.
Nesting: does not nest in Ohio.

Feeding: gleans buds, leaves and branches for aphids, mosquitoes, beetles, wasps, caterpillars and many other insects; often flycatches for insects.
Voice: song is an extremely high-pitched, uniform trill: *tsit tsit tsit;* call is a loud *chip.*
Similar Species: *Black-and-white Warbler* (p. 267): dark legs; black and white stripes on crown; male has black "chin," throat and "cheek" patch. *Bay-breasted Warbler* (p. 264): chestnut sides and flanks; buff undertail coverts; nonbreeding and immature lack dark streaking on underparts.
Best Sites: can be detected by song in almost every type of woody cover, including urban shade trees; best numbers in L. Erie shoreline habitats.

J F M A M J J A S O N D

265

CERULEAN WARBLER

Dendroica cerulea

The gorgeous, cerulean blue of this bird's upperparts is usually hard to see, as the Cerulean Warbler is a canopy specialist that rarely leaves the treetops. The smallest of the *Dendroica* warblers, the Cerulean is a persistent singer, with an upwardly spiraling, buzzy song that is easy to learn and often the only way these birds can be detected. • Many ornithologists are concerned about the rapid drop in Cerulean Warbler populations, attributable to habitat loss, forest fragmentation, overmanagement of woodlands for timber production and various arboreal diseases. Continued invasion of the nonnative gypsy moth, which attacks oaks, could further decrease Cerulean numbers. • Usually associated with mature bottomland forests along streams, Ohio birds seem most prevalent in older-growth, upland oak-hickory woodlands, probably because this alternate habitat is more common.

♀

breeding

♂

ID: white undertail coverts and wing bars. *Male:* blue upperparts; white throat and underparts; black "necklace" and streaking on sides. *Female:* blue green crown, nape and back; dark eye line; yellow "eyebrow," throat and breast; pale streaking on sides. *Immature:* pale yellow underparts and "eyebrow"; yellowish to white wing bars; pale olive green upperparts.

Size: *L* 4½–5 in; *W* 7 in.

Habitat: mature, oak-dominated hardwood forests and larger stands of mature bottomland forests along rivers and streams.

Nesting: high in a deciduous tree; female builds an open cup nest of bark strips, weeds, grass, lichen and spider silk and lines it with fur and moss; female incubates

3–5 brown-spotted, gray to creamy white eggs for 12–13 days.

Feeding: gleans insects from upper canopy foliage and branches; often hawks for insects.

Voice: song is a rapid, accelerating sequence of buzzy notes leading into a higher trilled note; call is a sharp *chip*.

Similar Species: adult male is unmistakable. *Blackburnian Warbler* (p. 259): immature can be confused with female and immature Cerulean, except Blackburnian has faint streaks on back, more prominent dusky "cheek," longer tail and paler, buffy underparts.

Best Sites: *Breeding:* Shawnee SF has largest breeding population in Ohio; also Clear Creek MP, Burr Oak SP, Lake Hope SP and MeadWestvaco Experimental Forest. *In migration:* uncommonly seen outside of breeding areas.

BLACK-AND-WHITE WARBLER

Mniotilta varia

The genus name *Mniotilta*, which means "moss-plucking," refers to this warbler's habit of creeping along tree trunks as it forages. In fact, it was once known as the "Pied Creeper" and was placed in the genus of creepers, *Certhia*. The Black-and-white Warbler is clearly related to the *Dendroica* genus of warblers, differing only in physiological adaptations to its feet that help it better grip bark. • Black-and-whites are fairly common migrants statewide, but as breeders they are mostly restricted to the large tracts of forest in southern and eastern Ohio. There, they favor mature hardwood forests of lower slopes and stream bottoms. • Very inconspicuous as they crawl along branches and tree trunks, Black-and-whites are more easily detected by their soft but easily recognized song. • The Black-and-white Warbler has hybridized with both the Cerulean Warbler and Blackburnian Warbler, and holds the longevity record among wood-warblers—a banded female was documented at more than 11 years old.

breeding

ID: black and white stripes on crown; white streaking on dark upperparts; 2 white wing bars; white underparts with black streaking on sides, flanks and undertail coverts; black legs. *Breeding male:* black "cheek" and throat. *Breeding female:* gray "cheek"; white throat.
Size: *L* 5 in; *W* 8 in.
Habitat: *Breeding:* mature deciduous forests, ranging from streamside bottomland woods to lower and midslopes. *In migration:* all types of woody vegetation.
Nesting: in a shallow scrape on the ground, often among a pile of dead leaves next to a tree, log or large rock; female builds a cup nest with grass, leaves, rootlets and pine needles and lines it with fur and

fine grasses; female incubates 5 brown-flecked, creamy white eggs for 10–12 days.
Feeding: gleans insect eggs, larval insects, beetles, spiders and other invertebrates while creeping along tree trunks and branches.
Voice: song is a series of high-pitched, thin, 2-syllable undulating notes: *weetsee weetsee weetsee weetsee weetsee weetsee;* call is a sharp *pit* and a soft, high *seat*.
Similar Species: *Blackpoll Warbler* (p. 265): breeding male has solid black "cap" and clean white undertail coverts.
Best Sites: *Breeding:* large tracts of southern forest, such as Shawnee, Zaleski and Tar Hollow state forests; also Clear Creek MP (entrance to Hemlock Trail). *In migration:* any woodland area.

J F M A M J J A S O N D

267

AMERICAN REDSTART

Setophaga ruticilla

American Redstarts are hyperactive, flashy warblers that consume large numbers of flying insects. They behave much like flycatchers and have broad, flat, flycatcher-like bills and well-developed rictal bristles (stiff hairs around the mouth) to aid in capturing winged bugs. • Redstarts frequently spread their wings and fan their tails, flashing the colored patterns in their plumage. It is thought that this display startles prey from foliage. • The brilliant orange patches on the wings and tail of adult males is replaced with yellow in females and first-year males. Females rarely if ever sing, so if a "Yellowstart" is seen singing, it's a young male. • Redstarts are quite common migrants and can be locally abundant breeders where appropriate habitat exists. They favor young, wet woods with a dense understory of shrubs, such as often occur along streams and around swamps.

ID: *Male:* black overall; red orange foreshoulder, wing and tail patches; white belly and undertail coverts. *Female* and *1st-year male:* olive brown upperparts; gray green head; yellow foreshoulder, wing and tail patches; clean white underparts.
Size: *L* 5 in; *W* 8½ in.
Habitat: shrubby woodland edges; open and semi-open deciduous and mixed forests with a regenerating deciduous understory of shrubs and saplings; often near water.
Nesting: in a shrub or sapling, 3–23 ft above the ground; female builds an open cup nest of plant down, bark shreds, grass and rootlets and lines it with feathers; female incubates 4 brown-marked, whitish eggs for 11–12 days.
Feeding: actively gleans foliage and hawks for insects and spiders on leaves, buds and branches; often hover-gleans.
Voice: song is a highly variable series of sharp, high-pitched *tseet* or *zee* notes, often given at different pitches; usually has a strongly downslurred ending; call is a sharp, sweet *chip*.
Similar Species: none in Ohio.
Best Sites: *Breeding:* Little Beaver SP; Shawnee SF; Clear Creek MP. *In migration:* common statewide; huge numbers along Magee Marsh WA bird trail in May.

J	F	M	A	M	J	J	A	S	O	N	D

PROTHONOTARY WARBLER

Protonotaria citrea

P rothonotaries are clergy in the Catholic Church, noted for garbing themselves in robes of brilliant yellow. Yellow-clad Prothonotary Warblers appear luminescent as they flit about their shady, swampy habitat, inspiring the colloquial name "Golden Swamp Warbler." • Two primary limiting factors make this species one of our more locally distributed breeding warblers. First, the Prothonotary is a cavity-nesting wood-warbler—one of only two, along with the western Lucy's Warbler (*Vermivora luciae*); and second, the Prothonotary requires swampy woods, stagnant river oxbows and mature riparian woodlands buffering large, sluggish streams. • As Ohio has lost over 90 percent of its original wetlands, species such as the Prothonotary have undoubtedly declined. Fortunately, this stunning bird takes readily to human-made nest boxes, and erecting these in suitable habitat can bolster populations.

breeding

ID: large, dark eyes; long bill; unmarked, yellow head; yellow underparts except for white undertail coverts; olive green back; unmarked, bluish gray wings and tail; dark gray legs.
Size: *L* 5½ in; *W* 8½ in.
Habitat: wooded deciduous swamps and riparian woods buffering sluggish streams.
Nesting: cavities in standing dead trees, rotten stumps, birdhouses or abandoned woodpecker nests, from water level to 10 ft above the ground; often returns to the same nest site; female chooses one of the male's moss-filled cavities, lines it with soft plant material and incubates 4–6 brown-spotted, creamy to pinkish eggs for 12–14 days.

Feeding: forages for a variety of insects and small mollusks; gleans vegetation; may hop on floating debris or creep along tree trunks.
Voice: song is a loud, ringing series of *sweet* or *zweet* notes issued on a single pitch; flight-song is *chewee chewee chee chee;* call is a brisk *tink.*
Similar Species: *Blue-winged Warbler* (p. 246): white wing bars; black eye line; yellowish white undertail coverts. *Yellow Warbler* (p. 252): dark wings and tail with yellow highlights; yellow undertail coverts; male has reddish breast streaking.
Best Sites: upper end of Hoover Reservoir; St. Marys R., which flows out of Grand Lake St. Marys (where US Rte. 127 crosses river); Killbuck WA; woods along Scioto R. just south of Greenlawn Dam.

J F M A M J J A S O N D

WORM-EATING WARBLER

Helmitheros vermivorus

P ossibly the most secretive and difficult to detect of our breeding warblers, the Worm-eating Warbler is not commonly seen in migration outside of nesting areas. Knowing this species' song—a dry, rather insectlike trill—will tremendously improve your chances of locating this species. • Worm-eaters are obligate slope species, occupying steep, wooded hillsides with a well-developed understory of shrubs; thus they are largely restricted to the unglaciated region of southeastern Ohio. • This bird's name stems from a major food item, caterpillars—once referred to as worms—that are gleaned from lower levels in the forest understory. This warbler forages in hanging clusters of dead leaves and will occasionally poke about in leaf litter on the forest floor. • This is one of six Ohio warblers that nest on the ground. If you inadvertently stumble into the vicinity of a Worm-eating Warbler nest, this normally retiring species will boldly advance within close range, scolding loudly with loud, sharp, *chip* call notes.

ID: black and buff orange head stripes; brownish olive upperparts; rich buff orange breast; whitish undertail coverts.
Size: *L* 5 in; *W* 8½ in.
Habitat: steep, deciduous woodland hillsides and ravines, and wooded slopes along rivers.
Nesting: usually on a hillside or ravine bank, often near water; on the ground hidden under leaf litter; female builds a cup nest of decaying leaves and lines it with fine grass, moss stems and hair; female incubate 3–5 brown-speckled, white eggs for about 13 days.

Feeding: forages on the ground and in trees and shrubs; eats mostly caterpillars and small insects.
Voice: song is a faster, thinner version of the Chipping Sparrow's chipping trill; call is a buzzy *zeep-zeep*.
Similar Species: sexes similar and unmistakable. *Swainson's Warbler* (p. 340): very rare; duller brown upperparts; lacks striped crown.
Best Sites: large contiguous tracts of forests in southeastern Ohio: Shawnee, Zaleski and Scioto Trail state forests; also Lake Katherine SNP and Lake Hope SP.

J F M A M J J A S O N D

OVENBIRD

Seiurus aurocapilla

The Ovenbird's loud, ringing song is common in woodlands throughout Ohio and is easily recognized. Robert Frost immortalized this species in his poem "Ovenbird," an ode to the distinctive voice of this odd warbler. • A more terrestrial version of the waterthrush, the Ovenbird resembles a small thrush as it walks through the leaf litter of the forest floor. It reaches peak numbers in large tracts of contiguous woodlands, but forest fragmentation has reduced its numbers. Always vulnerable to Brown-headed Cowbird parasitism, Ovenbirds forced to utilize smaller fragmented woodlands are probably heavily parasitized, leading to low nesting success. • The curious name stems from the style of this bird's nest, which is built on the ground and resembles an old-fashioned Dutch oven. • Ovenbirds can be surprisingly hardy, and there are at least 10 early winter records, with most of these birds observed at feeders.

ID: olive brown upperparts; white eye ring; heavy, dark streaking on white breast, sides and flanks; rufous crown bordered with black; pink legs; white undertail coverts; no wing bars.

Size: *L* 6 in; *W* 9½ in.

Habitat: *Breeding:* undisturbed, mature deciduous forests with a closed canopy and very little understory; often in ravines and slopes bordering riparian areas. *In migration:* dense riparian shrubbery and thickets and a variety of woodlands.

Nesting: on the ground; female builds an oven-shaped, domed nest of grass, weeds, bark, twigs and dead leaves and lines it with animal hair; female incubates 4–5 white eggs, spotted with gray and brown, for 11–13 days.

Feeding: gleans the ground for worms, snails, insects and occasionally seeds.

Voice: song is a loud, distinctive *tea-cher tea-cher tea-CHER tea-CHER,* increasing in speed and volume; rarely heard dusk song is an elaborate series of bubbly, warbled notes, often ending in *teacher-teacher;* call is a brisk *chip, cheep* or *chock.*

Similar Species: *Northern Waterthrush* (p. 272) and *Louisiana Waterthrush* (p. 273): bold, yellowish or white "eyebrows"; darker upperparts; lack rufous crown. *Thrushes* (pp. 233–39): larger; lack rufous crown outlined in black. *Kentucky Warbler* (p. 274): similar song, but more monotone and lacks upward inflection at end of each phrase.

Best Sites: *Breeding:* large woodlands such as Lake Katharine SNP. *In migration:* wherever woody cover is found.

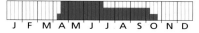

J F M A M J J A S O N D

NORTHERN WATERTHRUSH

Seiurus noveboracensis

The genus name *Seiurus*, which means "to wave tail," fits waterthrushes well, as they are chronic "bobber-butts." • Northern Waterthrushes are very unwarblerlike in their habits, as they forage almost exclusively on the ground. This trait, along with their distinctive plumage, makes waterthrushes unmistakable; however, the two waterthrush species are often confused, particularly in spring. Louisiana Waterthrushes nest along rocky, wooded streams and are only rarely detected as migrants outside this habitat. They also return several weeks earlier than Northerns. Any migrant waterthrush after late April is almost certain to be a Northern. • Waterthrushes sing frequently in migration, a characteristic that serves to clinch identification, as Louisiana and Northern songs are as different as night and day. • Northern Waterthrushes are common migrants, and prefer wet, swampy woodlands. Although they breed primarily north of Ohio, small numbers breed in swamp woods in four or five counties in extreme northeastern Ohio.

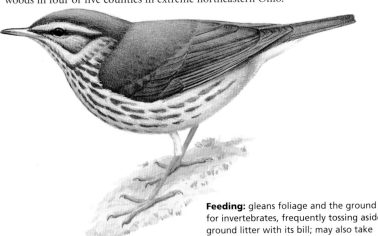

ID: pale yellowish to buff "eyebrow" and underparts (occasionally white); dark streaking on underparts; finely spotted throat; olive brown upperparts; pinkish legs; frequently bobs tail.

Size: *L* 5–6 in; *W* 9½ in.

Habitat: rarely seen away from water; prefers very wet, swampy woods; migrants use shorelines of ponds and riverbanks.

Nesting: on the ground, usually near water; female builds a cup nest of moss, leaves, bark shreds, twigs and pine needles and lines it with moss, hair and rootlets; female incubates 4–5 brown-speckled, whitish eggs for about 13 days.

Feeding: gleans foliage and the ground for invertebrates, frequently tossing aside ground litter with its bill; may also take aquatic invertebrates and small fish from shallow water.

Voice: song is a loud, 3-part *sweet sweet sweet, swee wee wee, chew chew chew chew;* call is a brisk *chip* or *chuck,* sharper and more metallic than Louisiana Waterthrush.

Similar Species: *Louisiana Waterthrush* (p. 273): broader, white "eyebrow"; unspotted, white throat; orange buff wash on flanks. *Ovenbird* (p. 271): rufous crown bordered with black stripe; white eye ring; unspotted throat; lacks pale "eyebrow."

Best Sites: low wet woodlands; borders of ponds and wetlands; Magee Marsh bird trail; pond at Green Lawn Cemetery; Calamus Swamp.

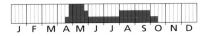

J F M A M J J A S O N D

LOUISIANA WATERTHRUSH

Seiurus motacilla

The first of the long-distance, neotropical migrant warblers to return in spring, the earliest Louisiana Waterthrushes appear in southern Ohio in the last week of March. Their arrival is heralded by their song—a loud, clear, ringing series of slurred whistles that ends with a delightfully rapid jumble of notes. They are also the first warblers to depart in fall. • Home for these birds is the narrow riparian zone of small, rocky creeks bordered by mature woodlands—habitat found mostly in the hill country of southern and eastern Ohio. • Louisiana Waterthrushes tend to forage on the ground along the banks of creeks, and are easily found by their song and their distinctive, explosive *chick* call note. • This is an amusing bird to watch; calling it a tail-bobber would be an understatement. It seems to involve its entire body in the action, pumping its hindquarters up and down, and bobbing about like a Spotted Sandpiper gone mad.

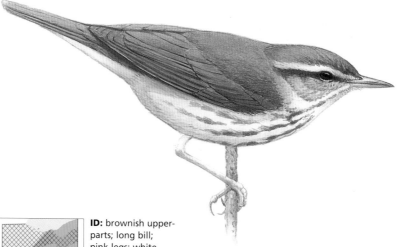

ID: brownish upper-parts; long bill; pink legs; white underparts; orange buff wash on flanks; long, dark streaks on breast and sides; bicolored, buff-and-white "eyebrow"; clean white throat.

Size: *L* 6 in; *W* 10 in.

Habitat: forested ravines alongside fast-flowing streams; rarely along wooded swamps.

Nesting: concealed within a rocky hollow or within a tangle of tree roots; both adults build a cup nest of leaves, bark strips, twigs and moss and line it with animal hair, ferns and rootlets; female incubates 3–6 finely speckled, creamy white eggs for about 14 days.

Feeding: gleans terrestrial and aquatic insects and crustaceans from rocks and debris in or near shallow water; dead leaves and other debris may be flipped and probed for food; occasionally catches flying insects over water.

Voice: song begins with 3–4 distinctive, shrill, slurred notes followed by a warbling twitter; call is a brisk *chick* or *chink*.

Similar Species: *Northern Waterthrush* (p. 272): narrower, off-white eye line; streaked throat; darker, duller legs; larger bill; birds with obvious yellow tones on eye line and underparts. *Ovenbird* (p. 271) and *thrushes* (pp. 233–39): lack broad, white "eyebrow."

Best Sites: Shawnee SF; Scioto Trail SF; Cuyahoga Valley NRC; Little Beaver Creek SP; Highbanks MP.

J F M A M J J A S O N D

KENTUCKY WARBLER

Oporornis formosus

The Kentucky Warbler is a striking species that is hard to spot, as it occupies forests with a dense understory of shrubs and tends to remain concealed in thick vegetation. It isn't hard to hear, though, delivering a loud, rich song—*chur-ree chur-ree*—suggestive of a Carolina Wren or Ovenbird, but more monotone and less musical. Becoming familiar with the song will greatly increase your chances of detecting this species. • The best views are often had when a birder happens into the vicinity of a nest. Then, these normally retiring birds will approach the intruder quite closely, scolding incessantly with a loud *chip*. • The Kentucky Warbler is common in Kentucky, but reaches the northern limits of its range in Ohio. It is common in the extensive woodlands of southern and eastern Ohio, becoming scattered and rare north of that area. • Increasing deer herds, with their attendant heavy browsing of forest understory plants, may have adversely affected Kentucky Warblers, as might the increased proliferation of exotic shrubs such as bush honeysuckle and privet.

ID: bright yellow "spectacles" and underparts; black crown, "sideburns" and "half mask"; olive green upperparts.
Size: *L* 5–5½ in; *W* 8½ in.

Habitat: moist deciduous and mixed woodlands with dense shrubby cover and herbaceous plant growth, such as wooded ravines and hillsides and creek bottomlands.

Nesting: on or close to the ground; both adults build a bulky cup nest of plant material and hair and line it with rootlets and hair; female incubates 4–5 cream-colored eggs, spotted or blotched with reddish brown, for 12–13 days.

Feeding: gleans insects while hopping along the ground and flipping over leaf litter or by snatching prey from the undersides of low foliage.

Voice: musical song is a series of 2-syllable notes: *chur-ree chur-ree;* call is a sharp *chick, chuck* or *chip*.

Similar Species: *Hooded Warbler* (p. 278): bright yellow face. *Common Yellowthroat* (p. 277): much smaller; thicker black "mask"; lacks yellow spectacles.

Best Sites: Clear Creek MP; MeadWestvaco Experimental Forest; Lake Katherine SNP; Mohican SF; Edge of Appalachia Preserve.

J F M A M J J A S O N D

CONNECTICUT WARBLER

Oporornis agilis

Of the 36 species of wood-warblers that are found annually in Ohio, the Connecticut Warbler is easily the most coveted and hardest to find. Many very good birders may have periods of several years without seeing one, and more than one seasoned birder has yet to add this one to their life list. There are two reasons that conspire to make a sighting a noteworthy event: Connecticuts are rare migrants with relatively few birds passing through, and they are late migrants prone to skulking in dense vegetation that is fully leafed out by the time of their passage. The Connecticut Warbler is probably somewhat more common than is thought, and if you know its loud but only infrequently given song, chances of detection rise considerably. • It wasn't until 1883 that the first nest was discovered—in wet, northern coniferous forests. This species also remains the least known of our warblers on its wintering grounds, which are thought to be mostly in the Amazon basin of South America.

breeding

ID: bold, white eye ring; yellow underparts; olive green upperparts; long undertail coverts make tail look short; pink legs; longish bill. *Breeding male:* blue gray "hood." *Female* and *immature:* gray brown "hood"; light gray throat.
Size: *L* 5–6 in; *W* 9 in.
Habitat: skulks in very dense growth, such as the understory of young woods, thickets and vine tangles.
Nesting: does not nest in Ohio.
Feeding: gleans caterpillars, beetles, spiders and other invertebrates from ground leaf litter; occasionally forages among low branches.

Voice: song is a loud, clear, explosive *chipity-chipity-chipity chuck* or *per-chipity-chipity-chipity choo;* call is a brisk, metallic *cheep* or *peak.*
Similar Species: relatively unmistakable; its habit of walking, not hopping, separates it from similar species. *Mourning Warbler* (p. 276): usually no eye ring; male has blackish breast patch. *Nashville Warbler* (p. 250): smaller; bright yellow throat; different behavior.
Best Sites: Magee Marsh WA bird trail; Green Lawn Cemetery; Mentor Headlands; Sheldon Marsh SNP.

J F M A M J J A S O N D

MOURNING WARBLER

Oporornis philadelphia

Spotting a Mourning Warbler is always a treat—they are relatively uncommon, and like their close relative the Connecticut Warbler, tend to skulk in very dense growth. On the plus side, Mournings forage on or near the ground, so at least observers won't contract "warbler neck" from peering up into the treetops. • Logging has aided this species by creating more suitable habitat, and the overall population is undoubtedly larger than before European settlement—Alexander Wilson, who discovered it in 1810, saw but one in his life. • Mourning Warblers primarily breed well north of Ohio and favor clearings of scruffy, successional tree growth, underlain with dense shrub zones. They are very rare breeders in Ohio, with small numbers nesting locally in a few counties of the extreme northeast, and possibly the Oak Openings and Mohican State Forest regions.

breeding

ID: blue gray "hood"; black upper breast patch; yellow underparts; olive green upperparts; short tail; pinkish legs. *Breeding male:* usually no eye ring, but may have broken eye ring. *Female:* gray "hood"; whitish "chin" and throat; may show thin eye ring.
Size: *L* 5–5½ in; *W* 7½ in.
Habitat: dense, shrubby thickets, tangles and brambles, often in moist areas of forest clearings and along edges of ponds, lakes and streams.
Nesting: on the ground at base of shrub or plant tussock or low in a small shrub or tree; bulky nest of leaves, weeds and grass is lined with fur and fine grass; female incubates 3–4 brown-blotched, creamy white eggs for about 12 days.

Feeding: forages in dense low shrubs for caterpillars, beetles, spiders and other invertebrates.
Voice: husky, 2-part song is variable and lower-pitched at the end: *churry, churry, churry, churry, chorry, chorry;* call is a loud, low *check.*
Similar Species: *Connecticut Warbler* (p. 275): bold, complete eye ring; long undertail coverts make tail look very short; lacks black breast patch; immature has light gray throat. *MacGillivray's Warbler:* western species, reported but not verified in Ohio; extremely similar; thick, white eye arcs; prominent, black lores through eyes.
Best Sites: *Breeding:* around scruffy openings in forests and shrubby perimeters of bogs and wetlands in Ashtabula, Geauga, Lake, Portage and Summit counties; also in vicinity of Irwin Prairie SNP and Mohican SF. *In migration:* irregular; dense shrubby growth at migrant hotspots.

COMMON YELLOWTHROAT

Geothlypis trichas

Famous Swedish biologist Carolus Linnaeus named the Common Yellowthroat in 1766, making it one of the first North American birds to be described. Linnaeus apparently improved his birding skills with time; his epithet *trichas* means "a thrush"—not even close. • Our most abundant and widespread breeding warbler, the yellowthroat's loud, ringing song is easily learned and can be heard in all manner of shrubby successional habitats, particularly in wet areas. Spotting this bird is harder, as it is a real skulker, but it can often be enticed to show itself by a loud bout of pishing. • Common Yellowthroats range throughout much of the continent, and have been split into 13 subspecies. • One of our hardier warblers, small numbers of Common Yellowthroats regularly attempt to overwinter, but most probably don't survive. These individuals are best sought in dense cattail stands.

ID: yellow throat, breast and undertail coverts; dingy white belly; olive green to olive brown upperparts; orangy legs. *Breeding male:* distinctive, broad, black "mask" with white upper border. *Female:* no "mask"; may show faint white eye ring.
Size: *L* 5 in; *W* 7 in.
Habitat: cattail marshes, riparian willow and alder clumps, sedge wetlands, beaver ponds and wet, overgrown meadows; sometimes in dry, abandoned brushy fields.
Nesting: on or near the ground, in a small shrub or among emergent aquatic vegetation; female builds a bulky open cup nest of weeds, grass, sedges and other materials and lines it with hair and soft plant fibers; female incubates 3–5 darkly marked, creamy white eggs for 12 days.

Feeding: gleans vegetation and hovers for adult and larval insects, including dragonflies, spiders and beetles; occasionally eats seeds.
Voice: song is a clear, oscillating *witchety witchety witchety-witch;* call is a sharp *tcheck* or *tchet.*
Similar Species: *Kentucky Warbler* (p. 274): yellow "spectacles"; all-yellow underparts; "half mask." *Mourning Warbler* (p. 276): immature is similar to plainer female and immature yellowthroats, but has bright yellow throughout underparts.
Best Sites: found in appropriate habitat statewide; rare in dense cattail stands in wetlands in winter.

J	F	M	A	M	J	J	A	S	O	N	D

HOODED WARBLER

Wilsonia citrina

The colorful Hooded Warbler is difficult to see, as it is an obligate forest understory bird of mature woodlands thickly overgrown with shrubs such as spice-
bush, viburnum, dogwood and various saplings. Fortunately, these persistent
singers normally stay low to the ground; by following their loud, clear, whistled song,
the birder can be rewarded with a view of one of our most spectacular wood-warblers.
• Females are promiscuous, and studies have suggested that as many as one-third
produce young fathered by males other than their mate; the reasons for this are as
yet unclear. • Life is fraught with danger for small birds, and some Hooded Warblers
face a particularly horrifying demise on their Central American wintering grounds.
An enormous tropical relative of our bullfrog, Vaillant's frog, which can grow to 7
inches in length, has been recorded snapping up and swallowing Hooded Warblers.
• To distinguish female and immature Hooded Warblers, note the very large, black eye
set off by the yellow face—this species has the largest eyes of any wood-warbler, an
adaptation for improved vision in shady habitats. Also,
the tail has large, white spots that are frequently
displayed via tail fanning.

grass, bark strips, dead leaves, animal hair,
spider webs and plant down; female incubates 4 brown-spotted, creamy white eggs
for about 12 days.

ID: bright yellow
underparts; olive
green upperparts;
white undertail; very
large, black eyes;
pinkish legs. *Male:*
black "hood"; bright
yellow face. *Female:*
yellow face and olive crown; may show
faint traces of black "hood."
Size: *L* 5½ in; *W* 7 in.
Habitat: dense, low shrubs in mature
deciduous forests; most common along
stream bottoms and moderately humid
lower slopes where woody understory
growth tends to be thicker.
Nesting: low in a deciduous shrub; mostly
the female builds an open cup nest of fine

Feeding: gleans insects and other forest
invertebrates from the ground or from
shrub branches; may flycatch or scramble
up tree trunks.
Voice: clear, whistling song is some variation of *whitta-witta-wit-tee-yo;* call note is
a metallic *tink, chink* or *chip.*
Similar Species: male is unmistakable.
Wilson's Warbler (p. 279) and *Yellow
Warbler* (p. 252): similar to female and
immature Hooded Warblers, but lack white
tail spots and dark lores.
Best Sites: Clear Creek MP; Lake Katherine
SNP; Scioto Trail SF; Shawnee SF; throughout Hocking Hills.

J F M A M J J A S O N D

WILSON'S WARBLER

Wilsonia pusilla

Wilson's Warbler is normally confiding and quite approachable, but don't expect one to sit still for detailed observation. This is perhaps the most hyperactive of warblers, constantly flitting from branch to branch, making frequent flycatching sallies, and continually flipping its tail and flicking its wings. • Spotting a Wilson's Warbler is always a treat, as they nest in wet shrub zones in the boreal region well north of Ohio and are generally uncommon migrants here. Even a good day might only yield a handful, or possibly a few dozen, along Lake Erie. This is also one of our latest spring migrants, often still passing through as late as early June. • Wilson's Warbler may turn up anywhere in thickets around wetlands; becoming familiar with the rather dry, rapid series of *chip* notes that comprise the song will lead you to many more birds than would otherwise be found. • This beautiful warbler was named in honor of Alexander Wilson, one of North America's preeminent ornithologists and the discoverer of many new species of birds, who lived a short 47 years from 1766 to 1813.

ID: yellow under-parts; yellow green upperparts; beady, black eyes; black bill; orange legs. *Male:* black "cap." *Female:* "cap" is very faint or absent.

Size: *L* 4½–5 in; *W* 7 in.

Habitat: prefers shrub zones, particularly around water; frequently seen in willows, dogwoods and buttonbush; sometimes in ornamental shrubbery and other land-scaped areas.

Nesting: does not nest in Ohio.

Feeding: hovers, flycatches and gleans vegetation for insects.

Voice: song is a rapid chatter that drops in pitch at the end: *chi chi chi chi chet chet;* call is a flat, low *chet* or *chuck.*

Similar Species: male's black "cap" is distinctive. *Yellow* (p. 252), *Hooded* (p. 278) or *Orange-crowned* (p. 249) *warblers:* females similar to Wilson's females, but lack one or more of tail spots, eye line, streaking on underparts or eye ring.

Best Sites: wet shrub zones bordering L. Erie, especially Magee Marsh WA bird trail; thickets around wetlands.

J F M A M J J A S O N D

CANADA WARBLER

Wilsonia canadensis

One of our rarest nesting warblers, the Canada Warbler breeding population in Ohio is quite disjunct from the primary nesting population. It occupies a very specialized breeding habitat—scruffy, overgrown successional openings caused by tree blowdowns in our largest and most pristine hemlock gorges. These hemlock communities occur on the steep slopes of narrow valleys, which possess a cooler microclimate that approximates the coniferous forests found well north of Ohio. Conservation of this rare and biologically significant habitat is vital. • Often found in association with Canada Warblers are other rare Ohio breeders, such as the Magnolia Warbler, Blue-headed Vireo and Hermit Thrush. • The male Canada Warbler, adorned with a black "necklace" that looks like it was painted on with a leaky fountain pen, is one of the showiest of any bird.

ID: yellow "spectacles"; yellow underparts (except white undertail coverts); blue gray upperparts; pale legs. *Male:* streaky, black "necklace"; dark, angular "half mask." *Female:* blue green back; faint "necklace."

Size: *L* 5–6 in; *W* 8 in.

Habitat: *In migration:* thickets and shrub zones, usually low to the ground. *Breeding:* restricted to hemlock gorges.

Nesting: on a mossy hummock or upturned root or stump; female builds a loose, bulky cup nest of leaves, grass, ferns, weeds and bark and lines it with animal hair and soft plant fibers; female incubates 4 brown-spotted, creamy white eggs for 10–14 days.

Feeding: gleans the ground and vegetation for beetles, flies, hairless caterpillars, mosquitoes and other insects; occasionally hovers.

Voice: song begins with 1 sharp *chip* note and continues with a rich, variable warble; call is a loud, quick *chick* or *chip;* knowing the song is essential to locating this species on its breeding grounds.

Similar Species: male is unmistakable. *Nashville Warbler* (p. 250) and *Magnolia Warbler* (p. 254): similar to dull female Canada, except Nashville has bright yellow undertail coverts and much greener wings; Magnolia has white wing bars, faint gray neck band and greenish back.

Best Sites: any good birding hotspot with shrubby thickets; best in L. Erie shoreline habitats such as Sheldon Marsh SNP, Mentor Headlands or Magee Marsh WA; breeders in hemlock gorges in the Hocking Hills, Mohican SF or select Lake County MPs.

J F M A M J J A S O N D

YELLOW-BREASTED CHAT

Icteria virens

The clown prince of the warbler family, the Yellow-breasted Chat bears little resemblance to any of the other warblers, and is quite different behaviorally. In fact, over the years some authorities have suggested that it shouldn't be grouped with the warblers—it was once placed in the genus *Turdus* along with the American Robin! However, recent genetic studies bear out the warbler relationship. • Prone to lurking about in dense thickets, our largest warbler is normally quite furtive and difficult to see, though the Yellow-breasted Chat's unwarblerlike song can't be missed. Hopping from branch to branch like a tipsy acrobat, the chat works its way to the summit of a densely leafed sapling, delivering a bizarre series of loud hoots, cackles, squawks and grunts. Pishing or squeaking loudly at the singer will often entice him briefly into the open. • A species of early successional habitats—old, brushy fields, regenerating clear-cuts and young red cedar groves—the chat is common and well distributed in Ohio wherever this habitat occurs.

♂

ID: white "spectacles"and jaw line; heavy, black bill; yellow breast; white undertail coverts; olive green upperparts; long tail; gray black legs.
Male: black lores. *Female:* gray lores.
Size: *L* 7½ in; *W* 9½ in.
Habitat: riparian thickets, brambles, overgrown fields, clear-cuts and shrubby tangles.
Nesting: low in a shrub or deciduous sapling; well-concealed, bulky base of leaves and weeds holds an inner woven cup nest made of vine bark and lined with fine grass and plant fibers; female

incubates 3–4 brown-spotted, creamy white eggs for about 11 days.
Feeding: gleans insects from low vegetation; eats many berries in fall.
Voice: song is an assorted series of whistles, "laughs," squeaks, grunts, rattles and mews; calls include a *whoit, chack* and *kook.*
Similar Species: none.
Best Sites: Crown City WA; Indian Creek WA; cedar thickets around Caesar Creek Reservoir, Chaparral Prairie SNP and Edge of Appalachia Preserve.

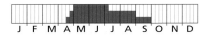

J F M A M J J A S O N D

SUMMER TANAGER

Piranga rubra

Far less common than the Scarlet Tanager in Ohio, Summer Tanagers are nonetheless locally common within a limited region. They are oak-hickory specialists, reaching greatest abundance in the largest tracts of woodlands of southern and eastern Ohio. There, they occupy ridge tops and other upland habitats, where their specialized forest community occurs. • Summer Tanagers feed heavily on bees and wasps, which they often pluck from the air in brief sallies and then beat against a branch until dead. • This species can be elusive, as the Summer Tanager tends to remain high in the crowns of leafy trees. Voice is by far the best way to detect its presence, and the call note—a mechanical *pit-i-tuck*—is easier to recognize than the song. • Early British naturalist Mark Catesby originally dubbed this species the "Summer Redbird." • Amazingly, there is one winter record for this highly migratory neotropical species, from Lorain County in January 2002.

ID: *Male:* rose red overall; pale bill; immature male has patchy red and greenish plumage. *Female:* grayish to greenish yellow upperparts; dusky yellow underparts; may have orange or reddish wash overall.
Size: *L* 7–7½ in; *W* 7¾ in.
Habitat: upland forests dominated by oak and hickory.
Nesting: about 12 feet above the ground in the fork of a horizontal limb, usually over an opening; nest is a ragged cup of vegetation, often so flimsy that eggs can be seen through the nest; female incubates 3–4 brown-spotted, pale blue to green eggs for 11–12 days.
Feeding: eats mainly insects; also takes berries and small fruits; gleans insects from the tree canopy; may hover-glean or hawk insects in midair; known to raid wasp nests.
Voice: song is a series of 3–5 sweet, clear, whistled phrases, like a faster version of the American Robin's song; call is *pit* or *pit-a-tuck*.
Similar Species: males are unmistakable. *Scarlet Tanager* (p. 283): female has smaller bill, darker wings, brighter underparts and uniformly olive upperparts.
Best Sites: Shawnee SF; Lake Katherine SNP; MeadWestvaco Experimental Forest; Burr Oak SP.

J F M A M J J A S O N D

SCARLET TANAGER

Piranga olivacea

One of our most colorful neotropical migrants, the Scarlet Tanager's spectacular plumage might even entice an "anti-birder" to join our ranks. • While Scarlet Tanagers favor oaks, they are far more common and wide-ranging, and much less restricted to oak-hickory habitats than are Summer Tanagers. They probably breed in every county and can be common in large forests. • The Scarlet Tanager is one of our longest-distance songbird migrants, with wintering grounds as far south as Bolivia. • The Scarlet Tanager, American Robin, Summer Tanager and Rose-breasted Grosbeak have similar songs. However, the Scarlet Tanager's song has a very raspy or burry quality—it sounds like a robin with a sore throat. Its oft-given *chip-burr* call also gives it away as it forages among the treetops. • Although very difficult to detect, Scarlet Tanagers have a small tooth on the upper mandible of the bill, a feature not shared by Summer Tanagers.

ID: *Breeding male:* bright red overall with pure black wings and tail; pale bill. *Female:* uniformly olive upperparts; yellow underparts; grayish brown wings. *Nonbreeding male:* patchy red and green yellow plumage; bright yellow underparts; olive upperparts; black wings and tail.
Size: *L* 6½–7½ in; *W* 11½ in.
Habitat: fairly mature, upland deciduous and mixed forests and large woodlots.
Nesting: high in a deciduous tree; female builds a flimsy, shallow cup of grass, weeds and twigs and lines it with rootlets and fine grass; female incubates 2–5 brown-spotted, pale blue green eggs for 12–14 days.
Feeding: gleans insects from the tree canopy; may hover-glean or hawk insects in midair; may forage at lower levels during cold weather; also takes some seasonally available berries.
Voice: song is a series of 4–5 raspy, whistled phrases, like a blurred version of the American Robin's song; call is a *chip-burrr* or *chip-churrr*.
Similar Species: *Summer Tanager* (p. 282): larger bill; male has red tail and wings; female has paler wings and is duskier overall, often with orange or reddish tinge.
Best Sites: good-sized woodlands; large forests of southern and eastern Ohio; migrant hotspots such as Green Lawn Cemetery or Magee Marsh WA bird trail; occasionally, early migrants caught in a severe cold snap forage for earthworms on lawns.

EASTERN TOWHEE

Pipilo erythrophthalmus

The "Rufous-sided Towhee" was reclassified as two species in 1995—our Eastern Towhee and the western Spotted Towhee. Few birds are as distinctive as the "Chewink," as the towhee is occasionally called, an imitative nickname of one of its common calls. • Scratching like little chickens in leaf litter under dense, brushy growth, males frequently sing a loud, ringing song that sounds like *drink your teeee!* • While towhees normally conceal themselves in thick shrubs, they are quite curious; making squeaking or pishing sounds will usually lure them into view, where their beautiful, bold markings can be admired. • Eastern Towhees remain common over much of the state and still breed in every county. The greatest densities occur in southern and eastern Ohio, where these birds regularly overwinter, though hardy individuals may be found anywhere in winter.

ID: rufous sides and flanks; white lower breast and belly; buff undertail coverts; white outer tail corners; red eyes; dark, conical bill. *Male:* black "hood," breast and upperparts. *Female:* brown "hood," breast and upperparts.

Size: *L* 7–8½ in; *W* 10½ in.

Habitat: dense, shrubby thickets, overgrown fields, red cedar glades, old clearcuts, woodland openings and edges.

Nesting: on the ground or low in a dense shrub; female builds a camouflaged cup nest of twigs, bark strips, grass, rootlets and animal hair; mostly the female incubates 3–4 brown-spotted, creamy white to pale gray eggs for 12–13 days.

Feeding: scratches at leaf litter for insects, seeds and berries; sometimes forages in low shrubs and saplings.

Voice: song is 2 high, whistled notes followed by a trill: *drink your teeeee;* call is a scratchy, slurred *cheweee!* or *chewink!*

Similar Species: *Spotted Towhee* (p. 341): very rare; heavily speckled with white on back and wings.

Best Sites: unglaciated and southern Ohio; hotspots such as Green Lawn Cemetery in migration.

J F M A M J J A S O N D

AMERICAN TREE SPARROW

Spizella arborea

A more appropriate name for this species might be "American Shrub Sparrow"—the American Tree Sparrow breeds in the High Arctic, mostly beyond the tree line and is a denizen of dwarf willow, alder and birch communities. • This sparrow was named by early European settlers who thought it resembled the Eurasian Tree Sparrow *(Passer montanus)*—a completely unrelated species in a different family. • American Tree Sparrows can be quite common in old fields and grassy meadows, where they associate in loose flocks that can number into the dozens or more. They are strictly winter visitors to Ohio. • These birds are prolific consumers of grass seeds, as well as the seeds of goldenrod, ironweed, wingstem and various other plants of successional habitats. They frequently visit feeders and are sometimes mistaken for Chipping Sparrows, which are extremely rare in winter. • Occasionally, toward the end of their stay, males will sing their beautiful, clear, warbling song—one of the more infrequently heard bird songs in Ohio.

ID: gray, unstreaked underparts; dark central breast spot; pale rufous "cap"; rufous stripe behind eye; gray face; mottled brown upperparts; notched tail; 2 white wing bars; dark legs; dark upper mandible; yellow lower mandible. *Nonbreeding:* gray central crown stripe. *Juvenile:* streaky breast and head.

Size: *L* 6–6½ in; *W* 9½ in.

Habitat: brushy thickets, roadside shrubs, grassy meadows, goldenrod fields and agricultural croplands.

Nesting: does not nest in Ohio.

Feeding: scratches exposed soil or snow for seeds in winter; eats mostly insects in summer; takes some berries and visits feeders.

Voice: song is a high, whistled *tseet-tseet* followed by a short, sweet, musical series of slurred whistles; song may be given in late winter and during spring migration; call is a 3-note *tsee-dle-eat*.

Similar Species: *Chipping Sparrow* (p. 286): clear, black eye line; white "eyebrow"; lacks dark breast spot. *Swamp Sparrow* (p. 299): white throat; lacks dark breast spot and white wing bars. *Field Sparrow* (p. 288): white eye ring; orange pink bill; lacks dark breast spot.

Best Sites: widespread in appropriate habitat throughout Ohio; often visits feeders.

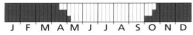

| J | F | M | A | M | J | J | A | S | O | N | D |

CHIPPING SPARROW

Spizella passerina

The Chipping Sparrow is most commonly found in areas carpeted by open lawns, shade trees and other landscaping—in fact, this is one of our few native birds to benefit from urban sprawl. • Soon after the first birds appear in mid-March, their monotonous, rather lengthy, trilling song becomes a common sound in appropriate habitats. This is one of a suite of "trillsters" that includes the Pine Warbler, Worm-eating Warbler, Swamp Sparrow and Dark-eyed Junco. These species are often heard singing within earshot of Chipping Sparrows, but their songs can be differentiated with practice. • Although Chipping Sparrows are highly migratory, a very few winter in Ohio, often appearing at feeders. These are either first-winter birds or nonbreeding adults that, lacking the bright rufous cap, more closely resemble the Clay-colored Sparrow rather than other rufous-capped species.

breeding

ID: *Breeding:* rufous "cap"; white "eyebrow"; black eye line; light gray, unstreaked underparts; mottled brown upperparts; all-dark bill; 2 faint wing bars; pale legs. *Nonbreeding:* paler crown with dark streaks; brown "eyebrow" and "cheek"; pale lower mandible. *Juvenile:* gray brown overall with dark brown streaking; pale lower mandible.
Size: *L* 5–6 in; *W* 8½ in.
Habitat: primarily in suburban areas with open expanses of lawn and scattered trees; also cemeteries, orchards, pastures and open woodlands.
Nesting: usually at midlevel in a conifer; female builds a compact cup nest of woven grass and rootlets, often lined with hair; female incubates 4 pale blue eggs for 11–12 days.

Feeding: gleans seeds from the ground and outer tree or shrub branches; prefers seeds from grass, dandelions and clovers; also eats adult and larval invertebrates; occasionally visits feeders.
Voice: song is a rapid, dry trill of *chip* notes; call is a high-pitched *chip*.
Similar Species: *American Tree Sparrow* (p. 285): dark central breast spot; rufous stripe behind eye; lacks bold, white "eyebrow." *Swamp Sparrow* (p. 299): lacks white "eyebrow," black eye line and white wing bars. *Field Sparrow* (p. 288): rufous stripe extends behind eye; white eye ring; gray throat; orange pink bill; lacks bold, white "eyebrow." *Clay-colored Sparrow* (p. 287): brown rump.
Best Sites: likely one of the first birds heard when you step outside your door.

J	F	M	A	M	J	J	A	S	O	N	D

CLAY-COLORED SPARROW

Spizella pallida

Primarily a breeder in brushlands of the vast Great Plains, where it is one of the most common and characteristic species, the Clay-colored Sparrow began an eastward range expansion in the early 1900s. As logging of the eastern deciduous forest created additional suitable habitat, this sparrow expanded as far east as Michigan and eastern Ontario. The first Ohio record was in 1940; since that time these sparrows have become rare but regular migrants. • Ohio has had a few records of summering territorial males, including birds in Holmes and Lucas counties. Our only documented nesting to date—in Franklin County in 1996—was destroyed by predators. Most records of Clay-colored Sparrows are of migrants in May, when males sing their curious, insectlike song with its distinctive series of buzzes. • Clay-colored Sparrows are one of North America's most migratory sparrows, with most birds wintering in Mexico and Central America.

breeding

ID: unstreaked, white underparts; buff breast wash; gray nape; light brown "cheek" edged with darker brown; brown crown with dark streak and pale, central stripe; white "eyebrow"; white jaw stripe bordered by brown; white throat; mostly pale bill. *Juvenile:* dark streaks on buff breast, sides and flanks.
Size: *L* 5–6 in; *W* 7½ in.
Habitat: brushy open areas along forest and woodland edges; forest openings, regenerating burn sites, abandoned fields, riparian thickets and young conifer farms.
Nesting: in a grassy tuft or small shrub; female builds an open cup nest of twigs, grass, weeds and rootlets and lines it with rootlets, fine grass and fur; mostly the female incubates 4 brown-speckled, bluish green eggs for 10–12 days.
Feeding: forages for seeds and insects on the ground and in low vegetation.
Voice: song is a series of 2–5 slow, low-pitched, insectlike buzzes; call is a soft *chip*.
Similar Species: *Chipping Sparrow* (p. 286): breeding adult has prominent rufous "cap," gray "cheek" and underparts, 2 faint, white wing bars and all-dark bill; juvenile lacks buff sides, flanks and gray nape.
Best Sites: along L. Erie in May at places such as Mentor Headlands and Ottawa NWR; grassy areas near parking lots at Magee Marsh WA bird trail; watch for summering individuals in northwestern Ohio, particularly Lucas, Ottawa and Williams counties.

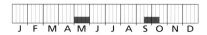

FIELD SPARROW

Spizella pusilla

This pink-billed sparrow is a denizen of overgrown fields, pastures and forest clearings, and is one of our most common breeding sparrows, nesting in every county. • The Field Sparrow is readily detected by its song, which carries long distances and is an accelerating series of short, liquid whistles, sounding like a table-tennis ball dropped on a table and bouncing to a stop. • The Field Sparrow can recognize unwelcome Brown-headed Cowbird eggs in its nest, but the eggs are usually too large for this small sparrow to eject, so the nest is simply abandoned. Affected pairs may make numerous nesting attempts in a single season. • Field Sparrows have declined at least locally, as trends toward "clean" agriculture and neatly manicured land have eliminated the scrubby, successional fields that these birds require, particularly in western Ohio. • While most Field Sparrows winter south of Ohio, they can be surprisingly common in brushy habitats from central Ohio southward. Making loud pishing sounds will often draw them out of the scrub to investigate.

ID: orange pink bill; gray face and throat; rusty crown with gray central stripe; rusty streak behind eye; white eye ring; 2 white wing bars; unstreaked gray underparts with buffy red wash on breast, sides and flanks; pinkish legs.
Size: *L* 5–6 in; *W* 8 in.
Habitat: abandoned or weedy and overgrown fields and pastures; woodland edges and clearings; extensive shrubby riparian areas; red cedar groves and young conifer plantations.
Nesting: on or near the ground, often sheltered by a grass clump, shrub or sapling; female weaves an open cup nest of grass and lines it with animal hair and soft plant material; female incubates 3–5 brown-spotted, whitish to pale bluish white eggs for 10–12 days.
Feeding: forages on the ground; takes mostly insects in summer and seeds in winter.
Voice: song is a series of woeful, musical, downslurred whistles accelerating into a trill; call is a *chip* or *tsee.*
Similar Species: *American Tree Sparrow* (p. 285): dark central breast spot; dark upper mandible; lacks white eye ring. *Chipping Sparrow* (p. 286): all-dark bill; white "eyebrow"; black eye line; lacks buffy red wash on underparts.
Best Sites: statewide in appropriate habitat; peak abundance in hill country of eastern and southern Ohio and red cedar-dominated areas of southwestern Ohio.

J F M A M J J A S O N D

VESPER SPARROW

Pooecetes gramineus

Along with Horned Larks, breeding Vesper Sparrows occupy some of the most desolate and bird-unfriendly habitat in Ohio, and thus are often overlooked. Barren agricultural lands of soybean, corn and wheat interspersed with the odd brushy fencerow and shade tree harbor the greatest number of Vesper Sparrows. Stopping along a country road that bisects this habitat will often give you the opportunity to hear the beautiful song of this species drifting across the wastelands. • Vesper Sparrows were probably quite rare if not absent from presettlement Ohio; the opening up of the original forest allowed their expansion into the state. They reached their heyday in the early 1900s, when breeding populations were found in every county; today Vesper Sparrows are most common in western and northern Ohio. • Most Vesper Sparrows winter south of Ohio, but very rarely one will attempt to overwinter. • This bird's name was derived from its habit of singing at twilight—vespers are evening prayers or services.

ID: chestnut brown shoulder patch; white outer tail feathers; pale yellow lores; weak flank streaking; white eye ring; dark upper mandible; lighter lower mandible; pale legs.

Size: *L* 5½–6½ in; *W* 10 in.

Habitat: large agricultural areas interspersed with fencerows and scattered trees, pastures, hayfields and young successional meadows.

Nesting: in a scrape on the ground, often under a canopy of grass or at the base of a shrub; loosely woven cup nest of grass is lined with rootlets, fine grass and hair; mostly the female incubates 3–5 brown-blotched, whitish to greenish white eggs for 11–13 days.

Feeding: walks and runs along the ground, picking up grasshoppers, beetles, cutworms, other invertebrates and seeds.

Voice: song is 4 characteristic, preliminary notes, with the second higher in pitch, followed by a bubbly trill: *here-here there-there, everybody-down-the-hill.*

Similar Species: *Other sparrows* (pp. 284–304): lack white outer tail feathers and chestnut shoulder patch. *Lark Sparrow* (p. 290): juvenile has streaking on breast and lacks chestnut shoulder patch. *American Pipit* (p. 244): thinner bill; grayer upperparts lack brown streaking; lacks chestnut shoulder patch. *Lapland Longspur* (p. 303): blackish or buff wash on upper breast; broad, pale "eyebrow" and reddish edges to wing feathers in nonbreeding plumage.

Best Sites: open agricultural lands; fields in vicinity of Killdeer Plains WA.

J F M A M J J A S O N D

289

LARK SPARROW

Chondestes grammacus

This beautiful bird has the unfortunate distinction of being our rarest breeding sparrow—only a few pairs are known to nest in the state. These breeders are limited to the Oak Openings ecosystem west of Toledo, which is characterized by sandy soil that once made up the beach ridges of preglacial Lake Warren, modern-day Lake Erie's larger predecessor. • Originally confined to the Great Plains, Lark Sparrows expanded eastward as settlers created suitable habitat in the process of clearing the vast eastern deciduous forest. Numbers peaked in Ohio in the early 1900s, when Lark Sparrows were widely scattered nesters throughout much of the glaciated region. Now that changing land use and development have greatly altered habitats, these sparrows are retreating back to their original range, illustrating the "boom and bust" cycle so typical of many birds that weren't originally part of our avifauna. • Thanks to the progressive management of Metroparks of the Toledo Area, which employs controlled burning to manage Oak Openings habitats, Lark Sparrows will hopefully remain in that area for some time to come.

ID: distinctive "helmet" of white throat, "eyebrow" and crown stripe and a few black lines break up otherwise chestnut red head; unstreaked, pale breast with dark central spot; black tail with white outer feathers; soft brown, mottled back and wings; pale legs.

Size: *L* 6 in; *W* 11 in.

Habitat: dry, sparsely vegetated sandy fields interspersed with scattered small trees; may appear in open pastures with occasional shrubs and trees.

Nesting: on the ground or in a low bush; bulky cup nest of grass and twigs is lined with finer material; occasionally reuses an abandoned nest, particularly of the Northern Mockingbird; female incubates 4–5 sparsely marked, white eggs for 11–12 days.

Feeding: walks or hops along the ground, gleaning seeds; also eats grasshoppers and other invertebrates.

Voice: melodious, variable song of short trills, buzzes, pauses and clear notes.

Similar Species: distinctive; juvenile vaguely resembles other streaked sparrows, but has prominent lateral throat stripe.

Best Sites: Oak Openings, especially near Girdham Road sand barrens, which border Girdham Rd. and are south of Monclova Rd. in Oak Openings MP.

J F M A M J J A S O N D

SAVANNAH SPARROW

Passerculus sandwichensis

A great many people have heard the Savannah Sparrow—for some reason its song is frequently used in movie soundtracks. Its pleasing, two-pitched, buzzy song is a common sound in Ohio farm country, but this wasn't always the case. This sparrow, named for the Georgia city where it was first discovered, is a relative newcomer here, with the first documented nestings in the 1920s. Today, it is locally common throughout much of Ohio, occupying hayfields, grasslands, reclaimed strip mines and even weedy soybean, wheat and corn stubble fields. Breeding Savannah Sparrows prefer sites with shorter grass interspersed with areas of barren soil. • This widely distributed species ranges nearly throughout North American and has been divided into 17 sub-species. • Savannah Sparrows can be very common but quite secretive migrants. For instance, 75 were captured in a few hours of banding in a 4-acre field near Chillicothe in October 2002.

ID: finely streaked breast, sides and flanks; light-colored, streaked underparts; mottled brown upperparts; yellow lores; pale jaw line; pinkish legs and bill; may show dark breast spot.

Size: *L* 4½–6½ in; *W* 6¾ in.

Habitat: agricultural fields (especially hay and alfalfa), moist sedge and grass meadows, pastures, reclaimed strip mines and grassy expanses at airports.

Nesting: on the ground; in a shallow scrape well concealed by grass or a shrub; female builds an open cup nest woven from and lined with grass; female incubates 3–6 brown-marked, whitish to greenish or pale tan eggs for 10–13 days.

Feeding: walks or runs along the ground gleaning insects and seeds; occasionally scratches.

Voice: song is a clear, high-pitched, buzzy *tea tea teeeeea today;* call is a high, thin *tsit.*

Similar Species: *Vesper Sparrow* (p. 289): white outer tail feathers; chestnut brown shoulder patches. *Lincoln's Sparrow* (p. 298): buff jaw line; buff wash across breast; broad, gray "eyebrow." *Grasshopper Sparrow* (p. 292): unstreaked breast. *Song Sparrow* (p. 297): triangular "mustache" stripes; pale central crown stripe; rounded tail; lacks yellow lores.

Best Sites: airports, alfalfa fields and most reclaimed strip mines.

J F M A M J J A S O N D

291

GRASSHOPPER SPARROW

Ammodramus savannarum

The passage of strip mine reclamation laws in 1974 forced coal companies to smooth the contours of former mine sites and to plant the soil with extensive stands of grass. These enormous grasslands in southeastern Ohio have proven to be a bonanza for Grasshopper Sparrows, which are often quite abundant on these sites. As farming practices become less bird-friendly, reclamation grasslands grow in importance as refugia for species such as this. • Grasshopper Sparrows can easily be overlooked until one becomes familiar with their weak, buzzy, insectlike trill. They probably still nest in every county, but are declining. Singing males can easily be found; they sing from fence posts or the tops of small saplings or tall weeds. After males cease singing, these sparrows becomes inconspicuous in the extreme—their migratory and wintering patterns are poorly understood.

ID: unstreaked, white underparts with buff wash on breast, sides and flanks; flattened head profile; dark crown with pale central stripe; buff "cheek"; mottled brown upperparts; beady, black eyes; sharp tail; pale legs; may show small yellow patch on edge of forewing. *Immature:* less buff on underparts; faint streaking across breast.
Size: *L* 4½–5½ in; *W* 7¾ in.
Habitat: grasslands, hayfields, sometimes fallow agricultural fields and pastures and reclaimed strip mine grasslands with little or no shrub or tree cover.
Nesting: in a shallow depression on the ground, usually concealed by grass; small cup nest woven from grass is lined with rootlets, fine grass and hair; female incubates 4–5 brown-spotted, creamy white eggs for 11–13 days.

Feeding: gleans insects and seeds from the ground and grass; eats many insects, including grasshoppers.
Voice: song is a high, faint, buzzy trill preceded by 1–3 high, thin, whistled notes: *tea-tea-tea zeeeeeeeeee;* also sings more complex song of erratic, jumbly buzzes.
Similar Species: *Le Conte's Sparrow* (p. 294): buff and black head stripes with white central crown stripe; gray "cheek"; dark streaking on sides and flanks. *Nelson's Sharp-tailed Sparrow* (p. 295): buff orange face and breast; gray central crown stripe; gray "cheek" and shoulders. *Henslow's Sparrow* (p. 293): similar to immature Grasshopper but has darker breast streaking, rusty wings, olive-toned head and small, dark ear and "whisker" marks.
Best Sites: grassy fields statewide; reclaimed strip mines; The Wilds; Crown City, Tri-Valley, Woodbury and Egypt Valley wildlife areas.

J	F	M	A	M	J	J	A	S	O	N	D

HENSLOW'S SPARROW

Ammodramus henslowii

Henslow's Sparrow has the dubious distinction of singing what may be the worst song of any North American songbird. The vocalizing male makes a production of throwing his head back, opening his bill to the sky, and letting loose with a sound akin to a cricket's hiccup. It is one of our easiest bird songs to overlook, and consequently this species is probably thought to be rarer than it actually is. Nevertheless, Henslow's Sparrow has shown the steepest population decline of any North American grassland bird. Like many such birds, changing agricultural practices have not been kind to the Henslow's Sparrow; by far the biggest populations now occur in reclaimed strip mine grasslands. • Knowing this bird's song is vital, as chances of detecting nonsinging birds is slim to none. Even when a singer is found, the sound is quite ventriloquial and is often delivered from within vegetation, so patience is required to locate the bird. • Henslow's Sparrows display an affinity for large patches of Chinese lespedeza *(Lespedeza cuneata)*, often planted on reclaimed strip mine sites.

ID: flattened head profile; olive green face, central crown stripe and nape; dark crown and "whisker" stripes; rusty tinge on back, wings and tail; white underparts with dark streaking on buff breast, sides and flanks; thick bill; deeply notched, sharp-edged tail. *Juvenile:* buff wash on underparts; faint streaking only on sides.
Size: *L* 4½–5½ in; *W* 6½ in.
Habitat: large, fallow fields and meadows with matted ground layer of dead vegetation and scattered shrub or herb perches; reclaimed strip mine grasslands.
Nesting: on the ground at the base of a grass clump or herbaceous plant; mostly the female builds an open cup nest of grass and weeds lined with fine grass and hair; female incubates 3–5 brown-spotted, whitish to pale greenish white eggs for about 11 days.

Feeding: gleans insects and seeds from the ground.
Voice: weak, liquidy, cricketlike *tse-lick* song is distinctive, often given during periods of rain or at night.
Similar Species: *Other sparrows* (pp. 284–304): lack greenish face, central crown strip and nape. *Grasshopper Sparrow* (p. 292): lacks dark "whisker" stripes and prominent streaking on breast and sides. *Savannah Sparrow* (p. 291): lacks buff breast.
Best Sites: reclaimed strip mines; Crown City WA; Tri-Valley WA; The Wilds; Killdeer Plains WA; large field at extreme western end of Toledo Express Airport, just west of U.S. Alt. Route 20.

LE CONTE'S SPARROW

Ammodramus leconteii

O f the 300 or so regular Ohio bird species, Le Conte's Sparrow may be the hardest to find. Like many sparrows, it is retiring and secretive; the difficulty in locating one is compounded by the fact that almost all occur in fall when males do not sing. Also, because Ohio is on the edge of this species' migratory corridor, relatively few individuals pass through. Spring records of Le Conte's Sparrow are few; October is the prime month and wetlands along Lake Erie are most likely to produce sightings. • The best way to find a Le Conte's is to walk through fields or wetlands with abundant herbaceous vegetation and carefully pick through the sometimes enormous numbers of sparrows that will be flushed. Even at a glimpse, Le Conte's Sparrow shows strong buff orange tones, a characteristic shared only by Nelson's Sharp-tailed Sparrow. However, Le Conte's are real skulkers, and getting one to leave the cover of vegetation long enough for a look can be frustrating.

Nesting: does not nest in Ohio.
Feeding: gleans the ground and low vegetation for insects, spiders and seeds.
Voice: song is a weak, short, raspy, insect-like buzz: *t-t-t-zeeee zee* or *take-it ea-zeee;* alarm call is a high-pitched whistle.
Similar Species: *Nelson's Sharp-tailed Sparrow* (p. 295): gray central crown stripe and nape; white streaks on dark back. *Grasshopper Sparrow* (p. 292): lacks buff orange face and streaking on underparts. *Henslow's Sparrow* (p. 293): darker overall; rufous-toned wings and back; olive head and face.
Best Sites: Gordon Park Impoundment (Dike 51) along L. Erie in Cleveland; weedy dredge-spoil impoundments in L. Erie; Lorain Harbor and Huron Municipal Pier; Big Island WA.

ID: buff orange face; gray "cheek"; black line behind eye; light central crown stripe bordered by black stripes; buff orange upper breast, sides and flanks; dark streaking on sides and flanks; white throat, lower breast and belly; mottled, brown black upperparts; buff streaks on back; pale legs. *Immature:* duller overall; more streaking on breast.
Size: *L* 4½–5 in; *W* 6½ in.
Habitat: grassy meadows with dense vegetation; drier edges of wet sedge and grass meadows; damp fields with dense herbaceous vegetation.

NELSON'S SHARP-TAILED SPARROW

Ammodramus nelsoni

Like Le Conte's Sparrow, this species is very difficult to detect. However, Nelson's Sharp-tailed Sparrow seems to be a bit bolder, and if a suspect is found in dense vegetation, a loud round of pishing is likely to draw the bird into view, if only briefly. • There are numerous spring records of this species, but it is a late migrant, with most birds passing through in mid- to late May, even early June. Most reports are from habitats along Lake Erie in September and October. • This sparrow is best found in the large stands of giant reed that occur in wetlands and dredge-spoil impoundments along Lake Erie. Unfortunately for birders, giant reed can grow to 12 feet in height and forms nearly impenetrable stands. Nevertheless, this plant provides good cover for this secretive sparrow, and no doubt hides many more than are ever detected. • In 1998, the "Sharp-tailed Sparrow" was split into two species—Nelson's Sharp-tailed Sparrow and the coastal Saltmarsh Sharp-tailed Sparrow.

ID: buff orange face, breast, sides and flanks; gray "cheek," central crown stripe and nape; dark line behind eye; light streaking on sides and flanks; white stripes on dark back; white to light buff throat; white belly.
Size: *L* 5–6 in; *W* 7 in.
Habitat: marshes, wet meadows and fields, weedy dredge-spoil impoundments and stands of giant reed.
Nesting: does not nest in Ohio.
Feeding: gleans ants, beetles, grasshoppers and other invertebrates from the ground and low vegetation; also eats seeds.

Voice: song is a short, raspy buzz: *ts tse-sheeeee* (rarely if ever heard here); call is a high *siss*.
Similar Species: *Le Conte's Sparrow* (p. 294): lacks gray nape and white stripes on dark back. *Grasshopper Sparrow* (p. 292): lacks streaking on underparts. *Savannah Sparrow* (p. 291): notched tail; heavily streaked underparts.
Best Sites: L. Erie coastal wetlands and dredge-spoil impoundments; also large interior wetlands such as Big Island WA, Killdeer Plains WA and Gilmore Ponds Preserve.

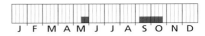

J F M A M J J A S O N D

FOX SPARROW

Passerella iliaca

If there was a beauty pageant for sparrows, the Fox Sparrow might be the winner. One of our largest sparrows, it is richly colored in hues of deep, rusty chestnut interspersed with patches of gray and heavily streaked with rufous below. Its song, which is heard occasionally in early spring migration, is a melodious warble of clear whistles, thought by many to be the most pleasing sparrow song of all. • One of the most geographically variable of all North American birds, the Fox Sparrow has been divided into four groups composed of 18 subspecies. Luckily, our birds are of the "Red" group and are the nominate subspecies *iliaca*, which is the most colorful of all. Odd-looking Fox Sparrows warrant study, as there is at least one Ohio record of the western "Sooty" subspecies. • Fox Sparrows often feed on the ground in dense thickets, where they kick-scratch the leaf litter in a manner reminiscent of the Eastern Towhee. • Fox Sparrows occasionally overwinter, most commonly in southern Ohio, and are likely to be detected by their call, a loud, sharp *smack*.

ID: whitish underparts; heavy, reddish brown spotting and streaking often converges into central breast spot; reddish brown wings, rump and tail; gray crown; brown-streaked back; gray "eyebrow" and nape; stubby, conical bill; pale legs.
Size: *L* 6½–7½ in; *W* 10½ in.
Habitat: thickets, brushy openings, dense shrubby landscaping and woodlands with thick understory.
Nesting: does not nest in Ohio.
Feeding: scratches the ground to uncover seeds, berries and invertebrates; visits feeders in migration and winter.

Voice: song is a variable, long series of melodic whistles: *All I have is what's here dear, won't you won't you take it?*; calls include an explosive *smack*.
Similar Species: *Song Sparrow* (p. 297): noticeably smaller; more obvious central crown stripe; dark "mustache"; dark brownish rather than reddish streaking and upperparts.
Best Sites: highly migratory; best detected at good birding hotspots such as Green Lawn, Spring Grove and Woodlawn cemeteries and Mentor Headlands; greatest numbers at Magee Marsh WA and Ottawa NWR.

J F M A M J J A S O N D

SONG SPARROW

Melospiza melodia

Ranging from Alaska's Aleutian Islands to Newfoundland and south through much of the continent, the Song Sparrow is one of the most successful and widely distributed sparrows. Our most common breeding sparrow, it occupies a broad range of open and semi-open habitats, including yards, parks and other urban areas. • This well-named sparrow stands among our great songsters for the complexity, rhythm and emotion of its springtime rhapsodies. • At a time when the role of women in science was often trivialized, Margaret Morse Nice conducted groundbreaking behavioral and ecological studies on the Song Sparrow, which were published in 1937 and 1943. Her field research was conducted along the Olentangy River in Columbus, Ohio, and to this day her work serves as a benchmark for avian ecology research. Among a wealth of information, Nice discovered that female Song Sparrows occasionally sing, and do so nearly as well as the males.

ID: whitish underparts with heavy brown streaking that converges into central breast spot; grayish face; dark line behind eye; white jaw line bordered by dark whisker and "mustache" stripes; dark crown with pale central stripe; mottled brown upperparts; rounded tail tip.
Size: *L* 5½–7 in; *W* 8¼ in.
Habitat: shrubby areas, often near water; riparian thickets, forest openings, woodland edges, densely landscaped parks and yards, fencerows and overgrown fields.
Nesting: usually on the ground or low in a shrub or small tree; female builds an open cup nest of grass, weeds, leaves and bark shreds and lines it with rootlets, fine grass and hair; female incubates 3–5 heavily spotted, greenish white eggs for 12–14 days; may raise 2–3 broods each summer.
Feeding: gleans the ground, shrubs and trees for cutworms, beetles, grasshoppers, ants, other invertebrates and seeds; also eats wild fruit and visits feeding stations.
Voice: song is 1–4 bright, distinctive introductory notes, such as *sweet, sweet, sweet,* followed by a buzzy *towee,* then a short, descending trill; calls include a short *tsip* and a nasal *tchep.*
Similar Species: *Fox Sparrow* (p. 296): larger; heavier breast spotting and streaking; reddish rather than dark brownish streaking and upperparts; lacks pale central crown stripe and dark "mustache." *Lincoln's Sparrow* (p. 298): lightly streaked breast with buff wash; buff jaw line. *Savannah Sparrow* (p. 291): lightly streaked breast lacks central spot; yellow lores; notched tail; lacks grayish face and dark, triangular "mustache."
Best Sites: ubiquitous; bird feeders; often one of the first birds you encounter upon leaving home.

J F M A M J J A S O N D

LINCOLN'S SPARROW

Melospiza lincolnii

Thomas Lincoln (1812–83) discovered this species while accompanying John James Audubon on his 1833 expedition to Labrador. In recognition of his accomplishment, Audubon named the sparrow after his young companion, making Lincoln, at the age of 21, perhaps the youngest person to have a North American bird named after him. Audubon noted "we found more wildness in this species than in any other inhabiting the same country." Those birders familiar with this painfully shy and retiring sparrow might identify with that remark; Lincoln's Sparrow seems to shun the proximity of humans. It breeds across the northern reaches of North America, inhabiting dense, wet, shrubby willow thickets and other boggy scrub. • The true migratory status of this species in Ohio is somewhat unclear owing to its secretive nature, but Lincoln's Sparrow is common in spring and fall. Occasional large flights occur, but generally, finding half a dozen in a day is doing well.

ID: buff breast band, sides and flanks with fine, dark streaking; buff jaw stripe; gray "eyebrow," face and "collar"; dark line behind eye; dark reddish "cap" with gray central stripe; white throat and belly; mottled gray brown to reddish brown upperparts; very faint, white eye ring.
Size: *L* 5½ in; *W* 7½ in.
Habitat: dense thickets and tangles, often in wet areas; also old fields, meadows and other dry, open habitats, particularly in fall.
Nesting: does not nest in Ohio.
Feeding: scratches at the ground, exposing invertebrates and seeds; very rarely visits feeders.

Voice: rich, musical, wrenlike mixture of buzzes, trills and warbled notes, not often heard here; calls include a buzzy *zeee* and *tsup*.
Similar Species: *Song Sparrow* (p. 297): heavier breast streaking; dark triangular "mustache"; lacks buff wash on breast, sides and flanks. *Savannah Sparrow* (p. 291): yellow lores; white "eyebrow" and jaw line. *Swamp Sparrow* (p. 299): more contrast between red and gray crown stripes; generally lacks streaking on breast
Best Sites: wet thickets, tangles and root clusters near water; wet thickets and buttonbush tangles along Magee Marsh WA bird trail; Green Lawn Cemetery pond; Calamus Swamp; Mentor Headlands; Springville Marsh SNP; Kelleys Island.

J	F	M	A	M	J	J	A	S	O	N	D

SWAMP SPARROW

Melospiza georgiana

Our most aquatic sparrow, the Swamp Sparrow is an obligate wetland species that breeds over much of the northern reaches of eastern North America and Canada, ranging south through the northern half of Ohio. Numbers swell significantly in migration, and migrants often turn up in parks, shrubby areas and fields far from water. Males often sing in migration and this species is most easily detected by its loud, musical trill emanating from marshes. • Swamp Sparrows are quite hardy and often overwinter throughout the state. Wintering birds are mostly silent, but occasionally deliver a loud, metallic *chink* call that is fairly easily learned and is the best way to detect them in the dense cattail stands they favor. Making pishing sounds near a cattail marsh in winter will often draw a surprising number of Swamp Sparrows into view. • There are no breeding records from the southern third of the state, but nesting Swamp Sparrows should be watched for in any large wetlands in the massive reclaimed strip mines of southeastern Ohio.

breeding

ID: gray face; reddish brown wings; brownish upperparts; dark streaking on back; dull gray breast; white throat and jaw line outlined by black stripes; dark line behind eye. *Breeding:* rusty "cap"; streaked buff sides and flanks. *Nonbreeding:* streaked brown "cap" with gray central stripe; more brownish sides. *Immature:* buffy "eyebrow" and nape; faint breast streaking.
Size: *L* 5–6 in; *W* 7¼ in.
Habitat: cattail marshes, open wetlands, wet meadows and open deciduous riparian thickets.
Nesting: in emergent aquatic vegetation or shoreline bushes; cup nest of coarse grass and marsh vegetation is lined with fine grass; usually has partial canopy and side entrance; female incubates 4–5 heavily marked, greenish white to pale green eggs for 12–15 days.

Feeding: gleans insects from the ground and from vegetation and the water's surface; takes seeds in late summer and fall.
Voice: song is a sharp, metallic, slow trill: *weet-weet-weet-weet;* call is a harsh *chink*.
Similar Species: *Chipping Sparrow* (p. 286): clean white "eyebrow"; full black eye line; uniformly gray underparts; white wing bars. *American Tree Sparrow* (p. 285): dark central breast spot; white wing bars; 2-tone bill. *Song Sparrow* (p. 297): heavily streaked underparts; lacks gray "collar." *Lincoln's Sparrow* (p. 298): fine breast streaking; less contrast between brown and gray crown stripes.
Best Sites: breeds in wetlands throughout northeastern Ohio; causeway at Magee Marsh WA; boardwalks at Irwin Prairie SNP and Maumee Bay SP.

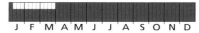

J F M A M J J A S O N D

WHITE-THROATED SPARROW

Zonotrichia albicollis

The mournful song of the White-throated Sparrow is a characteristic sound of the Canadian boreal forest and is one of the most easily learned of all bird songs. This species is an abundant migrant and a common winter resident throughout Ohio. However, nesters are quite rare, with only a handful of breeding records from the hemlock gorges of northeastern Ohio. • White-throated Sparrows come in two color types, or morphs—tan-striped and white-striped. Often, the uninitiated believe the tan-striped birds to be females and the white-striped ones to be males. Actually, the sexes of each morph look alike, so a tan-striped bird might be either male or female, and likewise for white-stripes. These color forms are maintained by a process known as "negative assortative mating," in which a morph of one type almost always mates with a morph of the other type. • The two morphs behave differently: tan-striped females provide more parental care than white-stripes, and white-striped males are more aggressive than tan-stripes. The genetics that dictate color morphology in White-throated Sparrows is apparently unique to this species.

tan-striped morph

ID: black and white (or brown and tan) head stripes; white throat; gray "cheek"; yellow lores; black eye line; unstreaked, gray underparts; mottled, brown upperparts; grayish bill.
Size: *L* 6½–7½ in; *W* 9 in.
Habitat: *Breeding* (very rare): large, undisturbed hemlock gorges. *In migration:* woodlots; wooded parks; overgrown fields and thickets; brushy areas.
Nesting: on or near the ground, often concealed by low shrubs or a fallen log; female builds an open cup nest of grass, weeds, twigs and conifer needles lined with rootlets, fine grass and hair; female incubates 4–5 variably marked, greenish blue to pale blue eggs for 11–14 days.
Feeding: scratches the ground to expose invertebrates, seeds and berries; also gleans insects from vegetation and while in flight; eats seeds from feeders in winter.
Voice: variable song is a clear and distinct whistled: *dear sweet Canada Canada Canada* or *Old Sam Peabody Peabody Peabody;* call is a sharp *chink.*
Similar Species: *White-crowned Sparrow* (p. 301): pinkish bill; gray "collar"; lacks bold, white throat and yellow lores.
Best Sites: *In migration:* easily found in appropriate habitat almost anywhere; huge numbers at hotspots such as Green Lawn Cemetery, Spring Grove Cemetery or Magee Marsh WA bird trail.

J	F	M	A	M	J	J	A	S	O	N	D

WHITE-CROWNED SPARROW

Zonotrichia leucophrys

Many sparrows are shrinking violets, but not White-crowned Sparrows. These birds incessantly sing their plaintive song—a series of variable buzzes and trills preceded by a whistle or two—and frequently perch in full view, where their stately, refined appearance can be admired. • Although a common spring and fall migrant, wintering birds, which are mostly found in the southern one-third of the state, are scarcer and are probably declining. Winter birds are often associated with the multiflora rose, an introduced shrub whose numbers are slowly diminishing, in part because of rose rosette disease. • Perhaps the most widely studied North American songbird, research on the White-crowned Sparrow has improved our understanding of song development and regional variation, and of songbird population biology. • This species is a close second to the Fox Sparrow for the title of largest sparrow. • Almost all of our birds are the nominate subspecies *leucophrys*. However, the western subspecies *gambelii* is a rare visitor—the adult can be recognized by its white, rather than black, lores.

ID: black and white head stripes; black eye line; pink orange bill; gray face; unstreaked, gray underparts; pale gray throat; mottled gray brown upperparts; 2 faint, white wing bars. *Immature:* broad gray "eyebrow" bordered by brown eye line and crown.
Size: *L* 5½–7 in; *W* 9½ in.
Habitat: brushy tangles; open fields and pastures, particularly those with multiflora rose; fencerows; shrubby areas of cemeteries and suburban parks.
Nesting: does not nest in Ohio.
Feeding: scratches the ground to expose insects and seeds; also eats berries, buds and moss spore capsules; may take seeds from bird feeders.
Voice: song is a frequently repeated variation of *I gotta go wee-wee now;* call is a high, thin *seet* or sharp *pink*.
Similar Species: *White-throated Sparrow* (p. 300): bold, white throat; grayish bill; yellow lores; browner overall; immature can be distinguished by cleanly marked plumage, large size, brown crown stripes and pinkish orange bill.
Best Sites: dry, open brushy fields with scattered shrubs; Killdeer Plains WA; Indian Creek WA; C.J. Brown Reservoir; Ottawa NWR; The Wilds.

J F M A M J J A S O N D

DARK-EYED JUNCO

Junco hyemalis

Common throughout their broad range—recent estimates put the total population at 630 million birds—Dark-eyed Juncos begin to appear here in late fall and remain as common winter residents. Often called "Snowbirds," these attractive, social sparrows are known to virtually everyone who feeds birds; juncos are easily lured to feeders, often in sizable numbers. • Most juncos depart by April's end, but there is a small breeding population in the hemlock gorges of extreme northeastern Ohio. • Prior to 1973, there were thought to be five species of "dark-eyed" juncos. Our birds were known as "Slate-colored Juncos," but these are now considered part of the species complex, which is split into about 15 subspecies. • At least one other subspecies, the "Oregon Junco," appears occasionally in Ohio. Adults are distinctive with their clearly demarcated, black to gray hoods and pinkish sides.

♂

"Oregon" Junco

"Slate-colored" Junco

ID: white outer tail feathers; pale bill. *Male:* dark slate gray overall, except for white lower breast, belly and undertail coverts. *Female:* brown rather than gray. *Juvenile:* brown similar to female, but streaked with darker brown.
Size: L 5½–7 in; W 9¼ in.
Habitat: occurs in all manner of habitats including urban yards, cemeteries, old fields, woods and brushy areas; breeders are confined to hemlock gorges.
Nesting: on the ground, usually concealed by a shrub, tree, root, log or rock; female builds a cup nest of twigs, bark shreds, grass and moss and lines it with fine grass and hair; female incubates 3–5 whitish to bluish white eggs, marked with gray and brown, for 12–13 days.
Feeding: scratches the ground for invertebrates; also eats berries and seeds.
Voice: song is a long, dry trill, very similar to the call of the Chipping Sparrow, but more musical; call is a smacking *chip* note, often given in series.
Similar Species: relatively unmistakable; be alert to other subspecies, particularly "Oregon Junco."
Best Sites: widespread from fall through spring; often seen at feeders.

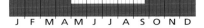

J F M A M J J A S O N D

LAPLAND LONGSPUR

Calcarius lapponicus

Flocks of Lapland Longspurs on the Great Plains—their primary wintering grounds—have been estimated to contain up to 4 million birds! These birds are not nearly so numerous here in Ohio, but are not difficult to find if you know where to look. • Breeding in the highest reaches of the Arctic, well beyond the tree line, longspurs shun forested areas in all seasons and are denizens of open, barren land. • Most common in spring migration in Ohio, flocks are best sought in large, freshly plowed farm fields from mid-March through mid-April. By this time, some of the males will have molted into their showy breeding plumage and may occasionally deliver their warbling flight song. • Wintering Lapland Longspurs are more uncommon and local, and are often found with Horned Larks and Snow Buntings. Becoming familiar with the longspur's distinctive, dry, rattling call will help you locate this bird. • Fields freshly doused with manure often attract longspurs, which extract seeds from the waste.

♀ nonbreeding

♂

ID: whitish outer tail feathers; pale yellowish bill. *Breeding male:* black crown, face and "bib"; chestnut nape; broad, white stripe curves down to shoulder from eye (may be tinged with buff behind eye). *Breeding female* and *nonbreeding:* often has rufous in wings; mottled brown-and-black upperparts; lightly streaked flanks; male has faint chestnut on nape and diffuse black breast; female has narrow, lightly streaked, buff breast band. *Immature:* grayish nape; broader, buff brown breast band.
Size: *L* 6½ in; *W* 11 in.
Habitat: large, barren agricultural fields; occasionally open lakeshores and mudflats.
Nesting: does not nest in Ohio.

Feeding: gleans the ground and snow for seeds and waste grain.
Voice: flight song is a rapid, slurred warble; musical calls; flight calls include a rattling *tri-di-dit* and a descending *teew*.
Similar Species: *Horned Lark* (p. 212): distinctive yellow-and-black head pattern; shallower, wispier flight pattern. *Snow Bunting* (p. 304): much more white in plumage. *American Pipit* (p. 244): much thinner bill; plain back.
Best Sites: in appropriate habitat throughout western Ohio. *Spring:* Maumee Bay SP (migrating flocks best seen from prominent hill near beach in morning); Killdeer Plains WA (fields along Washburn Rd. on south side of wildlife area).

J F M A M J J A S O N D

SNOW BUNTING

Plectrophenax nivalis

This is truly a bird that thrives in northern climates. In May 1987, a Snow Bunting was recorded near the North Pole, the most northerly record for any songbird! Snow Buntings generally don't begin to trickle into Ohio from their arctic breeding grounds until late October or early November, and they can become locally common winter residents. • These distinctive sparrows often fraternize with Horned Larks and Lapland Longspurs and frequent wide-open landscapes characterized by a lack of trees. They also display an affinity for a habitat not often used by any other songbird—the rocky riprap that lines the shorelines of reservoirs and parts of Lake Erie. • When snow blankets the ground, Snow Buntings are often seen along the verges of roads; driving country lanes is a good way to find them. • The genus name *Plectrophenax,* given by naturalist Leonhard Stejneger, is nonsensical in that it means "something to hit a cheat with." Stejneger bungled the spelling and thus altered the intended meaning of "showy claw," but rules of nomenclature require that the name stand as is.

nonbreeding

ID: black-and-white wings and tail; white underparts. *Breeding male:* black back; all-white head and rump; black bill. *Breeding female:* streaky, brown-and-whitish crown and back; dark bill. *Nonbreeding:* yellowish bill; golden brown crown and rump; female has blackish forecrown and golden back with dark streaks.
Size: *L* 6–7½ in; *W* 14 in.
Habitat: large, barren agricultural fields, mudflats and rocky riprap along lakeshores.
Nesting: does not nest in Ohio.

Feeding: gleans the ground and snow for seeds and waste grain; also takes insects when available, especially in summer.
Voice: spring song is a high-pitched, musical *chi-chi-churee;* call is a whistled *tew.*
Similar Species: distinctive. *Lapland Longspur* (p. 303): overall brownish upperparts; lacks black-and-white wing pattern.
Best Sites: throughout northwestern and northern Ohio; migrant flocks numerous in fall along L. Erie; rock riprap lining Findlay Reservoir; State Route 294 between Little Sandusky and Harpster in Wyandot Co. just northeast of Killdeer Plains WA (numerous birds along road when fields are snow covered).

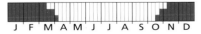

J F M A M J J A S O N D

NORTHERN CARDINAL

Cardinalis cardinalis

Of the 412 species of birds so far recorded in Ohio, the Northern Cardinal may be the best known among birders and nonbirders alike. Virtually everyone can instantly identify the showy males, in part because of their distinctive plumage, but also because they are common and widespread and are regular feeder visitors. The Northern Cardinal—named after highly ranked church officials who wear bright red robes—is Ohio's state bird, an honor that six other states have bestowed upon it. • Variations in Northern Cardinals throughout their range result from the 18 distinct subspecies, each slightly different in plumage. • Birds with missing head feathers are routinely seen. This condition is sometimes attributed to feather mites, but in most cases the cause is unknown. The feathers eventually regrow. • A 1993–95 Ohio study found that of 115 study nests, nearly 50 percent were parasitized by Brown-headed Cowbirds. • The oldest known wild cardinal lived to almost 16 years of age.

ID: *Male:* red overall; pointed crest; black "mask" and throat; red, conical bill. *Female:* brown buff to buff olive overall; red bill, crest, wings and tail. *Juvenile male:* similar to female but has dark bill and crest.

Size: *L* 7½–9 in; *W* 12 in.

Habitat: brushy thickets and shrubby tangles along forest and woodland edges; also backyards and suburban and urban parks.

Nesting: in dense shrubs, thickets, vine tangles or low in a coniferous tree; female builds an open cup nest of twigs, bark shreds, weeds, grass, leaves and rootlets and lines it with hair and fine grass; female incubates 3–4 brown-blotched, whitish to greenish white eggs for 12–13 days.

Feeding: gleans seeds, insects and berries from the ground or low shrubs.

Voice: song is sung year round and is a variable series of clear, bubbly, whistled notes: *What cheer! What cheer! birdie-birdie-birdie What cheer!;* call is a metallic *chip;* females can sing and sometimes do so from the nest, apparently to signal the male to bring food.

Similar Species: unmistakable.

Best Sites: statewide in suitable habitat; readily enticed to yards with appropriate landscaping and feeders.

J F M A M J J A S O N D

ROSE-BREASTED GROSBEAK

Pheucticus ludovicianus

The slurry, whistled song of the Rose-breasted Grosbeak is sometimes likened to a drunken American Robin, and is often the best way to find this sluggish treetop dweller. Its distinctive, easily learned call note is a loud, squeaky *chink*, suggestive of two trees rubbing together in the breeze. • A species of edge habitats and second-growth woodlands, Rose-breasted Grosbeaks have prospered in heavily logged Ohio and seem to be increasing in numbers. They are most conspicuous in early spring when migrants bolster numbers and the trees haven't yet leafed out, but they are also common nesters in many parts of the state. • Males assist females with incubating eggs, and the females sing nearly as well as the males, sometimes even while sitting on the nest. • There are at least a dozen winter records here, all of birds visiting feeding stations.

breeding

ID: pale, conical bill; dark wings with small white patches; dark tail. *Male:* black "hood" and back; red breast with "stem" down front; red inner under-wings; white underparts and rump. *Female:* bold, whitish "eyebrow"; thin crown stripe; brown upperparts; buff underparts with dark brown streaking. *1st-fall male:* similar to female, but has streaked orange breast and sides. *1st-spring male:* may show brown instead of black.
Size: *L* 7–8½ in; *W* 12½ in.
Habitat: younger second-growth deciduous and mixed forests, often near openings.
Nesting: fairly low in a tree or tall shrub, often near water; mostly the female builds a flimsy cup nest of twigs, bark strips, weeds, grass and leaves and lines it with rootlets and hair; pair incubates 3–5 brown-spotted, pale greenish blue eggs for 13–14 days.
Feeding: gleans vegetation for insects, seeds, buds, berries and some fruit; occasionally hover-gleans or catches flying insects on the wing; may also visit feeders.
Voice: song is a long, melodious series of whistled notes, much like a fast version of a robin's song; call is a distinctive squeak.
Similar Species: male is distinctive. *Black-headed Grosbeak:* accidental vagrant; female is similar but has bright lemon yellow underwings, less defined and less conspicuous streaking on underparts and bill often has dark upper mandible, at least in breeding season.
Best Sites: *In spring migration:* big numbers at any good migrant spot; anywhere there are shade trees. *Breeding:* Findley SP; Pymatuning SP; Mohican SF; Cuyahoga Valley NRC.

J F M A M J J A S O N D

BLUE GROSBEAK

Passerina caerulea

The Blue Grosbeak is one of the most coveted of Ohio's breeding song-birds—seen in good light, the male is truly a stunning sight. Many a birder has made the pilgrimage to Adams County, a traditional stronghold for this southern species. • Although uncommon and local, with a range that barely extends into southernmost Ohio, this species has been slowly but steadily expanding north-ward in recent decades. It can now be regarded as an uncommon breeder in many of the southernmost counties. Occasional birds even make it as far as northern Ohio in spring migration and sometimes stay into summer. • Males often sing their loud, rapid warble from exposed perches such as telephone wires and the tops of small trees. • In 1999, the mother lode of Blue Grosbeaks was discovered at Crown City Wildlife Area. This large, reclaimed strip mine supports a population density of this species unheard of elsewhere in Ohio.

ID: large, pale gray-ish, conical bill. *Male:* blue overall; 2 rusty wing bars; black around base of bill. *Female:* soft brown plumage overall; whitish throat; rusty wing bars; rump and shoulders faintly washed with blue. *1st-spring male:* similar to female, but has blue head.
Size: *L* 6–7½ in; *W* 11 in.
Habitat: thick brush, riparian thickets, shrubby areas, open fields, woodland edges and weedy fields near water.
Nesting: on ground or in a low bush; female builds a compact cup of grass, leaves and bark and lines it with roots and fine material; female incubates 3–4 pale blue eggs for 11–12 days; if conditions allow, may have 2 broods per year.

Feeding: gleans insects from the ground; occasionally takes seeds; rarely visits feeders.
Voice: sweet, melodious, warbling song with phrases that rise and fall; call is a loud *chink*.
Similar Species: *Indigo Bunting* (p. 308): smaller body and bill; lacks wing bars. *Brown-headed Cowbird* (p. 318): females are similar, but female cowbird has smaller bill and lacks wing bars.
Best Sites: Crown City WA (any location at peak of singing); country roads in Adams Co. (Waggoner Riffle and Vaughn Ridge roads) and in Lawrence, Gallia and Meigs counties.

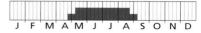

J F M A M J J A S O N D

INDIGO BUNTING

Passerina cyanea

Although one of the most common birds of rural open country, the Indigo Bunting can still be missed—in poor light, these birds can look black and will blend in among all the other roadside songbirds. However, in good light at close range, the Indigo Bunting displays a vivid electric blue, one of the most spectacular colors of any Ohio bird. • The male Indigo Bunting is one of the most persistent singers of any of our songbirds, vocalizing even throughout the heat of a summer day. • Females arrive a week or so later than the males in spring, both sexes having made the journey from Central and South American wintering grounds. • In some North American regions, such as Kentucky and Tennessee, the Indigo Bunting is thought to be the most numerous of any breeding songbird. Total population estimates for this species have ranged as high as 40 million pairs.

breeding

ID: stout, gray, conical bill; beady, black eyes; black legs; no wing bars. *Breeding male:* blue overall; black lores; wings and tail may show some black. *Female:* soft brown overall; brown streaks on breast; whitish throat; faint wing bars. *Fall male:* similar to female, but usually with some blue in wings and tail. *1st-spring male:* mottled blue and brown overall.
Size: *L* 5½ in; *W* 8 in.
Habitat: open deciduous forests and woodland edges, regenerating forest clearings, shrubby fields, orchards, abandoned pastures and hedgerows; occasionally along mixed woodland edges.
Nesting: in a small tree or shrub or within a vine tangle; female builds a cup nest of grass, leaves and bark strips and lines it with rootlets, hair and feathers; female

incubates 3–4 white to bluish white eggs for 12–13 days.
Feeding: gleans low vegetation and the ground for insects, especially grasshoppers, beetles, weevils, flies and larvae; also eats seeds of thistles, dandelions, goldenrods and other native plants.
Voice: song consists of paired, warbled whistles: *fire-fire, where-where, here-here, see-it see-it;* call is a quick *spit.*
Similar Species: *Blue Grosbeak* (p. 307): larger overall; larger, more robust bill; 2 rusty wing bars; male has black around base of bill; female lacks streaking on breast. *Brown-headed Cowbird* (p. 318): females are similar, but female cowbird is larger and darker and lacks wing bars.
Best Sites: red cedar-dominated habitats in southwestern Ohio; easy to find in rural areas, where males often sing from conspicuous spots.

DICKCISSEL

Spiza americana

Somewhat resembling little meadowlarks, Dickcissels are birds of the Great Plains and are at the extreme eastern edge of their range in Ohio. They are notorious cyclical irruptives—they may be common at a site one year and completely absent the next. Drought and subsequent food shortages in their core breeding range likely cause these eastward irruptions. • Prior to European settlement, Dickcissels probably inhabited the then 1000 square miles of prairie found in disjunct clusters in Ohio. Their continued presence is likely a relict of earlier prairie days, but they have learned to adapt to artificial, but superficially similar, habitats. • Dickcissels winter primarily in Venezuela, where some local farmers persecute them as agricultural pests. The birds forage in crops of rice and sorghum—one roost was estimated to contain over 2.3 million birds—and one farmer admitted to slaughtering over 1 million birds on his farm!

breeding

ID: yellow "eyebrow"; gray head, nape and sides of yellow breast; brown upperparts; pale grayish underparts; rufous shoulder patch; conical bill. *Male:* white "chin" and black "bib"; duller colors in nonbreeding plumage. *Female:* duller version of male; white throat.
Size: *L* 6–7 in; *W* 9¾ in.
Habitat: abandoned fields dominated by forbs; weedy meadows; strip mine reclamation grasslands; hayfields dominated by alfalfa or other legumes.
Nesting: on or near the ground, well concealed among tall, dense vegetation; female builds a bulky open cup nest of forbs, grass and leaves and lines it with rootlets, fine grass or hair; female incubates 4 pale blue eggs for 12–13 days.
Feeding: gleans insects and seeds from the ground and low vegetation; individuals

very rarely visit feeders in winter, often associating with House Sparrows.
Voice: song is 2–3 single notes followed by a trill, often paraphrased as *dick dick dickcissel;* flight call is a buzzerlike *bzrrrrt.*
Similar Species: *Eastern Meadowlark* (p. 312): much larger; long, pointed bill; yellow "chin" and throat with black "necklace." *House Sparrow* (p. 329): female resembles immature Dickcissel, which may appear at feeders in winter, but House Sparrow is slimmer, has prominent lateral throat stripe and longer, paler bill.
Best Sites: appropriate habitats throughout western Ohio (boom years); Crown City WA (plentiful in good years); Killdeer Plains WA and Big Island WA (most years); fields around Ottawa NWR (few).

J F M A M J J A S O N D

BOBOLINK

Dolichonyx oryzivorus

Generally as reliable as clockwork, the first returning Bobolinks appear the first week of May, and soon after, favored fields are awash with their joyful, bubbling songs. Like many prairie birds, Bobolinks benefited from the clearing of the eastern deciduous forest and expanded their range eastward from the Great Plains to exploit new habitats. Today, they occupy large grassy meadows, pastures and especially legume-dominated hayfields, particularly those planted with clover. Unfortunately, annual mortality is probably high because farmers' mowing schedules conflict with Bobolink nesting cycles.

• Our most migratory blackbird, the Bobolink makes a round-trip voyage of up to 9000 miles between northern North America and its wintering grounds in the pampas regions of Argentina and Brazil.

breeding

marked, grayish to light reddish brown eggs for 11–13 days.

ID: *Breeding male:* black bill, head, wings, tail and underparts; buff nape; white rump and wing patch. *Breeding female:* yellowish bill; brown buff overall; streaked back, sides, flanks and rump; pale "eyebrow"; dark eye line; light central crown stripe bordered by dark stripes; whitish throat. *Nonbreeding male:* similar to breeding female, but darker above and rich golden buff below.

Size: *L* 6–8 in; *W* 11½ in.

Habitat: tall, grassy meadows and pastures; hayfields; strip mine reclamation grasslands.

Nesting: on the ground, well-concealed cup nest of grass and forb stems is built in a shallow depression and lined with fine grass; female incubates 5–6 heavily

Feeding: gleans the ground and low vegetation for adult and larval invertebrates; eats many seeds.

Voice: song is a series of banjolike twangs: *bobolink bobolink spink spank spink,* often given in flight; also issues a distinctive musical *pink* call in flight.

Similar Species: breeding male is distinctive; female and nonbreeding somewhat resemble various sparrows, but are different in the combination of larger size, mostly unstreaked, buffy yellow underparts, and prominently striped back and crown.

Best Sites: along country roads near large, grassy pastures and hayfields, particularly those with clover; fields around Killdeer Plains WA and Big Island WA; Miami Whitewater Park; Tri-Valley WA (especially along Hall Rd. south of State Route 206).

RED-WINGED BLACKBIRD

Agelaius phoeniceus

One of the true avian harbingers of spring, the first Red-winged Blackbirds begin appearing with winter's first thaws, often in late February. Males precede females by a week or so and quickly become conspicuous as they establish territories, deliver their gurgling *konk a reee* song and flash their brilliant, red wing epaulets to ward off rival males. • The highly polygynous male will sometimes have up to 15 females nesting in his territory. Apparently females can play this game, too—recent genetic studies have shown that not all females within a territory bear offspring fathered by the resident harem-owner. • Red-winged Blackbirds are ubiquitous along Ohio roads, nesting in a wide variety of open habitats, even roadside ditches, but they are particularly attracted to wetlands. • This species is considered by some authorities to be the most abundant bird of any species in North America. Enormous hordes sometimes congregate in late fall; a flock estimated at 140,000 birds occurred in Ottawa County on November 7, 1989.

ID: *Male:* all black, except for large, occasionally concealed, red shoulder patch edged in yellow. *Female:* heavily streaked underparts; mottled brown upperparts; faint, red shoulder patch; pale "eyebrow."

Size: *L* 7–9½ in; *W* 13 in.

Habitat: cattail marshes, wet meadows, ditches and various wetlands, croplands and scrubby fields.

Nesting: colonial and polygynous; in cattails or shoreline bushes; female weaves an open cup nest of dried cattail leaves and grass and lines it with fine grass; female incubates 3–4 darkly marked, pale blue green eggs for 10–12 days.

Feeding: gleans the ground for seeds, waste grain and invertebrates; also gleans vegetation for seeds, insects and berries; occasionally catches insects in flight; may visit feeders.

Voice: song is a loud, raspy *konk a reee;* calls include a harsh *check* and a high

tseert; female may give a loud *che-che-che chee chee chee.*

Similar Species: male is distinctive when shoulder patch shows. *Brewer's Blackbird* (p. 316) and *Rusty Blackbird* (p. 315): females lack streaked underparts. *Brown-headed Cowbird* (p. 318): juvenile is smaller and has stubbier, conical bill. *Sparrows* (pp. 284–304): similar to female Red-wing, but smaller, with thicker bills and different behavior.

Best Sites: one of the most easily found Ohio birds; Springville Marsh SNP; Ottawa NWR; tough to find only during severe winters when most depart the state.

J F M A M J J A S O N D

EASTERN MEADOWLARK

Sturnella magna

One of the most adaptable of our original prairie birds, Eastern Meadowlarks can often be seen along highways on wires and fence posts, delivering their clear, ringing song. These birds probably got their name from English colonists who were reminded of the Sky Lark *(Alauda arvensis)* of the Old World, but meadowlarks are actually brightly colored members of the blackbird family. The scientific name means "large little starling," which is quite nonsensical and utterly inappropriate. • Eastern Meadowlarks are reminiscent of little quail, both in body structure and flight characteristics; this is a likely case of convergent evolution as both birds occupy similar habitats. • At their peak abundance in the 1930s, there were an estimated 91 meadowlarks per square mile in the Buckeye Lake region. Although they have declined since then, Eastern Meadowlarks remain common and can be found in every county. They winter in varying numbers depending on the severity of the winter and are more common toward the south.

ID: yellow underparts; broad, black breast band; mottled brown upperparts; short, wide tail with white outer tail feathers; long, pinkish legs; yellow lores; long, sharp bill; blackish crown stripes and eye line border pale "eyebrow" and median crown stripe; dark streaking on white sides and flanks.

Size: *L* 9–9½ in; *W* 14 in.

Habitat: grassy meadows and pastures; also in some croplands, weedy fields, grassy roadsides, reclaimed strip mines and old orchards.

Nesting: in a depression or scrape on the ground, concealed by dense grass; domed grass nest, with a side entrance, is woven into the surrounding vegetation; female incubates 3–7 heavily spotted, white eggs for 13–15 days.

Feeding: walks or runs along the ground gleaning grasshoppers, crickets, beetles and spiders from the ground and vegetation; probes the soil for grubs and worms; also eats seeds.

Voice: song is a rich series of 2–8 clear, slurred, melodic whistles: *see-you at school-today* or *this is the year;* gives a rattling flight call and a high, buzzy *dzeart.*

Similar Species: *Western Meadowlark* (p. 313): much rarer; virtually inseparable visually; paler gray back; whiter flanks; yellow on throat extends onto lower "cheek"; song and calls are different (best distinguishing feature).

Best Sites: extensive reclaimed strip mine grasslands such as The Wilds; Tri-Valley WA; Woodbury WA; easily found in appropriate habitat statewide.

J F M A M J J A S O N D

WESTERN MEADOWLARK

Sturnella neglecta

Never a common bird in Ohio, the Western Meadowlark now ranks as one of our rarest breeding birds; most years, no more than three or four territorial males are reported. This western species is at the extreme eastern edge of its range in Ohio, and the majority of birds are detected in former prairie regions in the central and western regions of the state.

• Vocalizations are by far the best way to separate the Western Meadowlark from the Eastern. The descending, sweet, gurgling melody of the Western Meadowlark is a staple of western movie soundtracks. • The Western Meadowlark was recognized and named by John James Audubon in 1844. He gave it the species name *neglecta* in recognition of the fact that early explorers of the western U.S., including Lewis and Clark, failed to recognize this species as distinct from the Eastern Meadowlark.

ID: yellow underparts; broad, black breast band; mottled brown upperparts; short, wide tail with white outer tail feathers; long, pinkish legs; yellow lores; brown crown stripes and eye line border pale "eyebrow" and median crown stripe; dark streaking on white sides and flanks; long, sharp bill; yellow on throat extends onto lower "cheek."
Size: *L* 9–9½ in; *W* 14½ in.
Habitat: grassy meadows and pastures; also in some croplands, weedy fields and grassy roadsides; most records coincide with former prairie regions.
Nesting: in a depression or scrape on the ground, concealed by dense grass, forbs or rarely low shrubs; domed grass nest, with a side entrance, is woven into the surrounding vegetation; female incubates 3–7 heavily spotted, white eggs for 13–15 days.

Feeding: walks or runs, gleaning grasshoppers, crickets, beetles, other insects and spiders from the ground and vegetation; probes the soil for grubs and worms; also eats seeds.
Voice: song is a rich, melodic series of bubbly, flutelike notes; calls include a low, loud *chuck* or *chup*, a rattling flight call or a few clear, whistled notes.
Similar Species: *Eastern Meadowlark* (p. 312): darker upperparts, especially crown stripes and eye line; yellow on throat does not extend onto lower "cheek"; different song and calls.
Best Sites: sites irregular from year to year; occasional birds at Maumee Bay SP, Killdeer Plains WA and in vicinity of Ottawa NWR; also former Darby Plains prairie in Madison and Union counties and throughout northwestern corner of state.

J F M A M J J A S O N D

YELLOW-HEADED BLACKBIRD

Xanthocephalus xanthocephalus

O hio lies on the extreme eastern fringe of the Yellow-headed Blackbird's range. This species is quite a rarity, too, as in most years only a few territorial males are found in the western Lake Erie marshes and most of these birds may be unmated nonbreeders. • The scientific name *Xanthocephalus* means "yellow head," which is appropriate for the brilliantly colored males that stand out like glowing jewels as they perch atop cattails or *Phragmites* in favored large marshes. Unfortunately, this blackbird's song fails to meet the standards set by the plumage. In the words of William Dawson in 1923, "he succeeds in pressing out a wail of despairing agony which would do credit to a dying catamount." • Yellow-headed Blackbirds occasionally turn up in large marshes away from Lake Erie or appear within large, mixed blackbird flocks. Rarely, birds will appear at feeders, usually in late fall or early winter.

ID: *Male:* yellow head and breast on otherwise black body; white wing patches; black lores; long tail; black bill. *Female:* dusky brown overall; yellow breast, throat and "eyebrow"; hints of yellow on face.
Size: *L* 8–11 in; *W* 15 in.
Habitat: usually in large, mixed-emergent marshes dominated by cattails interspersed with open water; occasionally in monocultures of giant reed, *Phragmites australis*.
Nesting: only a few nests have been found in Ohio; loosely colonial; female builds a deep, bulky basket of emergent aquatic plants and lines it with dry grass and other vegetation; nest is woven into emergent vegetation over water; female incubates 4 heavily speckled, pale green to pale gray eggs for 11–13 days.
Feeding: gleans the ground for seeds, beetles, snails, waterbugs and dragonflies; probes into cattail heads for larval invertebrates.
Voice: song is a strained, metallic, grating note followed by a descending buzz; call is a deep *krrt* or *ktuk;* low quacks and liquidy clucks may be given during breeding season.
Similar Species: none in Ohio.
Best Sites: large marshes bordering Magee Marsh WA causeway; sometimes seen in Mallard Marsh WA and nearby marshes; Metzger Marsh WA.

J F M A M J J A S O N D

RUSTY BLACKBIRD

Euphagus carolinus

Rusty Blackbirds nest well north of Ohio, but are much more common migrants than many people might think. These blackbirds are largely birds of swamp woods, particularly in spring, and can easily be overlooked as they forage on the ground in damp thickets. They are also occasional in winter, especially southward. At any time of year they favor wooded habitats much more than other blackbirds. • Once attuned to the Rusty's distinctive song—a rushed, squeaky gurgling often likened to the sound of rusty hinges on an opening door—the birder can easily find this species in wet woodlands statewide. • Rusty Blackbirds can be rather thuggish, and are known to kill and eat other birds on occasion. However, the Rusty may also have been eaten regularly at one time—the scientific name *Euphagus* means "good to eat."

breeding

ID: yellow eyes; dark legs; long, sharp bill. *Breeding:* dark plumage; male is darker with subtle green gloss on body and subtle bluish or greenish gloss on head. *Nonbreeding:* rusty wings, back and crown; male is darker; female has buffy underparts and rusty "cheek."
Size: *L* 9 in; *W* 14 in.
Habitat: all manner of wet woodlands; sometimes mixes with flocks of other blackbirds and feeds in agricultural fields.
Nesting: does not nest in Ohio.
Feeding: walks along the ground, gleaning waterbugs, beetles, dragonflies, snails, grasshoppers and occasionally small fish; also eats waste grain and seeds; known to kill and eat other birds occasionally.

Voice: song is a squeaky, creaking *kush-leeeh ksh-lay;* call is a harsh *chack.*
Similar Species: *Brewer's Blackbird* (p. 316): male has glossier, iridescent plumage, more purple on head, greener body and thicker base to bill; female has dark eyes; fall birds lack conspicuous rusty highlights. *Common Grackle* (p. 317): longer, keeled tail; larger body and bill; more iridescent plumage.
Best Sites: swampy woods; willow thickets along entrance road to Magee Marsh WA; Calamus Swamp; Funk Bottoms WA; Springville Marsh SNP.

J F M A M J J A S O N D

BREWER'S BLACKBIRD

Euphagus cyanocephalus

An uncommon to rare Ohio visitor, Brewer's Blackbird can easily be missed as it tends to consort with large, mixed-species blackbird flocks. Furthermore, it closely resembles the Rusty Blackbird, although the latter generally prefers wet woodlands, a habitat shunned by Brewer's Blackbird. The best way to find this species is to carefully pick through the large flocks of blackbirds that congregate in fields in early spring, particularly along Lake Erie from Ottawa County westward. Wintering blackbird flocks seen around barnyards, especially where cattle are present, should always be scrutinized, as Brewer's Blackbirds occasionally overwinter in this habitat. • Starting around 1920, this western species began an eastward range expansion, at a rate of about 40 miles per year. By 1950, these blackbirds had colonized Michigan's lower peninsula, and have nested within 20 to 30 miles of the Ohio border. Birders should be on the watch for possible breeders in northwestern Ohio in the tier of counties bordering Michigan, and especially Maumee State Forest and Williams County.

ID: *Male:* iridescent, green body and purplish head often look black; yellow eyes; some fall males may show faint, rusty feather edgings. *Female:* flat brown plumage; dark eyes.

Size: *L* 8–10 in; *W* 15½ in.

Habitat: open, human-modified habitats, usually with a nearby stream or wetland and interspersed with a patches of trees and shrubs.

Nesting: does not yet nest in Ohio.

Feeding: walks along shorelines and open areas, gleaning invertebrates and seeds.

Voice: song is a creaking, 2-note *k-shee;* call is a metallic *chick* or *check.*

Similar Species: *Rusty Blackbird* (p. 315): longer, more slender bill; iridescent plumage has subtler green gloss on body and subtle bluish or greenish gloss on head; female has yellow eyes. *Common Grackle* (p. 317): much longer, keeled tail; larger body and bill. *Brown-headed Cowbird* (p. 318): shorter tail; stubbier, thicker bill; male has dark eyes and brown head; female has paler, streaked underparts and very pale throat.

Best Sites: open fields in vicinity of Ottawa NWR, particularly in spring; in appropriate seasons, anywhere large blackbird flocks form, especially throughout glaciated western Ohio.

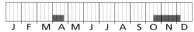

| J | F | M | A | M | J | J | A | S | O | N | D |

COMMON GRACKLE

Quiscalus quiscula

Anyone who feeds birds is probably familiar with our largest blackbird, the Common Grackle, and likely wishes it would ply its trade elsewhere. This exceedingly aggressive bird easily dominates even the loudmouthed Blue Jay. The Common Grackle is a major agricultural pest, being particularly fond of emerging shoots of corn. It also raids songbird nests and consumes eggs and nestlings. • Because of their opportunistic nature and generalized habitat preferences, Common Grackles are one of the most successful North American birds. Grackles occupy virtually every habitat except large tracts of unbroken forest, and are especially attracted to ornamental conifers in suburbia, in which they site their nests. • A true sign of spring, returning grackle flocks appear with the first spring thaws, often by late February. They often form massive winter roosts, usually in association with other blackbirds and European Starlings.

bronze morph

ID: iridescent plumage (purple blue head and breast, bronze back and sides, purple wings and tail) often looks blackish; long, keeled tail; yellow eyes; long, heavy bill; female is smaller, duller and browner than male. *Juvenile:* dull brown overall; dark eyes.
Size: *L* 11–13½ in; *W* 17 in.
Habitat: almost anywhere except large, unbroken forests.
Nesting: singly or in a small colony; in dense tree or shrub branches or emergent vegetation, often near water; female builds a bulky, open cup nest of twigs, grass, forbs and mud and lines it with fine grass or feathers; female incubates 4–5 brown-blotched, pale blue eggs for 12–14 days.

Feeding: slowly struts along the ground, gleaning, snatching and probing for insects, earthworms, seeds, waste grain and fruit; also catches insects in flight and eats small vertebrates; may take some bird eggs.
Voice: song is series of harsh, strained notes ending with a metallic squeak: *tssh-schleek* or *gri-de-leeek;* call is a quick, loud *swaaaack* or *chaack.*
Similar Species: *Rusty Blackbird* (p. 315) and *Brewer's Blackbird* (p. 316): smaller overall; lack heavy bill and keeled tail. *Red-winged Blackbird* (p. 311): shorter tail; male has red shoulder patch and dark eyes. *Great-tailed Grackle:* 1 Ohio record; much larger; heavier tail; less metallic sheen; much different vocalizations.
Best Sites: widespread and easily found, except perhaps in dead of winter.

J F M A M J J A S O N D

BROWN-HEADED COWBIRD

Molothrus ater

The Brown-headed Cowbird lays its eggs in the nests of other songbirds and parasitizes at least 140 species. In some cases the toll is high, as epitomized by the endangered Kirtland's Warbler, which was in jeopardy of disappearing until a concerted cowbird control program on the warbler's breeding grounds allowed the species to persist. • Brown-headed Cowbirds used to follow the movements of highly migratory bison herds. Their curious habit of brood parasitism likely evolved because they didn't remain in one area long enough to complete their own nesting cycle owing to the nomadic nature of bison. Although bison no longer range through Ohio, the Brown-headed Cowbirds now flock to cattle. • Cowbirds are a species of open country and woodland edges. As Ohio's forests were cleared, cowbird populations expanded tremendously, and this brought them into contact with many songbird species. These birds had not evolved a defense against the cowbird's brood parasitism, and so today, cowbirds can greatly reduce a species' nesting success.

ID: thick, conical bill; short, squared tail; dark eyes. *Male:* iridescent, blue green body plumage usually looks glossy black; dark brown head. *Female:* brown overall; faint streaking on light brown underparts; pale throat.
Size: *L* 6–8 in; *W* 12 in.
Habitat: open agricultural and residential areas including fields, woodland edges, suburban areas, parks, open woodlands, utility cutlines and pastures near cattle.
Nesting: does not build a nest; each female may lay up to 40 eggs a year in the nests of other birds, usually laying 1 egg per nest (larger numbers, up to 8 eggs in a single nest, are probably from several different cowbirds); brown-speckled, whitish eggs hatch after 10–13 days.

Feeding: gleans the ground for seeds, waste grain and invertebrates, especially grasshoppers, beetles and true bugs.
Voice: song is a high, liquidy gurgle: *glug-ahl-whee* or *bubbloozeee;* call is a squeaky, high-pitched *seep, psee* or *wee-tse-tse,* often given in flight; also a fast, chipping *ch-ch-ch-ch-ch-ch.*
Similar Species: *Rusty Blackbird* (p. 315) and *Brewer's Blackbird* (p. 316): slimmer, longer bills; longer tails; lack contrasting brown head and darker body; all have yellow eyes except female Brewer's Blackbird. *Common Grackle* (p. 317): much larger overall; longer, heavier bill; longer, keeled tail.
Best Sites: easily, and perhaps unfortunately, detected in all manner of habitats statewide.

J F M A M J J A S O N D

ORCHARD ORIOLE

Icterus spurius

O nce known as the "Bastard Baltimore Oriole," the Orchard Oriole's scientific name *spurious* means "illegitimate" and indicates an inferior relationship to the Baltimore Oriole. These inappropriate names resulted from a bungled identification—the relatively drab, female Baltimore was thought to be the male Orchard. However, the common name suits this bird, and though orchards are declining in Ohio, Orchard Orioles now occupy similar habitats of scattered, dense trees. • Orchard Orioles may spend less time here than any other neotropical breeder. They are largely finished with nesting their single broods by early to mid-July and most have departed by August. • The male takes two years to develop its distinctive, adult chestnut plumage; the first-year male resembles the female but is easily distinguished by its black throat. • The Orchard Oriole is the smallest of the nine species of orioles that regularly occur in the U.S. • As this species often skulks in dense growth, learning its distinctive, whistled song will help you to identify this bird.

Nesting: in a deciduous tree or shrub; female builds a hanging pouch nest from grass and fine plant fibers; female incubates 4–5 pale bluish white eggs, blotched with gray and brown, for 12–15 days.
Feeding: gleans insects and berries from trees and shrubs; sometimes probes flowers for nectar.
Voice: song is a loud, rapid, varied series of whistled notes ending in a downslurred note; call is a quick *chuck*.
Similar Species: *Baltimore Oriole* (p. 320): larger; male has brighter orange plumage with orange in tail; female and young male have orange overtones, are not as green overall and have more prominently patterned head than equivalent Orchard.
Best Sites: black locust thickets in reclaimed strip mine grasslands such as Crown City, Egypt Valley, Tri-Valley or Woodbury wildlife areas; Killdeer Plains WA.

ID: *Male:* black "hood" and tail; chestnut brown underparts, shoulder and rump; dark wings with white wing bar and feather edgings. *1st-spring male:* similar to adult female with blackish "bib." *Female* and *immature:* olive upperparts; yellow to olive yellow underparts; faint, white wing bars on dusky gray wings.
Size: *L* 6–7 in; *W* 9½ in.
Habitat: scattered shade trees, very open woodlands, suburban parklands, forest edges, hedgerows and groves of black locust trees in reclaimed strip mines.

J F M A M J J A S O N D

BALTIMORE ORIOLE

Icterus galbula

The uninitiated would scarcely believe that male Baltimore Orioles are members of the blackbird family, so resplendent are their black and orange colors. • Luckily for birders, this is an adaptive species that has adjusted to human-altered habitats. Prior to European settlement, Baltimore Orioles would have largely been confined to riparian woodlands. With development, Baltimores adopted large, urban shade trees, particularly the large American elms that were planted along many streets. Then, as Dutch elm disease decimated these trees starting in the 1930s, orioles switched over to other tree species and are still common suburban residents. • Baltimore Orioles are closely related to the western Bullock's Oriole (*I. bullockii*), and from 1983 to 1995 these species were lumped together under the colorless name "Northern Oriole."

ID: *Male:* black "hood," back, wings and central tail feathers; bright orange underparts, shoulder, rump and outer tail feathers; white wing patch and feather edgings; takes two years to reach adult plumage; young birds resemble females. *Female:* olive brown upperparts (darkest on head); dull, yellow orange underparts and rump; white wing bar.

Size: *L* 7–8 in; *W* 11½ in.

Habitat: open deciduous forests, particularly riparian woodlands, natural openings, roadsides, orchards, gardens and parklands; also scattered shade trees.

Nesting: high in a deciduous tree, suspended from a branch; female builds a hanging pouch nest of grass, bark shreds, plant stems and grapevines and lines it with fine grass, rootlets and fur; occasionally adds string and fishing line; female incubates 4–5 darkly marked, pale gray to bluish white eggs for 12–14 days.

Feeding: gleans canopy vegetation and shrubs for caterpillars, beetles, wasps and other invertebrates; also eats some fruit and nectar; may visit hummingbird feeders and feeding stations that offer orange halves.

Voice: song consists of slow, loud, clear whistles: *peter peter peter here peter;* calls include a 2-note *tea-too* and a rapid chatter: *ch-ch-ch-ch-ch.*

Similar Species: *Orchard Oriole* (p. 319): male has darker, chestnut brown plumage; female and young male are olive yellow and lack orange overtones.

Best Sites: along wooded streams such as at Blackhand Gorge SNP; urban parks and cemeteries such as Spring Grove Cemetery; open woodlands with large cottonwoods as at Sheldon Marsh SNP; as many as a dozen in a single tree at Magee Marsh bird trail in May.

J F M A M J J A S O N D

PURPLE FINCH

Carpodacus purpureus

Like many northern finches, Purple Finches are cyclically irruptive—in Ohio, big southward winter invasions occur perhaps once a decade. In most years, these birds are uncommon winter residents and occasionally virtually absent, but show variable yearly numbers in migration. • Purple Finches are particularly attracted to sunflower seeds and in good years will often visit feeders. When not at feeders, they can be surprisingly easy to overlook. To detect more birds, become familiar with their soft, slightly musical *pik* call note, which is given frequently. • In Ohio, nesting Purple Finches are rare and are confined to the natural and planted conifers of northern Ohio, with peak abundance in the northeastern corner of the state.

ID: *Male:* pale bill; raspberry red head, throat, breast and nape; brown and red streaks on back and flanks; reddish brown "cheek"; red rump; notched tail; pale, unstreaked belly and undertail coverts. *Female:* dark brown "cheek" and jaw line; white "eyebrow" and lower "cheek" stripe; heavily streaked underparts; unstreaked undertail coverts.

Size: *L* 5–6 in; *W* 10 in.

Habitat: *Breeding:* hemlock gorges, native white pine stands, plantations of spruce, scattered, large ornamental conifers and mixed white and red pine plantings. *In migration* and *winter:* coniferous, mixed and deciduous forests; shrubby open areas; successional fields and feeders with nearby tree cover.

Nesting: in a conifer; cup nest woven from twigs, grass and rootlets is lined with moss and hair; female incubates 4–5 darkly marked, pale greenish blue eggs for about 13 days.

Feeding: gleans the ground and vegetation for seeds, buds, berries and insects; readily visits table-style feeding stations.

Voice: song is a bubbly, continuous warble; call is a single metallic *pik*.

Similar Species: *House Finch* (p. 322): more prominent streaking on flanks; lacks reddish coloration on back; males are redder, often with yellow or orangy tones; females have longer, thinner streaks on underparts, lack obvious streaks on back and have no white eye line or "cheek" stripe.

Best Sites: feeders throughout state in invasion winters. *Breeding:* northeastern Ohio in areas with plenty of conifers such as Holden Arboretum. *In migration:* detected most easily in spring, when males are singing, at hotspots such as Magee Marsh WA bird trail or Green Lawn Cemetery.

J F M A M J J A S O N D

HOUSE FINCH

Carpodacus mexicanus

The rapid invasion of House Finches across the eastern U.S. is an excellent demonstration of the speed at which birds can exploit new habitats. In 1940, a small number of these finches was released in Long Island, New York, where they had been kept as cage birds. After a decade of local population expansion, they began spreading westward. First detected in Ohio in 1964, they had colonized the entire state by the mid-1980s. By the early 1990s, House Finches occupied the entire eastern U.S. and had reached the native western population, the original source of our birds. Today, House Finches are ubiquitous and abundant, and are particularly noticeable in suburbia. • Males vary in color from bright red to orange or yellowish. Bright red males have a selective advantage, as females tend to seek them out over their paler cohorts for breeding.

ID: streaked under-tail coverts; brown-streaked back; square tail. *Male:* brown "cap"; bright red "eyebrow," forecrown, throat and breast; heavily streaked flanks. *Female:* indistinct facial patterning; heavily streaked underparts.
Size: L 5–6 in; W 9½ in.
Habitat: primarily in cities, towns, suburbia and agricultural areas with an abundance of dense landscape plantings.
Nesting: in a cavity, building, dense foliage or abandoned bird nest; especially in evergreens and ornamental shrubs near buildings; mostly the female builds an open cup nest of grass, twigs, forbs, leaves, hair and feathers, often adding

other debris; female incubates 4–5 sparsely marked, pale blue eggs for 12–14 days.
Feeding: gleans vegetation and the ground for seeds; also takes berries, buds and some flower parts; often visits feeders.
Voice: song is a bright, disjointed warble lasting about 3 seconds, often ending with a harsh *jeeer* or *wheer;* flight call is a sweet *cheer,* given singly or in series.
Similar Species: *Purple Finch* (p. 321): scarcer and largely absent in summer; males are richer wine red, with red back and less prominent flank streaking; females have prominent white eye line, more patterned head, obvious "cheek" stripe and shorter, darker streaks on underpart.
Best Sites: statewide; at backyard feeders; males are prolific singers and easily detected.

J F M A M J J A S O N D

RED CROSSBILL

Loxia curvirostra

The last huge invasion of Red Crossbills, one of the rarest of our winter irruptives, occurred during the winter of 1972–73, and that summer a pair was found nesting in Hocking County. • Red Crossbills are notoriously nomadic in their search for food, which is almost exclusively conifer seeds. The crossed mandibles of their bills allow them to easily pop the protective scales off cones. Unfortunately, coniferous trees produce good cone crops only every two to four years on average. To compound the problem, a widespread boreal tree such as white spruce might have a prolific fruit crop in one region one year, while elsewhere production is poor. Consequently, Red Crossbills wander erratically far and wide in search of food. • Eight types, or tribes, of Red Crossbills have been identified. These types differ in cone preference, have different bill morphology and vocalizations, and apparently remain largely segregated from each other. They may eventually be split into separate species, and as many as four of them could occur in Ohio. Careful study should be given to Ohio Red Crossbills in an attempt to document types.

ID: bill has crossed tips. *Male:* dull, orange red to brick red plumage; dark wings and tail; always has color on throat. *Female:* olive gray to dusky yellow plumage; plain, dark wings. *Juvenile:* streaky brown overall.
Size: *L* 5–6½ in; *W* 11 in.
Habitat: almost exclusively conifers: pine plantations, ornamental conifers, natural stands of Virginia, pitch and white pine, and hemlock gorges.
Nesting: does not nest in Ohio (1 failed 1973 nesting attempt).
Feeding: primarily conifer seeds (especially pine); also eats buds, deciduous tree seeds and occasionally insects; often licks road salt or minerals in soil and along roadsides; rarely visits feeders.
Voice: distinctive *jip-jip* call note, often given in flight; song is a varied series of warbles, trills and *chips*.
Similar Species: unmistakable. *White-winged Crossbill* (p. 324): white wing bars in all plumages.
Best Sites: rare and irregular; absent some years; often not present for long at any given site; large conifers in cemeteries, especially Woodlawn Cemetery.

J F M A M J J A S O N D

WHITE-WINGED CROSSBILL

Loxia leucoptera

Crossbills are interesting birds to watch as they seek cones, clambering about branches in the manner of parrots, often pulling themselves along with their bills. White-winged Crossbills use the crossed mandibles of their distinctive bills to pop the scales off cones to access the seeds. • These birds are usually quite tame and approachable; unfortunately, they are also rare visitors to Ohio and can't be dependably found in most years. Crossbill irruptions, caused by conifer crop shortages in their principal range well north of Ohio, are very erratic. • White-wingeds are generally scarcer than Red Crossbills, though at least a few are seen most winters. They have smaller bills than Reds; consequently they prefer conifers with smaller cones, such as hemlock, spruce and tamarack. • Knowing the flight calls is helpful: crossbills are strictly diurnal migrants and are sometimes only observed in flight, but flocks constantly give their distinctive, rapid, dry, rattling calls.

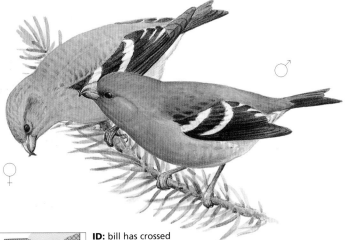

ID: bill has crossed tips; 2 bold, white wing bars. *Male:* pinkish red overall; black wings and tail. *Female:* streaked brown upperparts; dusky yellow underparts slightly streaked with brown; dark wings and tail. *Juvenile:* streaky brown overall with white wing bars.
Size: *L* 6–7 in; *W* 10½ in.
Habitat: attracted to large conifers, often in parks and cemeteries; sometimes feeds on the catkins of birch trees.
Nesting: does not nest in Ohio.
Feeding: prefers conifer seeds (mostly spruce and tamarack); also eats deciduous tree seeds and occasionally insects; often licks salt and minerals from roads when available.

Voice: song is a high-pitched series of warbles, trills and *chips;* call is a series of harsh, questioning *cheat* notes, often given in flight.
Similar Species: *Red Crossbill* (p. 323): lacks white wing bars in all plumages; male has pinkish coloration similar to Pine Grosbeak. *Pine Grosbeak* (p. 342): much rarer; larger and much more robust; lacks crossed bill tips.
Best Sites: cemeteries with large, ornamental conifers, especially Woodlawn Cemetery; also Green Lawn Cemetery and others; during invasion years, areas with large natural stands of hemlock such as Hocking Hills and Mohican SF.

J F M A M J J A S O N D

COMMON REDPOLL

Carduelis flammea

Just as crossbills are inextricably associated with conifer cones as a food source, so are redpolls with catkins of birch and alder. Given a choice, Common Redpolls will invariably forage in the boughs of these trees, where they acrobatically dangle on the outermost twigs as they pick apart catkins. Luckily, redpolls also visit backyard feeding stations, particularly thistle feeders. • Always most numerous along Lake Erie and in the northernmost counties (their primary wintering range is north of Ohio), Common Redpolls stage southward irruptions about every other year. Most species in the birch family produce bumper crops in alternate years, though our redpoll incursions are usually in lean years. In irruption years, flocks of several hundred have been noted, while off years may produce scarcely any reports. During big irruptions, large flocks often contain one or two Hoary Redpolls—a larger, paler species that breeds and winters even farther north than the Common Redpoll.

ID: red forecrown; black "chin"; yellowish bill; streaked upperparts, including rump; lightly streaked sides, flanks and undertail coverts; notched tail.
Male: pinkish red breast is brightest in breeding plumage. *Female:* whitish to pale gray breast.
Size: *L* 5 in; *W* 9 in.
Habitat: particularly attracted to birches; also found in old fields; visits feeders.
Nesting: does not nest in Ohio.
Feeding: gleans the ground, snow and vegetation in large flocks for seeds in winter; also forages in old fields for seeds; often visits feeders; takes some insects in summer.

Voice: song is a twittering series of trills; calls are a soft *chit-chit-chit-chit* and a faint *swe-eet;* indistinguishable from the Hoary Redpoll.
Similar Species: *Hoary Redpoll* (p. 342): generally paler and more plump; unstreaked or partly streaked rump; usually faint or no streaking on sides and flanks; bill may look stubbier; lacks streaking on undertail coverts. *Pine Siskin* (p. 326): heavily streaked overall; yellow highlights on wings and tail.
Best Sites: only reliably located during irruption years; tier of counties bordering L. Erie; ornamental birch trees and backyard feeders; Woodlawn Cemetery; fields around Maumee Bay SP.

J F M A M J J A S O N D

PINE SISKIN

Carduelis pinus

Gregarious and vociferous, nomadic flocks of Pine Siskins are much more likely to be heard before they are seen. Generally traveling in small groups, siskins constantly chatter and utter their diagnostic, ascending, buzzy *zeeeeee* calls. • Pine Siskins stage southward incursions in response to food shortages in the northern reaches of their range, not because of adverse weather. If siskins can maintain their metabolism through adequate diet, they can withstand temperatures colder than –40° F. Pine Siskins, on average, stage irruptions into Ohio every other year. Following large invasions, a few pairs often remain to nest, and breeding may take place in any part of the state. • Pine Siskins have become much more common and widespread since 1950; this increase may be attributable in part to the proliferation and maturation of ornamental conifers.

ID: heavily streaked underparts; darker, heavily streaked upperparts; yellow highlights at base of tail feathers and on wings (easily seen in flight); dull wing bars; slightly forked tail; indistinct facial pattern. *Immature:* similar to adult, but overall yellow tint fades through summer.

Size: *L* 4½–5½ in; *W* 9 in.

Habitat: *Breeding:* coniferous stands; urban and rural ornamental and shade conifers. *Winter:* coniferous and mixed forests, forest edges, backyards with feeders and sometimes overgrown fields.

Nesting: loosely colonial (Ohio nesters are usually single pairs); typically at midlevel in a conifer; female builds a loose cup nest of twigs, grass and rootlets and lines it with feathers, hair and fine plant fibers; female incubates 3–5 darkly dotted, pale blue eggs for about 13 days.

Feeding: gleans the ground and vegetation for seeds (especially thistle seeds), buds and some insects; attracted to road salts, mineral licks and ashes; regularly visits thistle feeders.

Voice: song is a variable, bubbly mix of squeaky, metallic, raspy notes, sometimes resembling a jerky laugh; call is a buzzy, rising *zzzreeeee*.

Similar Species: *Common Redpoll* (p. 325) and *Hoary Redpoll* (p. 342): red forecrowns; lack yellow on wings and tail. *Purple Finch* (p. 321) and *House Finch* (p. 322): females have thicker bills and no yellow on wings or tail. *Sparrows* (pp. 284–304): all lack yellow on wings and tail.

Best Sites: *Winter:* feeders during invasion years; sites with large conifers, such as Spring Grove, Green Lawn and Woodlawn cemeteries; also native stands of pitch and Virginia pines in Hocking Hills. *In migration:* anywhere, especially from mid-April to early May; hotspots such as Magee Marsh bird trail.

J F M A M J J A S O N D

AMERICAN GOLDFINCH

Carduelis tristis

American Goldfinches are familiar to feeder watchers, as these birds are easily lured to thistle seed feeders and are one of our most abundant native birds. They are unique among members of their genus in that males have very different breeding and nonbreeding plumages. These birds are often called "Wild Canaries," and the males are so colorful in breeding plumage that their late spring molt into summer colors may be the most noticed of any Ohio bird. • American Goldfinches generally do not commence nesting until well into July—the down of various thistles is almost always used to line the nest, and these plants don't mature until summer. • Although Brown-headed Cowbirds sometimes parasitize goldfinch nests, the young cowbirds never survive even though they do hatch. Apparently, they fail to receive adequate protein from the goldfinches' strictly vegetarian diet.

breeding

ID: *Breeding male:* black wings, tail and "cap" (extends onto forehead); bright yellow body; white wing bars, undertail coverts and tail base; orange bill and legs.
Nonbreeding male: olive brown back; yellow-tinged head; gray underparts.
Female: yellow green upperparts and belly; yellow throat and breast.
Size: *L* 4½–5½ in; *W* 9 in.
Habitat: weedy fields, woodland edges, meadows, riparian areas, parks and gardens.
Nesting: late summer and fall; in a deciduous shrub or tree, often in hawthorn, serviceberry or sapling maple; female builds a compact cup of plant fibers, grass and spider silk lined with plant down and hair; female incubates 4–6 pale bluish white eggs for 12–14 days.
Feeding: gleans vegetation for seeds, primarily thistles, sunflowers, goldenrods and others of the sunflower family; occasionally eats insects and berries; commonly visits feeders.

Voice: song is a long and varied series of trills, twitters, warbles and sibilant notes; calls include *po-ta-to-chip* or *per-chic-or-ee,* often delivered in flight, and a whistled *dear-me, see-me.*
Similar Species: breeding males, and even females, are distinctive. *Pine Siskin* (p. 326): similar to very dull nonbreeding American Goldfinch, but siskin body is streaked and face is not as plain.
Best Sites: widespread in open habitats; backyard thistle feeders.

J F M A M J J A S O N D

EVENING GROSBEAK

Coccothraustes vespertinus

A pinnacle of the bird feeding experience is the day that a flock of these utterly stunning finches descends on a backyard feeder and helps themselves to a meal. But the homeowner soon discovers what ravenous creatures Evening Grosbeaks are and may cringe as sunflower seed costs begin to soar. This grosbeak possesses the largest bill of any of our finches, and the scientific name *Coccothraustes* means "kernel crusher." The aptness of this name is quickly evident when these birds make short work of thick-hulled sunflower seeds. • Evening Grosbeaks are notoriously erratic and cyclical in their Ohio appearances. Some years, they are virtually absent and in others just a handful make brief showings. Larger than normal invasions occur about every two to three years. • Evening Grosbeaks have learned to recognize feeders as good food sources, and most birds are now seen at feeders. However, they are also fond of the fruit of box elder trees, which persists on the trees through winter; these trees are abundant along streams.

ID: massive, pale, conical bill; black wings and tail; broad, white wing patches. *Male:* black crown; bright yellow "eyebrow" and forehead band; dark brown head gradually fades into golden yellow belly and lower back. *Female:* gray head and upper back; yellow-tinged underparts; white undertail coverts.
Size: *L* 7–8½ in; *W* 14 in.
Habitat: mostly seen at feeders; also attracted to conifers and deciduous woods with an abundance of fruiting box elder and ash; also ornamental fruit-bearing trees.
Nesting: does not nest in Ohio.

Feeding: gleans the ground and vegetation for seeds, buds and berries; also eats insects and licks mineral-rich soil; often visits feeders for sunflower seeds.
Voice: song is a wandering, halting warble; call is a loud, sharp *clee-ip* or a ringing *peeer;* a distant calling flock sounds somewhat like House Sparrows.
Similar Species: unmistakable in all plumages.
Best Sites: platform feeders covered with sunflower seed; flocks tend to be erratic and irregular from year to year and even during a specific year; listen to the birding grapevine to learn where they've been seen.

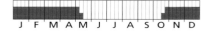

J F M A M J J A S O N D

HOUSE SPARROW

Passer domesticus

House Sparrows are not actually true sparrows, but belong to a small family of 37 species known as Old World Sparrows, none of which are native to North America. The presence of House Sparrows in North America is the result of one of the most spectacularly successful and ill-reasoned introductions of an animal ever undertaken. One hundred birds were released in New York City in the mid-1850s, as a cure for various insect crop pests. Subsequent releases occurred around the eastern U.S., including in Ohio. However, the avian rocket scientists responsible for the project failed to realize that House Sparrows are primarily seedeaters, and weren't at all effective in their intended purpose. In fact, House Sparrows have become agricultural pests in some areas by consuming large quantities of grains. • The House Sparrow is one of the most resilient and adaptive of the world's birds. Possibly the birds most affected by the House Sparrow are native cavity-nesting birds such as the Eastern Bluebird. The aggressive House Sparrow readily displaces the cavity nester to use the cavity for its own nest.

breeding

ID: *Breeding male:* gray crown; black "bib" and bill; chestnut brown nape; light gray "cheek"; white wing bar; dark, mottled upperparts; gray underparts. *Nonbreeding male:* smaller black "bib"; pale bill. *Female:* plain gray brown overall; buffy "eyebrow"; streaked upperparts; indistinct facial pattern; grayish, unstreaked underparts.
Size: *L* 5½–6½ in; *W* 9½ in.
Habitat: townsites, urban and suburban areas, farmyards and agricultural areas, railroad yards and other developed areas; largely absent from undeveloped areas.
Nesting: often communal; in a human-made structure, ornamental shrub or natural cavity; pair builds a large, dome-shaped nest of grass, twigs, plant fibers and litter

and often lines it with feathers; pair incubates 4–6 gray-speckled, whitish to greenish white eggs for 10–13 days.
Feeding: gleans the ground and vegetation for seeds, insects, fruit and human food waste; frequently visits feeders for seeds.
Voice: song is a familiar, plain *cheep-cheep-cheep-cheep;* call is a short *chill-up.*
Similar Species: male is distinctive. *Dickcissel* (p. 309): subadult is a rare winter feeder visitor often found with House Sparrow; similar to female House Sparrow, but slimmer, with prominent lateral throat stripe and longer, paler bill.
Best Sites: often first species encountered upon stepping outside your house.

J F M A M J J A S O N D

OCCASIONAL BIRD SPECIES

FULVOUS WHISTLING-DUCK
Dendrocygna bicolor

This large, somewhat gooselike duck of the southern U.S. and points south appears very rarely, with only nine records since the first report in 1962. When vagrants appear in the northern states, they are often in small flocks. The Black-bellied Whistling-Duck *(D. autumnalis)*, another member of this genus, could also appear in Ohio someday.

ROSS'S GOOSE
Chen rossii

The mallard-sized Ross's Goose was first recorded in Ohio in 1982; since then we've averaged about two or three reports annually. This remarkable spate of records is indicative of the overall population expansion of the species. Most Ross's Geese are detected in fall and winter with large concentrations of Canada Geese and Snow Geese.

CINNAMON TEAL
Anas cyanoptera

Adult male Cinnamon Teals in breeding plumage are stunning and unlikely to be overlooked or confused with other ducks. Females, however, may be confused with other teal and might go undetected. Since the first sighting in 1895, there have been only seven additional records.

KING EIDER
Somateria spectabilis

1st winter

This duck of the Far North winters in the northern reaches of the Atlantic Ocean, but small numbers pass through the Great Lakes. Ohio averages about one or two sightings annually, and all but four records come from Lake Erie. Females are similar to female Common Eiders.

BARROW'S GOLDENEYE
Bucephala islandica

The striking Barrow's Goldeneye has been documented about seven times in Ohio, with all records coming from Lake Erie between January and March. It is most likely to be found amid large flocks of migrant Common Goldeneyes. Females are very similar to female Commons.

PACIFIC LOON
Gavia pacifica

Pacific Loons could appear on just about any large water body, and small, atypical-looking loons should be carefully scrutinized. Ohio's first record was a bird found on Lake Erie at Huron in December 1985. There have been two other records since then.

nonbreeding

WESTERN GREBE
Aechmophorus occidentalis

An appearance in Ohio of this largest North American grebe is always an event. We have averaged only about one sighting per decade since the inaugural record in 1913. The similar-looking Clark's Grebe was split from this species in 1985. Ohio birds warrant careful scrutiny, as Clark's could possibly occur in Ohio.

NORTHERN GANNET
Morus bassanus

The northernmost of the boobies, Northern Gannets occur in Ohio about every three or four years. Occasionally, a small invasion happens and several birds are recorded. Almost all records are from Lake Erie, but rarely birds are found stranded or dead at inland locales. These giant birds are exciting to observe as they dive into the water from great heights like avian kamikazes.

juvenile

BROWN PELICAN
Pelecanus occidentalis

Several decades ago, Brown Pelicans nearly
disappeared from North America as a result
of pesticides in the food chain. Fortunately,
those poisons were banned, and the overall
pelican population has grown tremendously.
Ohio had its first record in 1990, followed by three
additional records over the next decade, with more to
come, no doubt.

TRICOLORED HERON
Egretta tricolor

Long known as "Louisiana Heron," this southern species regularly wanders northward to Ohio, particularly in summer.
We average about two records annually. Though most sightings occur in the Lake Erie marshes, these herons can
appear anywhere in the state. In recent years,
Pickerel Creek Wildlife Area and vicinity has
been a good location.

nonbreeding

WHITE IBIS
Eudocimus albus

First recorded here in 1964, this is one of the rarest
waders to visit Ohio, with about one sighting
every six years. All but one of these records has
been in July in postbreeding dispersals, a phenomenon of many
southern heron species. Most of our records are of the brownish
immatures.

GLOSSY IBIS
Plegadis falcinellus

The number of vagrants of this species in the northern
states is increasing, and Ohio records multiple sightings
annually. Separating this species from the similar White-faced
Ibis can be tricky, and subadults may not always be identifiable. The
Glossy Ibis can appear anywhere in the state, particularly in spring.

breeding

WHITE-FACED IBIS
Plegadis chihi

Like the Glossy Ibis, this species is on the increase, and more sightings can be expected here. Overall, though, there have been far fewer records of this species compared to the Glossy. Most records are from May, when the breeding-plumaged adults are easier to separate from the Glossy Ibis.

breeding

WOOD STORK
Mycteria americana

There are five Ohio records of this large southern wader, with four of them between 1909 and 1964. In 2001, a bird was spotted in Portage County. Wood Storks have declined precipitously in the U.S. in recent decades, but the population began to expand in the mid-1990s. This may lead to more Ohio records.

MISSISSIPPI KITE
Ictinia mississippiensis

Ohio averages about one sighting every other year of this kite. Most appear in May and June and are almost always one-day wonders or flyovers, making this one of the harder species to detect. Although it is almost the size of a Peregrine Falcon, the kite weighs only one-third as much. This light weight contributes to its oddly buoyant flight.

GYRFALCON
Falco rusticolus

These huge falcons generate tremendous excitement when they appear south of the Canadian border. Almost all records are from the Lake Erie shore. Of the few seen in Ohio, only one lingered and was widely seen by birders, in Paulding County in 1996. Strongly albinistic buteo hawks may be misidentified as Gyrfalcons.

gray morph

YELLOW RAIL
Coturnicops noveboracensis

It seems likely that many more of these incredibly secretive birds pass through Ohio than are detected. Yellow Rails are recorded about once every two or three years, and there is one nesting record from 1909. Most records come in spring, when birds are more likely to vocalize. The boardwalk at Irwin Prairie State Nature Preserve in April may be the best place to seek this species.

BLACK RAIL
Laterallus jamaicensis

As one of North America's most secretive and elusive species, the true status of the Black Rail in Ohio is poorly understood. The smallest North American rail, it probably regularly moves through our state in very small numbers, but has only been detected about every five years on average.

PURPLE GALLINULE
Porphyrio martinica

Finding one of these colorful birds of the southeastern U.S. in an Ohio wetland is always a shock, but they turn up with some regularity and may appear just about anywhere. We average about one sighting every other year, usually in April or May. Amazingly, there was one successful nesting in Franklin County in 1962.

breeding

SNOWY PLOVER
Charadrius alexandrinus

The years 1993–95 brought Ohio's first five records of this western species of salt flats and beaches. The overall population is shrinking, with only an estimated 21,000 birds in the entire U.S. Reasons for the Ohio mini-invasion are unclear, but this isn't a species to be expected with any regularity.

breeding

PIPING PLOVER
Charadrius melodus

Piping Plovers formerly occupied beaches on all of the Great Lakes; their numbers have now been greatly reduced in this region. The last Ohio breeders, on Lake Erie, disappeared in 1942. Today these birds are rare migrants, averaging about two sightings annually, with most records from sandy beaches along Lake Erie or inland lakes.

breeding

BLACK-NECKED STILT
Himantopus mexicanus

This striking and unmistakable stilt has graced Ohio wetlands fewer than 10 times, excluding anecdotal historic records. Most records come from the large marshes bordering western Lake Erie. Black-necked Stilts seem to be increasing overall, which might lead to more Ohio records.

CURLEW SANDPIPER
Calidris ferruginea

The first record of this sandpiper occurred in 1984, and only half a dozen sightings have followed. All but one of the Ohio birds have been the very distinctive breeding-plumaged adults. Nonbreeders closely resemble Dunlins, with which they sometimes fraternize, and may be overlooked.

molting

RUFF
Philomachus pugnax

To unexpectedly come upon a breeding-plumaged Ruff working a mudflat is a shock, as its ornate and colorful neck feathers are quite unique. Females, known as Reeves, are much less conspicuous; many Ohio birds are subadults or females. Ruffs are rare but regular, averaging about two sightings annually.

nonbreeding

PARASITIC JAEGER
Stercorarius parasiticus

Jaegers are notoriously difficult to identify, especially subadults, which is what almost all Ohio birds are. In fact, this species was thought for decades to be the most common Ohio jaeger, when in fact many early records were probably Pomarine Jaegers. We average about two to three reports a year; all modern records are from Lake Erie.

juvenile light morph

LONG-TAILED JAEGER
Stercorarius longicaudus

This is our rarest jaeger, with only about 15 records in total. Long-taileds are most likely to be seen in September and would be very unlikely after that month. Although almost all records are from Lake Erie—the most typical locale for any jaeger—one record is from the Hoover Reservoir.

juvenile intermediate morph

BLACK-HEADED GULL
Larus ridibundus

Not recorded in our state until 1965, Black-headed Gulls have been increasing in North America, and Ohio records reflect that. Since the inaugural bird, there have been at least 32 reports, all of them from Lake Erie. This species is most likely to be found with large concentrations of Bonaparte's Gulls.

nonbreeding

MEW GULL
Larus canus

This is our rarest semi-regularly occurring gull, with fewer than one dozen records to date, all from Lake Erie. Mew Gulls are not conspicuous, and they blend easily with the hordes of Ring-billed Gulls with which they normally associate, making detection difficult.

nonbreeding

CALIFORNIA GULL
Larus californicus

This gull saved the Mormons from a locust plague shortly after the group's arrival in Utah. Adult birds are not hard to identify, but often do not stand out from the other large gulls with which they fraternize. This species is a rare but increasing Ohio visitor, with more than 30 records since the first bird was seen here in 1979. All but two records are from Lake Erie.

nonbreeding

LEAST TERN
Sterna antillarum

The smallest tern in North America turns up about once each year in Ohio and may appear on any decent-sized water body. Interior U.S. populations have declined—this species was listed as endangered in 1985. Least Terns do not linger, but normally vanish within hours of being located.

breeding

RUFOUS HUMMINGBIRD
Selasphorus rufus

This hardy hummer of the western U.S. regularly wanders eastward, typically in late fall. Our first record was in 1985, and these tiny birds have been recorded almost annually since and are on the increase. Over a dozen individuals were reported in 2003. Leaving feeders up late into the fall might attract one of these vagrants.

BLACK-BACKED WOODPECKER
Picoides arcticus

There are only about one dozen Ohio records of this largely nonmigratory denizen of the boreal forest. Occasionally, and irregularly, small numbers move south of their principal range owing to food shortages. There has been only one sighting in Ohio in the last 40 years. Modern forest management practices may have reduced overall populations, limiting southward irruptions.

WESTERN KINGBIRD
Tyrannus verticalis

When discovered, Western Kingbirds are easy to observe, as they are bold, conspicuous birds that often hunt from roadside wires and treetops. Unfortunately, most don't linger in one spot for long. They are found about two out of every three years, mostly in the western half of the state. A pair even tried to nest in Lucas County in 1933.

SCISSOR-TAILED FLYCATCHER
Tyrannus forficatus

This western species is one of the flashiest and most unmistakable vagrants recorded in Ohio. There are only about 10 records, and May is the most likely month of occurrence. This bird could be mistaken for a Fork-tailed Flycatcher *(Tyrannus savana)*, a species that should appear here someday.

LOGGERHEAD SHRIKE
Lanius ludovicianus

Once common in rural Ohio, "Butcher Birds" were found breeding in most counties at their peak in the early 1900s. Shrikes began a precipitous decline in the 1940s. Today this species is exceedingly rare, with only about two nesting pairs reported per year. Most recent records are from extreme southern Ohio, and the occasional wintering bird is found, often at Killdeer Plains Wildlife Area.

BOREAL CHICKADEE
Poecile hudsonica

This species shares a very similar pattern to the Black-backed Woodpecker, with about the same number of records, the last from 1973. Fire suppression, plus insect control of species such as the spruce budworm, to maximize timber production in the boreal forest may have reduced fluctuations in food supply for birds such as this, reducing southward movements into Ohio.

BEWICK'S WREN
Thryomanes bewickii

Like the Loggerhead Shrike, this species was
a common and widespread breeder in the early
1900s. Wholesale habitat changes have caused a
population crash, not only in Ohio, but throughout
the range of the Appalachian subspecies *(T. b. altus),*
and today this species is imperiled. The last Ohio nesting record was in 1995 in Brown County.

TOWNSEND'S SOLITAIRE
Myadestes townsendi

This sleek, dapper-looking thrush has been recorded only six
times in Ohio; the first record is from 1938. When solitaires
wander east from their normal western North American
range, they are often attracted to ornamental fruit-bearing trees and shrubs. Often, these vagrants return
repeatedly to the same food source, such as the
widely seen bird at Holden Arboretum in 2001,
which gorged on the ornamental holly berries.

VARIED THRUSH
Ixoreus naevius

Ohio's first record of this gorgeous species of the
Pacific Northwest occurred in 1977. Since then,
one has turned up about every other year on
average. When this bird appears, it is almost
always at a backyard feeder and often stays
for an extended period.

BOHEMIAN WAXWING
Bombycilla garrulus

Irregular winter wanderers to the south and east of their
breeding range, Bohemian Waxwings are particularly
attracted to berries of native mountain-ash. When fruit
crops are poor, some waxwings drift as far south as Ohio
and are often attracted to the fruits of ornamental
trees. There have been only five records of this species
in the past 20 years.

BLACK-THROATED GRAY WARBLER
Dendroica nigrescens

This western warbler regularly wanders to the eastern states; Ohio has at least eight records. Most of these birds have occurred during the winter months, and they will visit feeders—particularly suet feeders—to survive. The small, yellow loral spot is diagnostic.

KIRTLAND'S WARBLER
Dendroica kirtlandii

In 2003, 1202 singing male Kirtland's were tallied on their northern Michigan breeding grounds, making this one of North America's rarest birds. Even though most of the entire population may pass through Ohio in migration, perhaps one bird per year is recorded on average. Although sightings might occur anywhere, the Magee Marsh Wildlife Area bird trail probably hosts the most records.

SWAINSON'S WARBLER
Limnothlypis swainsonii

The breeding range of this southern warbler of cane breaks, rhododendron thickets and wet woods nearly reaches Ohio, but there have been only 10 records here. Most are of spring overflights in April or May, but there are several instances of summering individuals near the Ohio River. Territorial individuals should be watched for in thick, overgrown tangles of young clear-cuts on very steep slopes in that region.

GREEN-TAILED TOWHEE
Pipilo chlorurus

This distinctive towhee of the western U.S. is very rare here in Ohio; there have been only four records. Three of the four birds frequented feeders, as is typical for eastern vagrants of this species, which usually appear in winter. Additional records will probably occur at someone's bird feeder.

SPOTTED TOWHEE
Pipilo maculatus

Until 1995, this species and the Eastern
Towhee were classified as the "Rufous-sided Towhee."
Spotted Towhees are birds of western North America that
are accidental in the East. Although there are only four Ohio
records, birders may notice and report more of these birds now
that this towhee is considered a distinct species. Unlike the
Green-tailed Towhee, none of the Ohio birds were seen at feeders.

BACHMAN'S SPARROW
Aimophila aestivalis

This secretive, southern sparrow exhibits a "boom and
bust" population cycle. Probably absent prior to
European settlement, when Ohio was 95 percent
forested, Bachman's Sparrow reached its heyday in the
early 1900s and was common in unglaciated Ohio.
Populations have steadily declined since then, with
the last record from 1978, and it is now extirpated.

LARK BUNTING
Calamospiza melanocorys

This Great Plains breeder irregularly wanders eastward,
often in late fall, and has been recorded 10 times in Ohio.
Breeding males are unmistakable, but females and
nonbreeding males are not nearly as conspicuous;
most Ohio records are of the latter. This bird
could appear almost anywhere, and individuals
may overwinter.

nonbreeding

HARRIS'S SPARROW
Zonotrichia querula

The largest North American sparrow, this species breeds to the
west of Ohio, but regularly appears here in migration and
sometimes overwinters. To date, there have been several
dozen records, and one or two birds appear most
years. Harris's Sparrow often visits feeders
and frequently associates with White-
crowned Sparrows.

nonbreeding

SMITH'S LONGSPUR
Calcarius pictus

From the 1940s through the early 1970s, small flocks of Smith's Longspurs were regularly detected in central and western Ohio, almost always in spring. Since then, they have essentially disappeared as an Ohio transient, with only a few records since 1980. This longspur should be sought in March and April, in large corn stubble fields interspersed with foxtail grass *(Setaria sp.)*.

breeding

PINE GROSBEAK
Pinicola enucleator

Although there have been many Ohio records of this large northern finch, reports have greatly dropped off. The last record dates to 1987. As with some other boreal forest birds that irregularly irrupt southward in winter, ecological changes in the northern forest may have reduced overall populations and limited southward invasions.

HOARY REDPOLL
Carduelis hornemanni

Large winter invasions of Common Redpolls, which happen every two to four years, may bring one or two of the more northern Hoary Redpolls. Separating these species is often not easy and differences between the two can be slight. Fortunately, these redpolls are often detected at feeders where they can be studied closely.

nonbreeding

SELECT REFERENCES

American Ornithologists' Union. 1998. *Checklist of North American Birds*. 7th ed. American Ornithologists' Union, Washington, D.C.

American Ornithologists' Union. 2003. *Forty-fourth supplement to the American Ornithologists' Union Check-list of North American Birds*. Auk 120: 923-932.

Anderson, M, E. Durbin, T. Kemp, S. Lauer, and E. Tramer. 2002. *Birds of the Toledo Area*. Ohio Biological Survey, Columbus, OH.

Choate, E.A. 1985. *The Dictionary of American Bird Names, Revised Edition*. The Harvard Common Press, Boston, MA.

Dister, D.C., J.W. Hammond, R. Harlan, B.F. Master, and B. Whan. 2002. *Ohio Bird Records Committee Checklist of the Birds of Ohio*. Ohio Department of Natural Resources, Division of Natural Areas and Preserves, Columbus, OH.

Dunn, J. and K. Garrett. 1997. *A Field Guide to the Warblers of North America*. Houghton Mifflin Co., Boston, MA.

Hayman, P., J. Marchant, T. Prater. *Shorebirds: An Identification Guide*. Houghton Mifflin Co., Boston, MA.

Jones, J. O. 1990. *Where The Birds Are*. William Morrow and Company, Inc., New York.

Kaufman, K. 1996. *Lives of North American Birds*. Houghton Mifflin Co., Boston, MA.

Kaufman, K. 2000. *Birds of North America*. Houghton Mifflin Co., New York.

National Geographic Society. 1999. *Field Guide to the Birds of North America*. 3rd ed. National Geographic Society, Washington, D.C.

Peterjohn, B.G. 2001. *The Birds of Ohio*. The Wooster Book Company, Wooster, OH.

Peterjohn, B.G., and D.L. Rice. 1991. *The Ohio Breeding Bird Atlas*. Ohio Department of Natural Resources, Division of Natural Areas and Preserves, Columbus, OH.

Peterson, R.T. 1996. *A Field Guide to the Birds: Including all species found in Eastern North America*. Houghton Mifflin Co., Boston, MA.

Rosche, L. 1988. *A Field Book of the Birds of the Cleveland Region*. 2nd ed. Cleveland Museum of Natural History, Cleveland, OH.

Sauer, J.R., J.E. Hines, I. Thomas, J. Fallon, and G. Gough. 1999. *The North American Breeding Bird Survey, Results and Analysis 1966-1998*. Version 98.1. USGS Patuxent Wildlife Research Center, Laurel, MD.

Sibley, D.A. 2000. *The Sibley Guide to Birds*. Alfred A. Knopf, Inc., New York.

Stokes, D., and L. Stokes. 1996. *Stokes Field Guide to Birds: Eastern Region*. Little, Brown and Co., Toronto, Canada.

Terres, J.K. 1995. *The Audubon Society Encyclopedia of North American Birds*. Wings Books, New York.

Thomson, T. 1994. *Birding in Ohio*. 2nd ed. Indiana University Press, Bloomington, IN.

Thompson, B. 1997. *Bird Watching for Dummies*. IDG Books Worldwide, Foster City, CA.

Wheeler, B.K. 2003. *Raptors of Eastern North America: The Wheeler Guides*. Princeton University Press, Princeton, NJ.

American Redstart

GLOSSARY

avifauna: the community of birds found in a specific region or environment.

borrow pit: in construction, an area of land where materials have been excavated for use as fill at another site.

cavity nester: a bird that builds its nest in a tree hollow or nest box.

cere: bare skin joining the forehead and the base of the upper bill.

corvid: any member of the family *Corvidae*, including crows, jays, magpies and ravens.

cryptic plumage: a coloration pattern that helps to conceal the bird.

diagnostic: the distinguishing characteristics of a bird.

disjunct population: two groups of the same species found in widely separated regions.

diurnal: active primarily during the day.

extirpated: a species that no longer exists in the wild in a particular region but occurs elsewhere.

fecundity: fertility.

feral: a bird that is living in the wild but was once domesticated.

flush: a behavior in which frightened birds explode into flight in response to a disturbance.

forb: a herb other than grass.

gizzard shad: a freshwater fish from the herring family.

grub: to dig up from the roots.

hybridism: the mating of two species to produce an offspring, often infertile, with characteristics of both species.

irruptive: a sporadic, mass migration of birds into an unusual range.

lead: a small channel of water that runs between ice floes.

lek: a place where males gather to display for females in the spring.

lores: the area on a bird between its eye and upper bill.

mantle: the area that includes the back and uppersides of the wings.

mast: food that is high in fiber, including tree nuts, conifer seeds and acorns.

migrant trap: an oasis of vegetation that attracts large groups of migrating birds.

mixed-emergent marshes: a wetland with a mixed community of vegetation, including bulrushes, cattails, wild rice and water lilies.

morph: one of several alternate color phases displayed by a species; may change seasonally or from region to region.

neotropical: the biogeographic region that includes southern Mexico, Central and South America and the West Indies.

nominate subspecies: the subspecies for which the scientific species name and subspecies name are identical; for example, the lineatus subspecies of the Red-shouldered Hawk (*Buteo lineatus lineatus*).

obligate: a habitat, feeding style or other factor that is essential to the survival of a species.

peeps: small, similar-looking sandpipers of the *Calidris* genus.

pelagic: open ocean habitat far from land.

Phragmites: reeds found in the wetlands of temperate or tropical regions.

pigeon milk: not true milk; a nutritious secretion produced in the crop of members of the pigeon family that is fed to their young.

pishing: a repeated, sibilant sound made especially to attract birds.

polygynous: a mating strategy in which one male breeds with several females.

raft: a gathering of birds.

relict: the remaining small population of a species in a region where it is otherwise extirpated.

riparian: habitat along riverbanks.

riprap: a pile of rocks and concrete chunks used as a supporting wall on an embankment near water.

scapulars: feathers of the shoulder, seeming to join the wing and back.

sexual dimorphism: a difference in plumage, size or other characteristics between males and females of the same species.

silviculture: division of forestry that deals with the care and cultivation of forests.

speculum: a brightly colored patch on the wings of many dabbling ducks.

squeaking: making a sound to attract birds by loudly kissing the back of the hand, or by using a specially designed, squeaky bird call.

stage: to gather in one place during migration, usually when birds are flightless or partly flightless during molting.

stoop: a steep dive through the air, usually performed by birds of prey while foraging or during courtship displays.

successional woodlands: sequence of vegetation that grows after a major disturbance such as a fire.

swale: a low-lying, marshy tract of land.

tertials: innermost feathers under the wing.

understory: the shrub or thicket layer beneath a canopy of trees.

vent: the single opening for excretion of uric acid and other wastes and for sexual reproduction; also known as the "cloaca."

zygodactyl feet: feet that have two toes pointing forward and two pointing backward; found in osprey, owls and woodpeckers.

eyebrow/supercilium
crown
eye line
lore
wing bars
chin
greater coverts
throat
rump
breast
tips of primary feathers
flank
tail feathers
belly
undertail coverts
secondary feathers

CHECKLIST

The following checklist contains 412 species of birds that have been officially recorded in Ohio. Species are grouped by family and listed in taxonomic order in accordance with the A.O.U. *Check-list of North American Birds* (7th ed.) and its supplements.

Accidental and casual species (those that are not seen on a yearly basis) are listed in *italics*. In addition, the following risk categories are noted: extinct or extirpated (ex), endangered (en) and threatened (th).

We wish to thank the Ohio Birds Records Committee for their kind assistance in providing the information for this checklist.

Waterfowl (Anatidae)

- ☐ *Fulvous Whistling-Duck*
- ☐ Greater White-fronted Goose
- ☐ Snow Goose
- ☐ *Ross's Goose*
- ☐ Canada Goose
- ☐ Brant
- ☐ Mute Swan
- ☐ *Trumpeter Swan*
- ☐ Tundra Swan
- ☐ Wood Duck
- ☐ Gadwall
- ☐ Eurasian Wigeon
- ☐ American Wigeon
- ☐ American Black Duck
- ☐ Mallard
- ☐ Blue-winged Teal
- ☐ *Cinnamon Teal*
- ☐ Northern Shoveler
- ☐ Northern Pintail
- ☐ *Garganey*
- ☐ Green-winged Teal
- ☐ Canvasback
- ☐ Redhead
- ☐ Ring-necked Duck
- ☐ *Tufted Duck*
- ☐ Greater Scaup
- ☐ Lesser Scaup
- ☐ *King Eider*

- ☐ *Common Eider*
- ☐ Harlequin Duck
- ☐ Surf Scoter
- ☐ White-winged Scoter
- ☐ Black Scoter
- ☐ Long-tailed Duck
- ☐ Bufflehead
- ☐ Common Goldeneye
- ☐ *Barrow's Goldeneye*
- ☐ Hooded Merganser
- ☐ Common Merganser
- ☐ Red-breasted Merganser
- ☐ Ruddy Duck

Grouse & Allies (Phasianidae)

- ☐ *Gray Partridge* (ex)
- ☐ Ring-necked Pheasant
- ☐ Ruffed Grouse
- ☐ *Greater Prairie-Chicken* (ex)
- ☐ Wild Turkey

New World Quail (Odontophoridae)

- ☐ Northern Bobwhite

Loons (Gaviidae)

- ☐ Red-throated Loon
- ☐ *Pacific Loon*
- ☐ Common Loon

Grebes (Podicipedidae)

- ☐ Pied-billed Grebe
- ☐ Horned Grebe
- ☐ Red-necked Grebe
- ☐ Eared Grebe
- ☐ *Western Grebe*

Petrels & Shearwaters (Procellariidae)

- ☐ *Black-capped Petrel*

Storm-Petrels (Hydrobatidae)

- ☐ *Leach's Storm-Petrel*

Gannets (Sulidae)

- ☐ *Northern Gannet*

Pelicans (Pelecanidae)

- ☐ American White Pelican
- ☐ *Brown Pelican*

Cormorants (Phalacrocoracidae)

- ☐ Double-crested Cormorant

Darters (Anhingidae)

- ☐ *Anhinga*

Frigatebirds (Fregatidae)
☐ *Magnificent Frigatebird*

Herons (Ardeidae)
☐ American Bittern (en)
☐ Least Bittern (th)
☐ Great Blue Heron
☐ Great Egret
☐ Snowy Egret (en)
☐ Little Blue Heron
☐ *Tricolored Heron*
☐ Cattle Egret (en)
☐ Green Heron
☐ Black-crowned Night-Heron (th)
☐ Yellow-crowned Night-Heron (th)

Ibises (Threskiornithidae)
☐ *White Ibis*
☐ *Glossy Ibis*
☐ *White-face Ibis*
☐ *Roseate Spoonbill*

Storks (Ciconiidae)
☐ *Wood Stork*

Vultures (Cathartidae)
☐ Black Vulture
☐ Turkey Vulture

Kites, Hawks & Eagles (Accipetridae)
☐ Osprey (en)
☐ *Swallow-tailed Kite*
☐ *Mississippi Kite*
☐ Bald Eagle (en)
☐ Northern Harrier (en)
☐ Sharp-shinned Hawk
☐ Cooper's Hawk
☐ Northern Goshawk
☐ *Harris's Hawk*
☐ Red-shouldered Hawk
☐ Broad-winged Hawk
☐ *Swainson's Hawk*
☐ Red-tailed Hawk
☐ Rough-legged Hawk

☐ Golden Eagle

Falcons (Falconidae)
☐ American Kestrel
☐ Merlin
☐ *Gyrfalcon*
☐ Peregrine Falcon (en)
☐ *Prairie Falcon*

Rails, Gallinules & Coots (Rallidae)
☐ *Yellow Rail*
☐ *Black Rail*
☐ King Rail (en)
☐ *Virginia Rail*
☐ Sora
☐ *Purple Gallinule*
☐ Common Moorhen
☐ American Coot

Cranes (Gruidae)
☐ Sandhill Crane (en)

Plovers (Charadriidae)
☐ *Northern Lapwing*
☐ Black-bellied Plover
☐ American Golden-Plover
☐ *Snowy Plover*
☐ *Wilson's Plover*
☐ Semipalmated Plover
☐ *Piping Plover* (en)
☐ Killdeer

Stilts & Avocets (Recurvirostridae)
☐ *Black-necked Stilt*
☐ American Avocet

Sandpipers & Allies (Scolopacidae)
☐ Greater Yellowlegs
☐ Lesser Yellowlegs
☐ *Spotted Redshank*
☐ Solitary Sandpiper
☐ Willet
☐ Spotted Sandpiper
☐ Upland Sandpiper (th)

☐ *Eskimo Curlew*
☐ Whimbrel
☐ *Long-billed Curlew*
☐ Hudsonian Godwit
☐ Marbled Godwit
☐ Ruddy Turnstone
☐ Red Knot
☐ Sanderling
☐ Semipalmated Sandpiper
☐ Western Sandpiper
☐ *Red-necked Stint*
☐ Least Sandpiper
☐ White-rumped Sandpiper
☐ Baird's Sandpiper
☐ Pectoral Sandpiper
☐ *Sharp-tailed Sandpiper*
☐ Purple Sandpiper
☐ Dunlin
☐ *Curlew Sandpiper*
☐ Stilt Sandpiper
☐ Buff-breasted Sandpiper
☐ *Ruff*
☐ Short-billed Dowitcher
☐ Long-billed Dowitcher
☐ Wilson's Snipe
☐ *Eurasian Woodcock*
☐ American Woodcock
☐ Wilson's Phalarope
☐ Red-necked Phalarope
☐ Red Phalarope

Gulls & Allies (Laridae)
☐ Pomarine Jaeger
☐ *Parasitic Jaeger*
☐ *Long-tailed Jaeger*
☐ Laughing Gull
☐ Franklin's Gull
☐ Little Gull
☐ *Black-headed Gull*
☐ Bonaparte's Gull
☐ *Heermann's Gull*
☐ *Mew Gull*
☐ Ring-billed Gull
☐ *California Gull*

☐ Herring Gull
☐ Thayer's Gull
☐ Iceland Gull
☐ Lesser Black-backed
   Gull
☐ Glaucous Gull
☐ Great Black-backed
   Gull
☐ Sabine's Gull
☐ Black-legged Kittiwake
☐ *Ross's Gull*
☐ *Ivory Gull*
☐ Caspian Tern
☐ *Royal Tern*
☐ Common Tern (en)
☐ *Arctic Tern*
☐ Forster's Tern
☐ *Least Tern*
☐ *Large-billed Tern*
☐ Black Tern (en)

Alcids (Alcidae)
☐ *Thick-billed Murre*
☐ *Black Guillemot*
☐ *Long-billed Murrelet*
☐ *Ancient Murrelet*
☐ *Atlantic Puffin*

**Pigeons & Doves
(Columbidae)**
☐ Rock Pigeon
☐ *Eurasian Collared-
   Dove*
☐ *White-winged Dove*
☐ Mourning Dove
☐ *Passenger Pigeon* (ex)
☐ *Common Ground-Dove*

Parrots (Psittacidae)
☐ *Carolina Parakeet* (ex)

**Cuckoos & Anis
(Cuculidae)**
☐ Black-billed Cuckoo
☐ Yellow-billed Cuckoo
☐ *Smooth-billed Ani*
☐ Groove-billed Ani

**Barn Owls
(Tytonidae)**
☐ Barn Owl (th)

Owls (Strigidae)
☐ Eastern Screech-Owl
☐ Great Horned Owl
☐ Snowy Owl
☐ *Northern Hawk Owl*
☐ *Burrowing Owl*
☐ Barred Owl
☐ *Great Gray Owl*
☐ Long-eared Owl
☐ Short-eared Owl
☐ *Boreal Owl*
☐ Northern Saw-Whet
   Owl

**Nightjars
(Caprimulgidae)**
☐ Common Nighthawk
☐ Chuck-will's-widow
☐ Whip-poor-will

Swifts (Apodidae)
☐ Chimney Swift

**Hummingbirds
(Trochilidae)**
☐ Ruby-throated
   Hummingbird
☐ *Calliope Hummingbird*
☐ *Rufous Hummingbird*

**Kingfishers
(Alcedinidae)**
☐ Belted Kingfisher

**Woodpeckers
(Picidae)**
☐ Red-headed
   Woodpecker
☐ Red-bellied
   Woodpecker
☐ Yellow-bellied
   Sapsucker (en)
☐ Downy Woodpecker
☐ Hairy Woodpecker

☐ *Red-cockaded
   Woodpecker*
☐ *Black-backed
   Woodpecker*
☐ Northern Flicker
☐ Pileated Woodpecker
☐ *Ivory-billed
   Woodpecker* (ex)

**Flycatchers
(Tyrannidae)**
☐ Olive-sided Flycatcher
☐ Eastern Wood-Pewee
☐ Yellow-bellied
   Flycatcher
☐ Acadian Flycatcher
☐ Alder Flycatcher
☐ Willow Flycatcher
☐ Least Flycatcher (th)
☐ *Gray Flycatcher*
☐ Eastern Phoebe
☐ *Say's Phoebe*
☐ *Vermilion Flycatcher*
☐ Great Crested
   Flycatcher
☐ *Western Kingbird*
☐ Eastern Kingbird
☐ *Scissor-tailed
   Flycatcher*

Shrikes (Laniidae)
☐ *Loggerhead Shrike* (en)
☐ Northern Shrike

Vireos (Vireonidae)
☐ White-eyed Vireo
☐ Bell's Vireo
☐ Yellow-throated Vireo
☐ Blue-headed Vireo
☐ Warbling Vireo
☐ Philadelphia Vireo
☐ Red-eyed Vireo

**Jays & Crows
(Corvidae)**
☐ Blue Jay
☐ *Black-billed Magpie*
☐ American Crow
☐ *Common Raven*

Larks (Alaudidae)
- ☐ Horned Lark

Swallows (Hirundinidae)
- ☐ Purple Martin
- ☐ Tree Swallow
- ☐ *Violet-green Swallow*
- ☐ Northern Rough-winged Swallow
- ☐ Bank Swallow
- ☐ Cliff Swallow
- ☐ Barn Swallow

Chickadees and Titmice (Paridae)
- ☐ Carolina Chickadee
- ☐ Black-capped Chickadee
- ☐ Boreal Chickadee
- ☐ Tufted Titmouse

Nuthatches (Sittidae)
- ☐ Red-breasted Nuthatch
- ☐ White-breasted Nuthatch
- ☐ *Brown-headed Nuthatch*

Creepers (Certhiidae)
- ☐ Brown Creeper

Wrens (Troglodytidae)
- ☐ *Rock Wren*
- ☐ Carolina Wren
- ☐ *Bewick's Wren* (en)
- ☐ House Wren
- ☐ Winter Wren
- ☐ Sedge Wren
- ☐ Marsh Wren

Kinglets (Regulidae)
- ☐ Golden-crowned Kinglet
- ☐ Ruby-crowned Kinglet

Gnatcatchers (Sylviidae)
- ☐ Blue-gray Gnatcatcher

Thrushes (Turdidae)
- ☐ *Northern Wheatear*
- ☐ Eastern Bluebird
- ☐ *Mountain Bluebird*
- ☐ *Townsend's Solitaire*
- ☐ Veery
- ☐ Gray-cheeked Thrush
- ☐ Swainson's Thrush
- ☐ Hermit Thrush (th)
- ☐ Wood Thrush
- ☐ American Robin
- ☐ *Varied Thrush*

Mockingbirds & Thrashers (Mimidae)
- ☐ Gray Catbird
- ☐ Northern Mockingbird
- ☐ Brown Thrasher

Starlings (Sturnidae)
- ☐ European Starling

Wagtails & Pipits (Motacillidae)
- ☐ American Pipit
- ☐ *Sprague's Pipit*

Waxwings (Bombycillidae)
- ☐ *Bohemian Waxwing*
- ☐ Cedar Waxwing

Wood-Warblers (Parulidae)
- ☐ Blue-winged Warbler
- ☐ Golden-winged Warbler (en)
- ☐ Tennessee Warbler
- ☐ Orange-crowned Warbler
- ☐ Nashville Warbler
- ☐ Northern Parula
- ☐ Yellow Warbler

- ☐ Chestnut-sided Warbler
- ☐ Magnolia Warbler
- ☐ Cape May Warbler
- ☐ Black-throated Blue Warbler
- ☐ Yellow-rumped Warbler
- ☐ *Black-throated Gray Warbler*
- ☐ Black-throated Green Warbler
- ☐ *Townsend's Warbler*
- ☐ Blackburnian Warbler
- ☐ Yellow-throated Warbler
- ☐ Pine Warbler
- ☐ *Kirtland's Warbler*
- ☐ Prairie Warbler
- ☐ Palm Warbler
- ☐ Bay-breasted Warbler
- ☐ Blackpoll Warbler
- ☐ Cerulean Warbler
- ☐ Black-and-white Warbler
- ☐ American Redstart
- ☐ Prothonotary Warbler
- ☐ Worm-eating Warbler
- ☐ *Swainson's Warbler*
- ☐ Ovenbird
- ☐ Northern Waterthrush
- ☐ Louisiana Waterthrush
- ☐ Kentucky Warbler
- ☐ Connecticut Warbler
- ☐ Mourning Warbler
- ☐ Common Yellowthroat
- ☐ Hooded Warbler
- ☐ Wilson's Warbler
- ☐ Canada Warbler
- ☐ *Painted Redstart*
- ☐ Yellow-breasted Chat

Tanagers (Thraupidae)
- ☐ Summer Tanager
- ☐ Scarlet Tanager
- ☐ *Western Tanager*

Sparrows & Allies (Emberizidae)

- ☐ *Green-tailed Towhee*
- ☐ *Spotted Towhee*
- ☐ Eastern Towhee
- ☐ *Bachman's Sparrow*
- ☐ American Tree Sparrow
- ☐ Chipping Sparrow
- ☐ Clay-colored Sparrow
- ☐ Field Sparrow
- ☐ Vesper Sparrow
- ☐ Lark Sparrow (en)
- ☐ *Black-throated Sparrow*
- ☐ *Lark Bunting*
- ☐ Savannah Sparrow
- ☐ Grasshopper Sparrow
- ☐ *Baird's Sparrow*
- ☐ Henslow's Sparrow
- ☐ Le Conte's Sparrow
- ☐ Nelson's Sharp-tailed Sparrow
- ☐ Fox Sparrow
- ☐ Song Sparrow
- ☐ Lincoln's Sparrow
- ☐ Swamp Sparrow
- ☐ White-throated Sparrow
- ☐ *Harris's Sparrow*
- ☐ White-crowned Sparrow
- ☐ Dark-eyed Junco (th)
- ☐ Lapland Longspur
- ☐ *Smith's Longspur*
- ☐ Snow Bunting

Grosbeaks & Buntings (Cardinalidae)

- ☐ Northern Cardinal
- ☐ Rose-breasted Grosbeak
- ☐ *Black-headed Grosbeak*
- ☐ Blue Grosbeak
- ☐ Indigo Bunting
- ☐ *Painted Bunting*
- ☐ Dickcissel

Blackbirds & Allies (Icteridae)

- ☐ Bobolink
- ☐ Red-winged Blackbird
- ☐ Eastern Meadowlark
- ☐ Western Meadowlark
- ☐ Yellow-headed Blackbird
- ☐ Rusty Blackbird
- ☐ Brewer's Blackbird
- ☐ Common Grackle
- ☐ *Great-tailed Grackle*
- ☐ Brown-headed Cowbird
- ☐ Orchard Oriole
- ☐ *Bullock's Oriole*
- ☐ Baltimore Oriole

Finches (Fringillidae)

- ☐ *Brambling*
- ☐ *Gray-crowned Rosy-Finch*
- ☐ *Pine Grosbeak*
- ☐ Purple Finch
- ☐ House Finch
- ☐ Red Crossbill
- ☐ White-winged Crossbill
- ☐ Common Redpoll
- ☐ *Hoary Redpoll*
- ☐ Pine Siskin
- ☐ American Goldfinch
- ☐ Evening Grosbeak

Old World Sparrows (Passeridae)

- ☐ House Sparrow

INDEX OF SCIENTIFIC NAMES

This index references only the primary species accounts.

INDEX

INDEX OF COMMON NAMES

Page numbers in **boldface** type refer to the primary, illustrated species accounts.

ABOUT THE AUTHORS

James S. McCormac

A lifelong Ohioan, James S. McCormac became fascinated with birds at a very young age, and was actively seeking them in the field by the age of ten. Thirty years later, his Ohio list stands at 344. He has also birded in almost every state and Canadian province. For the past six years, he has served as Secretary of the Ohio Bird Records Committee, a task suited to his interest in rarities and changing bird distribution. He has published numerous papers and articles on birds, and conducted research on Ohio's grassland species. James is also the inaugural President of the Ohio Ornithological Society, a statewide network of birders organized to promote education, conservation, research and interaction among people interested in birding.

McCormac is also keenly interested in plants, and is employed as a botanist for the Ohio Department of Natural Resources, Division of Natural Areas and Preserves. As with birds, he has a strong interest in rare species and their distribution, ecology and conservation. In the course of botanizing, he's discovered several new native species for Ohio, and published numerous scientific papers and general interest articles on plants. He was a founder and is a member of the Ohio Rare Plant Advisory Committee.

Gregory Kennedy

Gregory Kennedy has been an active naturalist since he was very young. He is the author of many books on natural history and has also produced film and television shows on environmental issues and indigenous concerns in Southeast Asia, New Guinea, South and Central America, the high Arctic and elsewhere. He has also been involved in numerous research projects around the world ranging from studies in the upper canopy of tropical and temperate rainforests to deepwater marine investigations.

Peregrine Falcon